Programming Heterogeneous MPSoCs

Jerónimo Castrillón Mazo
Rainer Leupers

Programming Heterogeneous MPSoCs

Tool Flows to Close the Software Productivity Gap

Springer

Jerónimo Castrillón Mazo
Rainer Leupers
Chair for Software for Systems on Silicon
RWTH Aachen University
Aachen
Germany

ISBN 978-3-319-00674-1 ISBN 978-3-319-00675-8 (eBook)
DOI 10.1007/978-3-319-00675-8
Springer Cham Heidelberg New York Dordrecht London

Library of Congress Control Number: 2013940298

Printed on acid-free paper

Springer is part of Springer Science+Business Media (www.springer.com)

Dedicated to my extended family

Acknowledgments

This book is based on my doctoral dissertation, written at the Chair for Software for Systems on Silicon at the RWTH Aachen University. The work presented here is the result of 6 years of research, filled with interactions with people that enriched my life, both personally and academically. I am glad to start this book by thanking them.

I would like to start with a sincere *danke schön* to my advisor Professor Rainer Leupers. Beyond his excellent guidance in the role of an advisor, I learned many things by working with him, more notably professionalism and pragmatism. I learned to focus on building useful solutions to problems that matter, rather than spending time on problems that are only interesting from a theoretical point of view. I would also like to thank Professor Gerd Ascheid and Professor Heinrich Meyr for their inspiring discussions in the context of joint projects I had the fortune to work in.

It would have been impossible to work out the solutions presented in this book without the support of the MAPS team, including previous doctoral works by Jianjiang Ceng and Hanno Scharwächter and contemporary efforts by my colleagues Weihua Sheng, Anastasia Stulova, Stefan Schürmans and Maximilian Odendahl. I was also lucky to count with high quality students that worked on parts of the solutions presented in this book. Special thanks go to Christian Klein for helping me put together the sequential programming flow, to Ricardo Velásquez for his efforts on earlier versions of the parallel programming flow, to Andreas Tretter for helping me extend the parallel flow to new hardware architectures and to Aamer Shah for his work on multiple applications.

I am grateful to Filippo Borlenghi, Jan Weinstock, Felix Engel, Maximilan Odendahl, Juan Eusse, Luis Murillo and Jovana Jovic for proof-reading parts of this book. Their feedback was decisive to bring this work into its final shape.

Finally, I would like to thank those people that contributed to the completion of this book in a *non technical* form—my friends and my family. Thanks to my good *old* friends Jorge and Sebas with whom I shared the *ups and downs*, typical of a doctoral path. Thanks to my *new* friends Filippo, Luis Gabriel, Max and Eusse for making me feel like at home while being thousands of kilometers away from it. Thanks to my *very old* friends Daniel, Ana María, María Adelaida, Andrés and Juan Pablo for their support from the distance.

My biggest thanks go to my family and my girlfriend Hera. It is difficult to express how thankful I am to my parents for their unconditional support throughout my life—they made me what I am today. It was also very important for me to have my siblings to count on, Juli, Jaco, Sebas and Esteban. With my family so far away, it is impossible to measure what it meant to me to count with Hera during the last 5 years. My most sincere thanks for being there with me *Herita*.

Contents

1 **Introduction** 1
1.1 Evolution and Trends in Embedded Systems 2
1.1.1 Embedded Hardware 3
1.1.2 Embedded Software 5
1.1.3 Design Gaps 6
1.2 Tool Flow Requirements: Closing the SW Gap 7
1.3 MAPS: A Framework for MPSoC Programming 10
1.4 Programming Tool Flows 11
1.5 Synopsis and Outline 13

2 **Background and Problem Definition** 15
2.1 Motivational Example 15
2.2 Preliminaries 17
2.2.1 Mapping and Scheduling 17
2.2.2 Sequential Code Performance Estimation 19
2.2.3 Notation 21
2.3 Multi-Application Problem 21
2.3.1 Definitions 21
2.3.2 Problem Statement 28
2.4 Sequential Code Problem 28
2.4.1 Compilers and Parallelism Extraction 29
2.4.2 Problem Statement 38
2.5 Parallel Code Problem 39
2.5.1 Process Networks and Dataflow Models 39
2.5.2 The C for Process Networks Language 44
2.5.3 KPN Mapping and Scheduling 44
2.5.4 Problem Statement 46
2.6 Software Defined Radio Problem 47
2.6.1 The Nucleus Project 48
2.6.2 Extensions to the Parallel Problem 49
2.6.3 Problem Statement 52
2.7 Synopsis 52

3 Related Work 53
3.1 Runtime Management 53
3.1.1 Survey of Runtime Managers 53
3.1.2 This Book in Perspective 55
3.2 Parallelism Extraction from Sequential Code 56
3.2.1 General-Purpose Domain 56
3.2.2 Embedded Domain 57
3.2.3 Commercial Solutions 59
3.2.4 This Book in Perspective 59
3.3 Synthesis of Parallel Specifications 60
3.3.1 General-Purpose Parallel Programming Models 60
3.3.2 Embedded Parallel Programming 61
3.3.3 This Book in Perspective 65
3.4 Software Defined Radio 66
3.4.1 SDR Approaches 66
3.4.2 This Book in Perspective 68
3.5 Multiple Applications 68
3.5.1 Handling Multiple Applications 69
3.5.2 This Book in Perspective 71

4 MPSoC Runtime Management 73
4.1 OSIP Solution 73
4.1.1 Mapping and Scheduling Approach 74
4.1.2 OSIP Architecture 76
4.1.3 Implementation Results 78
4.1.4 OSIP Firmware 79
4.2 Platform Integration 80
4.2.1 Hardware Integration 80
4.2.2 Software Integration 80
4.2.3 System Level Modeling 81
4.3 Benchmarking 84
4.3.1 Experimental Setup 84
4.3.2 Synthetic Benchmarking 85
4.3.3 H.264 Application 86
4.4 Synopsis 87

5 Sequential Code Flow 89
5.1 Tool Flow Overview 89
5.1.1 Clang and LLVM 90
5.1.2 Tool Flow Components 91
5.2 Application Analysis 93
5.2.1 Application Tracing 93
5.2.2 Building the Application Model 95
5.2.3 Sequential Performance Estimation Revisited 97
5.2.4 Analysis Results 98

5.3 Partitioning and Parallelism Extraction . . . 99
5.3.1 Graph Clustering . . . 99
5.3.2 Discovering Parallelism Patterns . . . 100
5.3.3 Global Analysis . . . 105
5.4 Backend . . . 108
5.4.1 C Backend . . . 108
5.4.2 CPN Hints . . . 111
5.5 Case Study . . . 113
5.5.1 Synthetic Example . . . 113
5.5.2 Audio Filter Application . . . 116
5.6 Limitations and Outlook . . . 121
5.7 Synopsis . . . 122

6 Parallel Code Flow . . . 123
6.1 Tool Flow Overview . . . 123
6.1.1 The CPN C Compiler . . . 124
6.1.2 Tool Flow Components . . . 125
6.1.3 Input and Output Modeling . . . 126
6.2 Token Logging and KPN Tracing . . . 130
6.2.1 KPN Traces . . . 130
6.2.2 Sequential Performance Estimation Revisited . . . 131
6.3 Building and Evaluating the Model . . . 132
6.3.1 Model Construction . . . 132
6.3.2 Model Evaluation . . . 137
6.4 Best-Effort Mapping and Scheduling . . . 139
6.4.1 Trace Graph . . . 139
6.4.2 Graph-Based Buffer Sizing . . . 141
6.4.3 Heuristics for Setting Scheduler Parameters . . . 142
6.4.4 Heuristics for Decoupled Process and Channel Mapping . . . 144
6.4.5 Joint Process and Channel Mapping: The GBM Algorithm . . . 147
6.4.6 Post-Processing Phase . . . 151
6.5 Mapping and Scheduling with Timing Constraints . . . 151
6.6 OSIP Backend . . . 154
6.6.1 OSIP Generated Code . . . 154
6.6.2 Debugging Environment . . . 155
6.7 Case Study . . . 156
6.7.1 Experimental Setup . . . 156
6.7.2 Best-Effort Results . . . 159
6.7.3 Results for Real-Time Applications . . . 160
6.7.4 Validation on TI's Keystone Platform . . . 161
6.8 Limitations and Outlook . . . 164
6.9 Synopsis . . . 164

7 Extensions for Software Defined Radio 165
7.1 Tool Flow Overview 165
7.1.1 Tool Flow Components 166
7.1.2 Input and Output Modeling 167
7.2 Tracing for Software Defined Radio 169
7.2.1 Flavor Trace Generation 169
7.2.2 Sequential Performance Estimation Revisited 170
7.3 Configuration Matching 172
7.3.1 From Nucleus to Flavors 172
7.3.2 Interface Matching 173
7.4 Mapping and Code Generation 175
7.4.1 Testing Timing Constraints 175
7.4.2 Code Generator Extensions 176
7.5 Case Study 178
7.5.1 Target Application: MIMO OFDM Transceiver 178
7.5.2 Target Platforms and Flavor Libraries 179
7.5.3 Results of the SDR Flow 181
7.5.4 Execution on the Virtual Platform 183
7.6 Limitations and Outlook 184
7.7 Synopsis 185

8 Multi-Application Flow 187
8.1 Tool Flow Overview 187
8.1.1 Tool Flow Components 188
8.1.2 Input and Output Modeling 189
8.2 Scenario Analysis 191
8.2.1 Generating Configurations for Single Applications 192
8.2.2 Pre-Processing 193
8.2.3 Composability Analysis 195
8.2.4 Results Export 198
8.3 Case Study 198
8.3.1 Experimental Setup 199
8.3.2 Scenario Analysis Results 199
8.4 Limitations and Outlook 201
8.5 Synopsis 203

9 Conclusions and Outlook 205

References 209

Glossary 225

Index 229

Figures

Fig. 1.1 SoC Trends: Processor count
Fig. 1.2 SoC Trends: Processing elements in OMAP and Snapdragon
Fig. 1.3 Software complexity trends
Fig. 1.4 Design Gaps
Fig. 1.5 MPSoC programming flow
Fig. 1.6 Sample target MPSoCs
Fig. 1.7 Tool flows overview
Fig. 2.1 Motivational example
Fig. 2.2 Example of mapping and scheduling for a DAG
Fig. 2.3 Phases in compilation flow
Fig. 2.4 Motivational example for the sequential code problem
Fig. 2.5 3AC example
Fig. 2.6 CDFG Example
Fig. 2.7 Forms of parallelism
Fig. 2.8 MoC examples
Fig. 2.9 MoC expressiveness
Fig. 2.10 Example of mapping for a KPN application
Fig. 2.11 The Nucleus waveform development concept
Fig. 4.1 Hierarchical mapping and scheduling
Fig. 4.2 Profiling results for the OSIP application
Fig. 4.3 OSIP Architecture
Fig. 4.4 Generic OSIP-based platform
Fig. 4.5 Example of OSIP tCEFSM model
Fig. 4.6 OSIP's synthetic benchmark results
Fig. 4.7 OSIP operational ranges
Fig. 4.8 OSIP's results for H.264 decoding
Fig. 5.1 Sequential tool flow
Fig. 5.2 Trace file example
Fig. 5.3 Flow for sequential application model
Fig. 5.4 Sample analysis results
Fig. 5.5 Parallelism patterns in a DFG
Fig. 5.6 Efficiency-speedup tradeoff

Fig. 5.7 CPN hints in the MAPS IDE
Fig. 5.8 The Pitaya platform
Fig. 5.9 Synthetic example results
Fig. 5.10 Gantt Charts from the Pitaya simulator
Fig. 5.11 LP-AF profile
Fig. 5.12 Dynamic call graph of the LP-AF application
Fig. 5.13 Parallelism extraction for the LP-AF application
Fig. 6.1 Parallel tool flow
Fig. 6.2 Conceptual view of a schedule descriptor
Fig. 6.3 Process trace example
Fig. 6.4 Platform graph construction
Fig. 6.5 Measurement of timing constraints
Fig. 6.6 Trace Replay Module
Fig. 6.7 Evaluation results for the JPEG application
Fig. 6.8 Flow of the mapping and scheduling phase
Fig. 6.9 Example of a trace graph
Fig. 6.10 OSIP configuration for code generation
Fig. 6.11 OSIP state from debugger
Fig. 6.12 Experimental setup
Fig. 6.13 Sample randomly generated KPN graphs
Fig. 6.14 Mapping results for platform DCP
Fig. 6.15 Real-time mapping results
Fig. 6.16 Real-time iterative mapping
Fig. 6.17 Keystone results
Fig. 7.1 SDR tool flow
Fig. 7.2 Example of flavor traces
Fig. 7.3 Example of time annotation for an irregular nuclei
Fig. 7.4 Code generation flow SDR applications
Fig. 7.5 MIMO OFDM transceiver diagram
Fig. 7.6 SDR target heterogeneous platforms
Fig. 7.7 Gantt charts examples in the SDR flow
Fig. 7.8 Best rates obtained by the SDR flow
Fig. 7.9 Execution traces for TX-RX obtained from VP
Fig. 8.1 Multi-application tool flow
Fig. 8.2 Example of multi-application description
Fig. 8.3 Scenario analysis flow
Fig. 8.4 Illustration of utilization functions
Fig. 8.5 Pre-processing example
Fig. 8.6 Example of combined processor utilization
Fig. 8.7 Multi-application configurations
Fig. 8.8 Composability results

Tables

Table 3.1 Comparison of runtime management approaches
Table 3.2 Comparison of parallelism extraction approaches
Table 3.3 Comparison of parallel code synthesis approaches
Table 3.4 Comparison of SDR approaches
Table 3.5 Comparison of multi-application approaches
Table 4.1 OSIP synthesis results
Table 5.1 Instrumentation functions
Table 6.1 Test results for 800 random KPNs on platform SCP
Table 7.1 Trace generation in the SDR flow
Table 7.2 Configuration matching for buffer/flag interfaces
Table 7.3 Configuration matching for a register interface
Table 7.4 Overview of latency values (in cycles) for all flavors used in the case study
Table 7.5 SDR flow: Results summary
Table 7.6 Comparison of rates
Table 8.1 Results of the scenario analysis phase

Chapter 1
Introduction

This book is concerned with the improvement of the *programming experience* of today's and upcoming embedded systems in the multimedia and wireless communication domains. In these domains, two main interrelated trends have made programming a daunting task for embedded software developers. On the one hand, the underlying computing engines of new embedded devices have evolved to become complex *Systems-on-Chip* (SoCs) that integrate several processors of different types as well as dedicated hardware accelerators. These heterogeneous *Multi-Processor Systems-on-Chip* (MPSoCs) are nowadays commonplace in portable consumer electronics. On the other hand, the complexity of the software being deployed on multimedia and wireless terminals keeps rising exponentially. This software is no longer characterized by a single small application, but by a set of individually complex applications, each with constraints that have to be respected regardless of the presence of other applications in the system. As a result, programmers have to implement applications with ever increasing requirements, on platforms with ever increasing complexity. Additionally, they have to meet project schedules that shrink due to tight time-to-market constraints. This situation has given rise to the *software productivity gap*.

This book aims at narrowing the gap, thereby improving the programming experience of heterogeneous MPSoCs. To that end, a set of methodologies together with their underlying algorithms are proposed and evaluated. They are embedded in four different tool flows in an integrated programming environment. The first two flows are solutions for sequential and parallel programming of heterogeneous MPSoCs. The *sequential flow* provides support for parallelizing applications written using the C programming language, offering a solution for existing legacy code bases. The *parallel flow* determines a mapping of an explicitly parallel application onto the target platform. It supports a parallel programming language that describes applications in form of *Process Networks* (PNs). A third flow, an extension of the parallel flow, is presented that targets baseband processing applications in the context of *Software Defined Radio* (SDR). This *SDR flow* includes means to exploit potential hardware acceleration in the platform, without changing the original application source code.

J. Castrillón Mazo and R. Leupers, *Programming Heterogeneous MPSoCs*, DOI: 10.1007/978-3-319-00675-8_1,

A fourth and last tool flow is then presented that incorporates the previous ones into a unified methodology for deploying multiple applications onto heterogeneous MPSoCs. This *multi-application flow* helps the user to find a mapping for a group of applications which ensures that each application meets its constraints. Each tool flow is evaluated on Virtual Platforms (VPs), which mimic different characteristics of state-of-the-art embedded MPSoCs. Multi-application and multi-tasking support in the VPs is enabled by a novel resource manager.

This chapter is organized as follows. Section 1.1 gives an overview of the landscape of today's embedded systems. Based on this landscape, Sect. 1.2 lists the requirements for a toolset aiming at narrowing the aforementioned software productivity gap. Thereafter, the *MAPS Framework*, a research prototype of such a toolset, is introduced in Sect. 1.3. Finally, Sect. 1.4 gives an overview of the programming tool flows presented in this book. This chapter ends with a summary and an outline of the rest of the book in Sect. 1.5.

1.1 Evolution and Trends in Embedded Systems

Today, embedded systems are much more widespread than other computing systems. Already by 1997, the amount of embedded processors shipped equaled that of Personal Computers (PC) at approximately 110 million units [110]. By the year 2000, twice as many embedded processors were sold compared to PC processors. Recent reports show that the number of PCs sold per year keeps decreasing, while the embedded processors market continues to grow. As an example, the PC market had a negative growth of 18.9 % between years 2010 and 2011 in Western Europe [90]. During the same period, the wireless market was foreseen to grow by 35.8 % [54]. Today, it is fair to estimate that over 98 % of the two billion multiprocessor chips sold every year are integrated into embedded devices [86, 125, 181]. This overwhelming growth, together with the myriad of challenges associated with each embedded application niche, have motivated plenty of research.

Traditionally, an embedded system has been considered an electronic system with one or several processors designed to perform a specific function within a larger system [285], as opposed to mainstream desktop computing or High Performance Computing (HPC). Example of embedded systems include multimedia players, set-top boxes and cryptographic chips in radio-frequency identification cards. Over the last few years, however, the boundary between embedded and other computing systems began to blur. Several systems, still considered embedded, started to outgrow the traditional definition adding functionalities commonly associated to PCs. A prominent example are today's smartphones which often feature a fully-fledged Operating System (OS) and thus are not restricted to a single specific function. Nonetheless, these systems include components that are embedded in the traditional sense, like the baseband processing subsystem in a smartphone. In the coming years, even such *deeply* specialized systems will have to support a small set of applications to improve hardware reuse.

It is important to take a look into the hardware and software evolution that made embedded systems what they are today. This will allow to get a grasp of the complexity of the problems faced in this book.

1.1.1 Embedded Hardware

Embedded hardware has undergone a tremendous change. Some years ago it consisted of a single processor, a small number of peripherals and some hardware accelerators. Product differentiation was achieved by hardware, with *Application-Specific Integrated Circuits* (ASICs) designed with maximum performance as the primary goal. With advances in technology, increased engineering costs made dedicated hardware solutions affordable only for a few. Already in 2001, the *International Technology Roadmap for Semiconductors* (ITRS) recognized the *design cost* as the greatest threat to the semiconductor roadmap [125]. Beside the cost factor, shrinking project schedules, down to few months, made it impossible to design products from scratch. *Component reuse* became the key concept for reducing *Non Recurring Engineering* (NRE) costs and meeting project deadlines. In fact, the ITRS foresees an impressive reuse ratio of 98 % by 2026 compared to 46 % in 2009 [124]. This motivated the so-called *Platform-based design* paradigm [37, 228], in use since the early 2000s. New design paradigms helped engineers to meet project deadlines for a competitive time-to-market. However, wireless communications and multimedia standards kept changing rapidly, so products had to be *flexible* to maximize the time-in-market and therefore amortize NRE costs. This produced a migration to software-dominated solutions, where previously hardwired functionalities were moved to programmable processors. With increasingly demanding applications, embedded processors faced the *power wall*, similar to what happended in desktop computing around the year 2000. As a result, the originally simple embedded hardware evolved into a complex multiprocessor platform.

Figure 1.1 shows the ITRS 2011 prediction for the number of Processing Elements (PEs) that will be integrated in future portable consumer electronics SoCs. According to the ITRS, platforms may contain over a thousand PEs by 2019. Although predictions are often wrong, e.g., in 2005 the ITRS predicted a 15 GHz on-chip local clock by 2010, embedded MPSoCs are already following this trend, though considerably more slowly. It is arguable that this slowdown can be attributed to the lack of programming tools, considering that recent examples have shown that the semiconductor integration is not the main issue. Prominent examples are Intel's *Single-chip Cloud Computer* (SCC) [114, 284] with 48 cores and Tilera's TILE-Gx100 [223] with 100 cores.

Within the application domain relevant to this work, the PEs in Fig. 1.1 will not be of the same type, as is the SCC and the TILE-Gx100 processors [148]. While a homogeneous array offers advantages for devising programming methodologies, it cannot achieve the energy efficiency required by battery-powered portable devices. When allowed by the applications, some hardware components will expose *just*

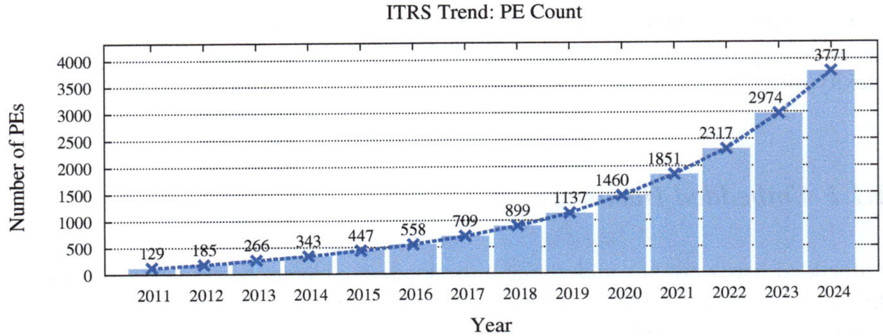

Fig. 1.1 SoC trends: Predicted processor count in portable consumer electronics designs (adapted from [124])

enough flexibility (e.g., in form of a restricted instruction set). For that reason, domain and application specific processors such as *Digital Signal Processors* (DSPs) and *Application Specific Instruction-set Processors* (ASIPs) [120] will continue to be common in embedded systems [25, 137]. For high end systems, e.g., for high data rate communication, using hardware accelerators will continue to be unavoidable [247]. These accelerators will have a flexibility reduced to a small set of hardware configuration parameters, without any software programmability. The work in [291] illustrates the cost of flexibility in terms of energy efficiency for such systems.

To get an idea of the evolution of embedded platforms, consider the actual processor counts of two commercial product families illustrated in Fig. 1.2. The first curve (in solid red) contains sample processors from the *Open Multimedia Application Platform* (OMAP) family from Texas Instruments (TI) [258]. This family displays an initial slow increase in the PE count from the early 2-processor OMAP1 to the first 3-processor OMAP3 SoCs. Thereafter, the curve for the OMAP family features a steep increase. The OMAP5430, available from the third quartal of 2012, contains already 14 PEs. It consists of a dual ARM Cortex-A15 MPCore (with 4 cores each), two ARM Cortex-M4, a graphic processor (PowerVR), a C64x DSP, an *Image/Video/Audio accelerator* (IVA) and an image signal processor. A similar trend can be observed in the Snapdragon family from Qualcomm [220]. The initial S1 SoC included a Central Processing Unit (CPU), a Hexagon DSP and an Adreno Graphics Processing Unit (GPU). The latest S4 processor, announced for the end of 2012, includes 11 PEs: a quad Krait Multicore, an Adreno 320 GPU, 3 Hexagon DSPs, a multimedia processor, audio/video acceleration and an LTE modem. Note that Fig. 1.2 only shows the processing elements that are visible to the user. Such platforms typically include several other smaller processors, for example, inside the modem or in the interconnect fabric.

The OMAP and Snapdragon families are typical examples of how hardware progresses nowadays. Due to high NRE costs, new platforms will more often be an evolution of a previous SoC rather than an entirely new design. In fact, the amount

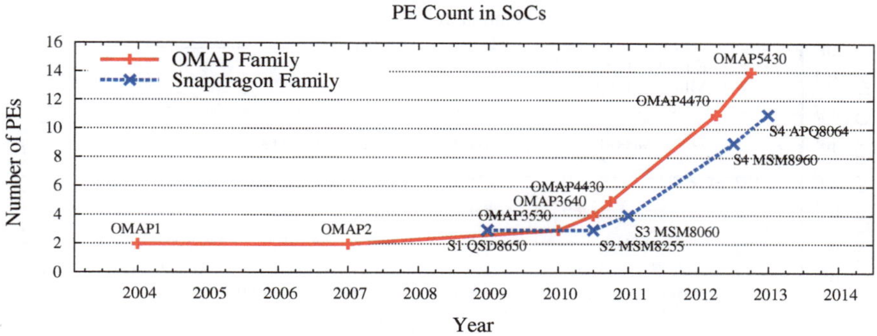

Fig. 1.2 SoC trends: Processing elements in TI OMAP processors [258, 287, 293] and qualcomm snapdragon processors [221, 288]

of new designs stopped growing worldwide by 2005 and has since then decreased [63, 248].

1.1.2 Embedded Software

The tremendous change undergone by embedded hardware is overshadowed by the evolution of embedded software. Advances in computing systems are produced by a combination of the technology *push* and the market *pull*. The former used to dominate advances in the early times, where semiconductor companies produced faster processors and applications followed. Today, it is the market pull in form of end user application demands that drives advances. This phenomenon was true for the PC industry in the late 1990s [280], and is even stronger in the embedded domain, due to its inherent application-specific nature. The embedded hardware advances described in Sect. 1.1 can therefore be explained in part by the software evolution.

In the past, embedded software consisted of simple routines within a single control loop that were often programmed in assembly. Current embedded devices run much more complex applications such as video conferencing and voice recognition (see for example [137]). Additionally, the recent trend towards software-defined communication platforms suddenly increased the amount of software in wireless embedded devices. Such baseband SDR applications have stringent real-time requirements that make them extremely difficult to program.

This impressive change in the kinds of applications that run on embedded devices has been driven by the market pull, where software requirements double every ten months [72]. Increasing requirements translate into an exponential growth in application complexity, measured in *Operations Per Second* (OPS). This trend is shown in Fig. 1.3 for applications from the domains of interest for this book. The complexity of baseband algorithms increases at an impressive rate of an order of magnitude every five years [22], with a similar trend displayed by multimedia algorithms [63,

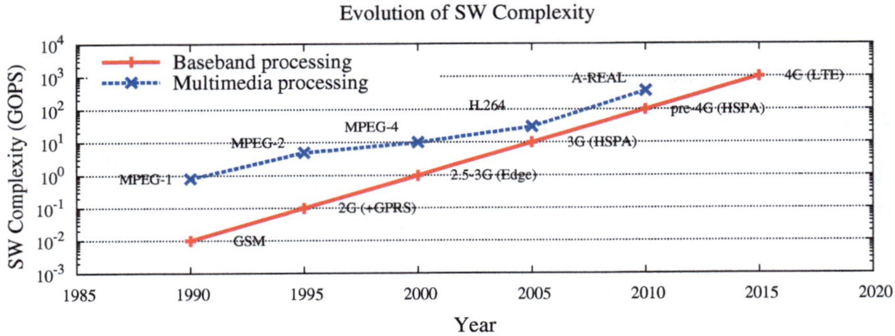

Fig. 1.3 Software complexity trends for baseband and multimedia processing (data from [22, 63, 170, 297]). MPEG-X and H.264 are video standards. A-REAL stands for the augmented reality implementation in [297]

170, 297]. Protocols, baseband and multimedia processing account for above 60 % of the power and over 90 % of the available performance in a 3G phone [22].

Today, the software complexity situation is aggravated by the presence of multiple applications. Since embedded platforms are no longer meant to serve a single application, programmers have to ensure that every application meets its constraints in a multi-application environment. In current platforms, such as the Snapdragon S4, different application classes are by design mapped to different hardware subsytems: *Modem*, *Multimedia* and *Multicore* (for general applications) [220]. It is arguable that this separation in hardware is a consequence of the lack of *multi-application-aware* programming tools that ease the design-time verification effort. Note that a more aggressive resource sharing leads to smaller, more energy-efficient platforms.

With hardware differentiation becoming unaffordable as discussed in Sect. 1.1.1, software is the key product differentiator today. That is the reason why software developers outnumber hardware designers [190] and also why software aspects account for 80 % of the design costs [123]. It is therefore not surprising that today, among all software and hardware design tools, the compiler and the debugger are the most important for embedded designers [199]. Surprisingly, there is no true programming support for heterogeneous MPSoCs.

1.1.3 Design Gaps

The complexity of both embedded hardware and software has given rise to several design gaps, illustrated in Fig. 1.4. On the one hand, hardware designers are not able to exploit all the transistors that can be potentially put into a chip by technology. This trait grows bigger over the years and is denoted as the *hardware productivity gap*. On the other hand, the *software productivity gap* describes the fact that engineers are not able to keep pace with increasing software requirements. As shown in Fig. 1.4, the

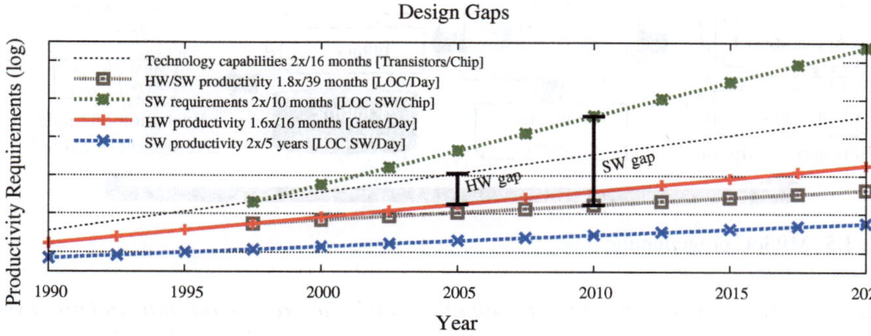

Fig. 1.4 Design gaps (adapted from [72]), with software productivity defined as *Lines of Code* (LOCs) over time. Actual products are a combination of hardware and software, so that the actual HW/SW productivity is better than the pure software productivity

software gap enlarges at a higher pace than the hardware gap. This phenomenon can be observed in recent design reports from semiconductor companies. As an example, the hardware and software complexity increase in television sets from 1992 to 2004 was of 43x and 900x respectively [226]. Similarly, according to [225], the hardware of a 4G modem will be 500 times more complex than that of a 2G modem, whereas the software complexity increase will be of 10000x as shown in Fig. 1.3.

Design gaps have been in the semiconductor industry for decades, driving innovation in the *Electronic Design Automation* (EDA) community. The hardware gap, for instance, would be several orders of magnitude bigger if it were not for *Hardware Design Language* (HDL) compilers. Recent design methodologies and tools, known collectively under the term *Electronic System Level* (ESL), continue to improve productivity and have contributed to narrow the gaps. According to the ITRS, hardware-oriented methodologies would have reduced design costs from 900 to 18 million USD by 2005 [122]. Examples of ESL tools include processor designer kits from *Architecture Description Languages* (ADLs) [252, 255], customizable processor templates [256] and Virtual Platforms (VPs) [250].

The software gap, however, has not been addressed that intensively by today ESL tools [182, 183]. This is especially critical since software has become a dominant design aspect as discussed in Sect. 1.1.2. As a consequence of this lack of efficient programming tools, today only 13% of large software projects are finished on time [226]. This need for a software productivity boost is what motivated the work presented in this book.

1.2 Tool Flow Requirements: Closing the SW Gap

The discussion in the previous section made clear that embedded software has become a bottleneck for most designs. In fact, improving programming of parallel architectures is the matter of the next two major items in the ITRS agenda for the coming five

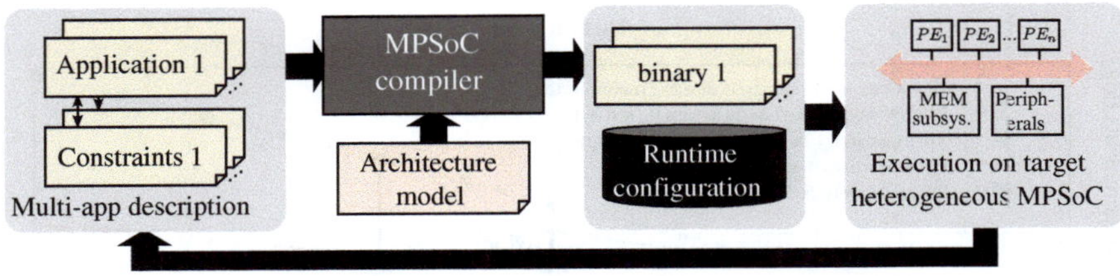

Fig. 1.5 MPSoC programming flow

years, namely *"concurrent software compiler"* and *"heterogeneous parallel processing"* [123]. There are many aspects to consider when attempting to close the software productivity gap, including programming, testing, verification and debugging. This section identifies concrete requirements for an ideal programming methodology that is up to the challenge in the multimedia and baseband domains.

A generic MPSoC programming flow is shown in Fig. 1.5. In this idealized flow, a compiler takes as input a set of applications with their constraints and produces binary images that can be executed on the target platform. The architecture model is an explicit input to the compiler, which implies that the compiler is retargetable, at least to some extent. This allows a single compiler to be used across different generations of a given MPSoC. Apart from the binary images, the compiler also exports a configuration database intended for the runtime system. This database contains configurations for different runtime situations or *scenarios*, where a scenario describes a particular combination of applications running simultaneously. The compiler must ensure that the constraints of each application are respected for every possible scenario. The feedback arrow at the bottom of the figure suggests that the user can modify the applications after observing its behavior, in the same way it is done in a traditional compilation flow.

The programming flow in Fig. 1.5 shares similarities with the well-known *Y-chart* approach [146] used in the system synthesis community. Notice however that there is no feedback to the architecture in the flow. This is because, as discussed in Sect. 1.1.1, with increasing hardware production costs, it will become less common to optimize a complete MPSoC for a given application or application set. This would only become feasible by a breakthrough in platform design methodologies and production techniques. Since it is unlikely for this breakthrough to happen in the next decade, a fixed platform is assumed in the rest of this book.

Not every programming flow with the structure shown in Fig. 1.5 will effectively help to close the software gap. There are several general and practical considerations that have to be taken into account, as discussed in the following.

Applications and Constraints: The MPSoC compiler must support and understand applications of different classes (hard, soft and non real-time). Besides time constraints, it should be possible to specify resource constraints, allowing the programmer to fix a resource type for a given task.

Programming Model: *Abstraction* is one of the best ways of improving productivity, for which the programming flow must support high-level languages. However, notwithstanding how appealing new parallel programming models are for MPSoCs, a compiler must support sequential legacy code, since C remains the most widespread language in the embedded domain [75].[1] For the same reason, new *Domain Specific Languages* (DSLs) should be close to C to ensure acceptability and an easy code migration path. The language should avoid having calls to specific Application Programming Interfaces (APIs), e.g., the *POSIX threads* (Pthreads) API [201, 240] or the *Message Passing Interface* (MPI) [238]. These calls would reduce code portability, and hence software reuse. Implicit parallel programming languages [117] would therefore be preferred.

MPSoC Compiler: As expected from a compiler, functionally correct code generation is its foremost requirement. In the presence of non functional constraints, e.g., real-time, the compiler should provide guarantees, in terms of whether or not the constraints are met, or with which degree of certainty they will be. As a second requirement, the compiler should generate *efficient* code. In traditional compilers, two main metrics have been used to measure efficiency, namely performance and code size. In this context, *efficient execution* refers to providing just enough performance to meet the constraints while minimizing the energy consumption, thereby optimizing the battery lifetime. Additionally, the MPSoC compiler must be equipped with methodologies that retrieve the performance that is typically lost when raising the programming abstraction level. This is specially challenging for high end systems which make heavy use of hardware acceleration. In such cases, a traditional compilation approach would ignore the accelerators and produce inefficient code.

Runtime-awareness: With an architecture-agnostic programming model, the compiler must be aware of the runtime support or even generate the runtime for bare-metal systems. Such *Hardware-dependent Software* (HdS) component is one of the most error-prone, thus, automatic HdS generation would greatly contribute to increase productivity.

Target Architectures: This work is concerned with architectures for multimedia and baseband processing. These architectures do not follow the regular mesh approach of the aforementioned Intel's SCC or Tilera's TILE-Gx100, but are more irregular in terms of processing elements, interconnect and memory architectures.

Some sample MPSoC templates are shown in Fig. 1.6. A generic abstract model can be seen in Fig. 1.6a, with several processors of different types, i.e., *Reduced Instruction Set Computers* (RISCs), DSPs, ASIPs and *Very Large Instruction Word* (VLIW) processors. Two more specific templates are illustrated in Fig. 1.6b, c for multimedia and baseband processing respectively. The multimedia template is a sim-

[1] A recent survey reported that 85 % of the embedded projects in year 2012 were programmed either in C (65 %) or C++ (20 %) [269]. The next two preferred languages were assembly (5 %) and Java (2 %).

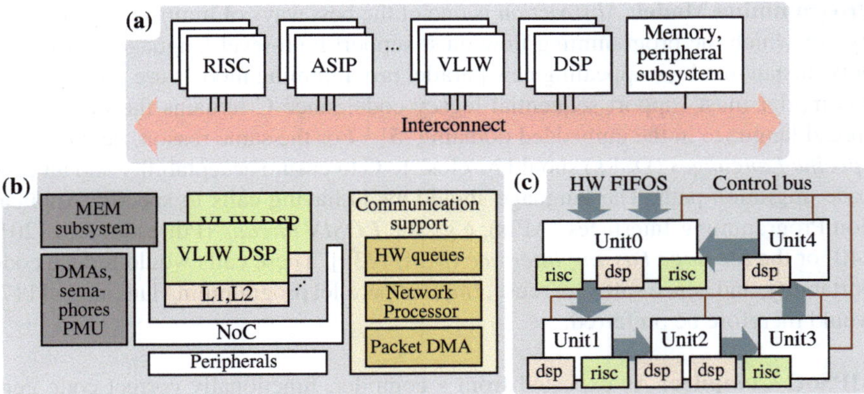

Fig. 1.6 Sample target MPSoCs. **a** Generic MPSoC. **b** MPSoC for multimedia (see Keystone [257]). **c** MPSoC for baseband processing (see BWC200 [67])

plified view of TI's Keystone architecture [257], whereas the baseband template is inspired by BlueWonder's BWC200 [67]. These architectures represent new challenges for the tool flow in Fig. 1.5 since both the processors and the communication architecture are heterogeneous. The template in Fig. 1.6b has a memory hierarchy and hardware support for communication, e.g., *Direct Memory Access* (DMA) peripherals and hardware queues. All elements are interconnected by a *Network on Chip* (NoC) that enables symmetric communication among the VLIW DSPs. The template in Fig. 1.6c is a data streaming architecture with dedicated communication links for data streaming and a control bus for small control messages. This platform features asymmetric data communication among *units*, for the sake of energy efficiency.

ESL Environment: A flawless collaboration of several tools is required to improve the overall programming experience. An MPSoC programming flow must therefore be well integrated with other ESL tools, including frameworks for performance estimation and virtual platforms for software simulation, analysis and debugging.

1.3 MAPS: A Framework for MPSoC Programming

The work presented in this book forms part of a project that attempts to fulfill the requirements presented in Sect. 1.2. The outcome of this project is the so-called *MPSoC Application Programming Studio* (MAPS) [47, 48, 173], developed at the chair for Software for Systems on Silicon (SSS) and partly founded by the German Excellence Cluster for *Ultra-high speed Mobile Information and Communication* (UMIC) [271]. However, most of the methodologies and algorithms in this book are not bound to any specific framework.

MAPS started as a framework for semi-automatic parallelism extraction from sequential legacy code [48]. The goal was to help programmers parallelize legacy C code applications for the relatively new multi-processor platforms of 2007. Later on, with dataflow programming models gaining momentum in the DSP community [239], MAPS was extended to support a programming model based on the *Kahn Process Networks* (KPN) [135] *Model of Computation* (MoC). The parallel language evolved from a rudimentary pragma-based approach, briefly introduced in [173], into what is now called the *C for Process Networks* (CPN) [236]. With this language, MAPS aimed at providing a smooth software migration path for algorithm designers, who are used to write specifications as block diagrams. Being a C-based language, CPN offers a steep learning curve for C programmers. Finally, around the year 2010, initial design considerations were made in order to analyze multiple applications at design time [47].

From the beginning, MAPS followed a holistic approach, covering programming models, simulation (see [49]), performance estimation, code profiling, code analysis, optimization and code generation. MAPS tools are all accessible from an Integrated Development Environment (IDE) based on Eclipse [73].

Today, MAPS attempts to fullfill the requirements listed in the previous section by (1) supporting different kinds of *applications and constraints* as discussed in Chap. 2, (2) providing two fundamentally different *programming models*, namely the C and the CPN languages, (3) including an optimizing *MPSoC compiler* that features the methodologies described in Chaps. 5–8, (4) integrating target-specific backends that are *runtime aware* for the *target architectures* shown in Fig. 1.6 and (5) interoperating with the *ESL environment*, by exporting execution models at different abstraction levels for commercial ESL tools, including Synopsys Platform Architect (PA) [250] and Synopsys MultiCore Optimization technology (MCO) [7].

1.4 Programming Tool Flows

Figure 1.7 shows an overview of the tool flows presented in this book, three for single applications and one for multiple applications. For the sake of clarity, the single application flows are subdivided into three phases: *frontend*, *middle-end* and *backend*. As in traditional compilers, the frontend's task is to produce an abstract representation of the application with rich information to be used in the analysis and optimization phase that takes place in the middle-end. The backend is the phase where a new representation of the input application is generated, usually, in a form that is closer to the target architecture. The main highlights of the tool flows are described in the following.

Sequential Flow: The sequential flow follows a semi-automatic approach for application parallelization. The parallelized version may use the CPN programming model and can be later fed to the parallel flow. The frontend of the flow is based on the *Low Level Virtual Machine* (LLVM) compiler infrastructure [161]. This frontend adds support to ANSI C99 and robustness compared to the previous MAPS frontend,

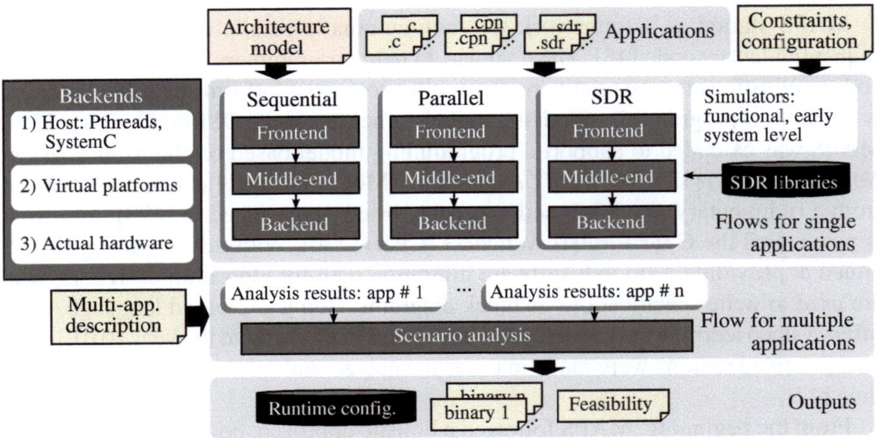

Fig. 1.7 Tool flows overview

based on the LANCE compiler [172]. The middle-end contains parallelism extraction algorithms based on graph clustering techniques and pattern recognition for known parallelism constructs (see [43]). Finally, the backend of the sequential flow produces a parallel representation of the application using Pthreads and MPI APIs. This backend makes the flow more general, compared to the previous MAPS sequential backend which was restricted to the so-called *Tightly Coupled Thread* (TCT) MPSoC [272].

Parallel Flow: The parallel flow determines a mapping of a KPN application onto the target platform. As frontend, the parallel flow uses the CPN compiler of the MAPS framework. Due to the dynamic nature of KPN applications, the middle-end uses *application tracing* to derive a mapping and a schedule (see [40, 44, 45]). The flow has a retargetable backend, with implementations for several commercial MPSoCs. It also includes an infrastructure for virtual platforms that includes generation of debugging scripts (see [42]).

SDR Flow: This book presents a new methodology and a tool flow for demanding applications that require hardware acceleration or optimized software routines. Available routines and accelerators are known to the tool via platform-specific libraries (see right-hand side of Fig. 1.7). In this way, the application code remains portable while retaining the efficiency that would otherwise be lost if a traditional compilation flow is used. The tool flow is applied to applications in the SDR domain (see [41, 46]).

Multi-application Flow: The last tool flow in this book provides support for multiple applications (see [40, 45]). It uses several mapping results from the parallel flow (or SDR flow) for each single application. The scenario analysis phase uses these results to help the user to define a mapping for a set of applications, so that each application in the group meets its constraints.

The programming flows above address software optimization for a given MPSoC. Notice however that the efficiency of the runtime system on the target greatly influences the execution time of a parallel implementation, and hence limits the possibilities of the software optimizer. This book therefore introduces an ASIP for lightweight task management in heterogeneous MPSoCs called OSIP—*Operating System.*

1.5 Synopsis and Outline

This chapter explained the process that made embedded systems what they are today. It showed how programming tools lag behind the rapid software and hardware evolution, opening the software productivity gap. The chapter then listed several aspects that would contribute to narrow the gap and described how they are handled in the MAPS framework. Finally, an overview of the tool flows presented in this book was given.

The remainder of this book is organized as follows. First, a formal definition of the problem and background knowledge are presented in Chap. 2. Chapter 3 provides a survey of other works that aim at solving similar problems to the ones presented in this book. The OSIP runtime manager and its system-level model are described in Chap. 4. The sequential, parallel, SDR and multi-application flows are presented in Chaps. 5–8 respectively. Finally, conclusions are drawn and an outlook is given in Chap. 9.

Chapter 2
Background and Problem Definition

The previous chapter motivated and gave an intuitive description of the problem of mapping multiple applications onto heterogeneous MPSoCs. Three sub-problems were identified, corresponding to three different application types and tool flows: sequential, parallel and SDR. Each of these problems is broad enough to allow for different formulations, interpretations and solutions. To avoid ambiguity, this chapter presents a formal description of the main problem and its sub-problems. The notation and the definitions introduced in this chapter are used throughout this book.

This chapter is organized as follows. Section 2.1 presents a motivational design that serves as running example in this book. Basic terminology and notation are introduced in Sect. 2.2. The multi-application mapping problem and its sub-problems are treated in Sects. 2.3–2.6. The chapter ends with a summary in Sect. 2.7.

2.1 Motivational Example

Consider the problem setup illustrated in Fig. 2.1, in which an embedded software engineer is asked to implement several applications on a heterogeneous MPSoC. Together with these applications, the input specification also provides a list of *use cases* or scenarios. Each use case consist of a list of applications that may run simultaneously, e.g., JPEG and LP-AF in the figure. The designer must then ensure that every use case executes on the target platform without violating any of the constraints.

The applications in Fig. 2.1 are real-life applications, which are used later in the case studies of this book. They are specified in different formats and have different constraints. The following four applications are used for test purposes:

Low-Pass Audio Filter (LP-AF): This is a C implementation of a low pass stereo audio filter in the frequency domain. It reads an input audio signal, transforms it into the frequency domain and applies a digital *Finite Impulse Response* (FIR) filter. The resulting signal is transformed back to the time domain and is forwarded to the

J. Castrillón Mazo and R. Leupers, *Programming Heterogeneous MPSoCs*, DOI: 10.1007/978-3-319-00675-8_2,

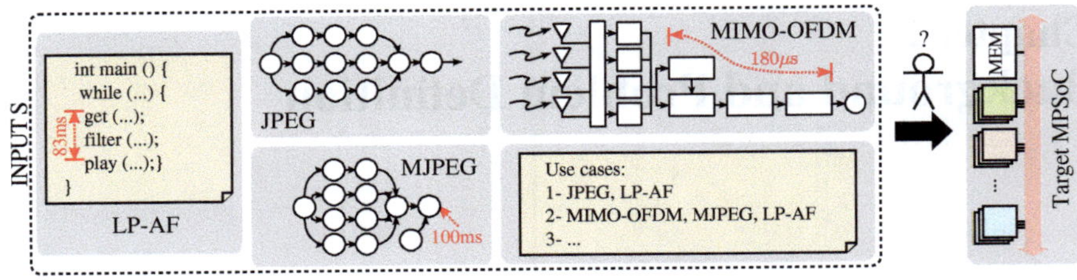

Fig. 2.1 Motivational example

loudspeakers. The application must produce filtered samples every 83 ms for a rate of 192 kbit/s.

JPEG Image Encoding/Decoding: This is a best-effort KPN implementation of the *Joint Photographic Experts Group* (JPEG) encoding and decoding standard [277].

MJPEG Video Decoder: This is a KPN implementation of *Motion JPEG* (MJPEG), which requires a minimum frame-rate of 10 frames-per-second.

MIMO-OFDM Transceiver: This is a generic algorithmic specification of a *Multiple-Input Multiple-Output* (MIMO) *Orthogonal Frequency-Division Multiplexing* (OFDM) digital wireless transceiver [188]. The specification includes a path latency constraint of 180 μs.

Today, provided with this input specification, a designer would follow a manual solution approach, due to the lack of automatic tools. He would start by implementing every single application individually and tuning its performance. Only after studying the behavior of every single application, a designer can start making multi-application design decisions. The following items describe a typical design process.

1. LP-AF (C application): A designer would first look for the best-suited processor to run the application on, which may include code optimizations. If the constraints are not met, the code must be parallelized and mapped to the platform. Several iterations may be required to finally meet the constraints.
2. JPEG (KPN application): For an already parallel application, the designer would start by analyzing and profiling the processes to identify bottlenecks. Typically, after obtaining an initial mapping of the application, several iterations are needed to maximize application performance.
3. MJPEG (KPN application): For this real-time application, instead of maximizing performance, different mappings must be explored until the constraints are met.
4. MIMO-OFDM (algorithmic application): An algorithmic description, e.g., in MATLAB, must be first understood in order to identify potential hardware acceleration or DSP routines. Algorithmic blocks that are not directly supported by the platform must be then implemented, e.g., in assembly or C. Once an implementation is available, the design is similar to that of MJPEG.
5. Multiple applications: With available single-application implementations, every use case must be analyzed individually. If the constraints are violated, the designer must *understand* the reasons and modify the mapping of some of the applications

accordingly. Usually, the configuration of less critical applications is modified first. For example, the best performing mapping of JPEG may block resources that are needed by the LP-AF application.

The steps above represent time-consuming *trial-and-error* tasks, e.g., application understanding, re-coding, debugging, profiling, simulation and testing. The steps change depending on the application type, which motivated the introduction of different tool flows. These intuitive steps underpin the problem definitions given in this chapter and the semi-automatic solutions presented in this book.

2.2 Preliminaries

Before introducing the formal problem definition, this section introduces the fundamentals of mapping and scheduling, basic concepts of source code performance estimation and the notation used throughout this book.

2.2.1 Mapping and Scheduling

Mapping and scheduling are recurring terms in this book. It is therefore important to first intuitively understand the differences between these two processes.

Consider an application composed of several tasks to be executed on an MPSoC. *Scheduling* refers to the process of deciding at which time instant a given task is to be run on a given PE. In *static scheduling*, the time instants are decided at design time, whereas in *dynamic scheduling*, they are decided at runtime. The latter is implemented by *scheduling policies*, which are used during program execution to arbitrate resource sharing. Examples of such policies include First-Come First-Served (FCFS) and priority-based scheduling. In the latter, a numerical value is assigned to every computation. At runtime, the scheduler selects the computation that is ready to execute and has the highest value. In FCFS, in turn, the scheduler executes the computations in the order they were issued.

Mapping refers to the process of assigning a PE to a task. In static or *fixed* mapping the allocation of PEs to tasks is performed at design time, whereas in *dynamic* mapping, tasks are dispatched to PEs at runtime according to a *mapping policy*. Examples of mapping policies include Round-Robin (RR) and Earliest Finishing Time (EFT). In a RR mapper, PEs are selected in a circular fashion, usually for the purposes of load balancing. In EFT, the dispatcher assigns the PE to the task that would execute it the earliest. The mapping and scheduling processes can be either coupled or decoupled. A decoupled mapper dispatches tasks to local schedulers, which then decide when to execute the task. A fixed mapping strategy is an example of such an approach. In a coupled process, both the *where* (mapping) and the *when* (scheduling)

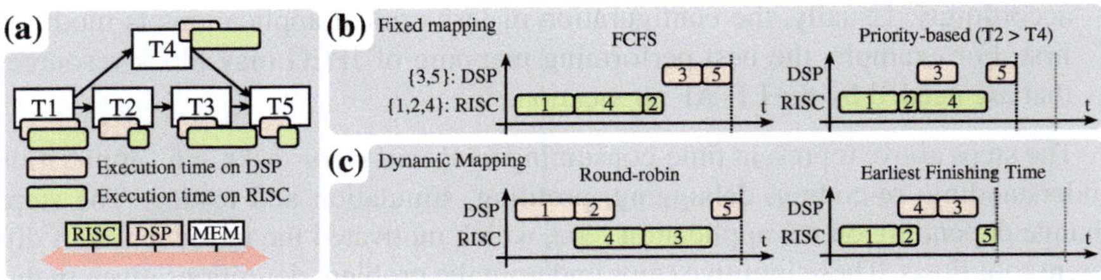

Fig. 2.2 Mapping and scheduling for the DAG problem. **a** Sample application. **b** Fixed mapping. **c** Dynamic mapping

are decided jointly. The EFT mapper is an example of a coupled process. Therefore, if not expressed explicitly, the term mapping refers to mapping *and* scheduling.

Figure 2.2 shows an example to illustrate the mapping and scheduling concepts introduced here. For the sake of clarity, a simple model is used in which an application is represented as a *Directed Acyclic Graph* (DAG), shown in Fig. 2.2a. In the graph, nodes represent tasks and arrows represent precedence constraints, i.e., tasks T2 and T4 can only execute after T1 has finished. The colored bars attached to the nodes represent the execution time of each task on each processor type (RISC and DSP). In this example, the communication cost is ignored. Figures 2.2b and c show the results of the mapping and scheduling process for different configurations in a *Gantt Chart*, where the time is plotted on the horizontal axis and the PEs on the vertical axis. Dotted lines are used to mark the total application execution time, often denoted *makespan*. Figure 2.2b shows possible results of a fixed mapping approach for the two different scheduling policies mentioned earlier. In both configurations, tasks T1, T2 and T4 are mapped to the RISC and tasks T3 and T5 to the DSP. It is assumed that after task T1 finishes, the FCFS scheduler selects task T4 before T2. The priority-based scheduler, in turn, selects T2 due to its priority. The results of a dynamic mapping approach are shown in Fig. 2.2c. In the RR example, it is assumed that the mapper assigns tasks to PEs in a strict order, even if they are busy. It is further assumed that the first assignment is T1 to DSP. Following the same order used for FCFS, the next assignments are T4 to RISC, T2 to DSP, T3 to RISC and T5 to DSP. The EFT mapper, instead, assigns a task to the PE that reports the earliest finishing time. As a result, T4 is assigned to the DSP and T2 to the RISC. This example shows how important the *runtime configuration* is for reducing the application makespan.

In general the problem of determining a static schedule for a DAG that minimizes the makespan on limited resources is NP-Complete [270]. Several heuristics have been therefore proposed to solve it, with a comprehensive survey in [157]. The parallel code problem introduced in Sect. 2.5 is in general more complex than the DAG problem.

There are other flavors of the mapping and scheduling processes which lie in between static and dynamic. For example, in *Quasi Static Scheduling* (QSS) several static schedules are computed for a single application. At runtime, different static scheduling configurations are switched depending on some execution conditions or *scenarios* (see for example [95]). More generally, in *Quasi Dynamic Mapping* (QDM), several configurations are computed at design time which are switched dynamically at runtime. Each configuration may describe either a static or a dynamic mapping and scheduling. This is the approach followed in the multi-application problem.

2.2.2 Sequential Code Performance Estimation

In the DAG example in Fig. 2.2, the timing information of the tasks was assumed to be known. It is represented by colored bars with a length that expresses the sequential execution time of each one of the tasks on each of the two different processor types. It is clear that without this information it would be impossible to obtain the Gantt Charts shown in Fig. 2.2. In the multi-application problem, such information is not assumed to be known, but has to be determined by means of sequential code performance estimation.

Performance estimation is orthogonal to the problems addressed in this book. It is however very relevant both for the quality of the results and the applicability of the framework. While inaccurate estimates may invalidate the results, very accurate ones may be an overkill in early design phases. In such phases, many iterations are needed, hence fast, slightly inaccurate estimates may be preferred. In order to account for different use cases, the following methods for source code performance estimation can be used:

Annotation-based: This is an approach in which the programmer manually specifies the time of a given task on every given processor type. This simple technique is applicable for coarse-grained estimation, in cases where the programmer knows both the application and the target architecture. Within this book, this technique is only used for describing the timing of low-level assembly routines and hardware accelerators (see Chap. 7). For arbitrary source code, this technique could otherwise prove too inaccurate.

Table-based: Table-based performance estimation is a profiling technique based on source instrumentation, similar to the well-known GNU tool *gprof* [101]. For cross-platform estimation, the gprof approach is too coarse-grained since it estimates at the level of a line of code in the original program. In fine-grained approaches, the source code is lowered to a level of elementary operations, e.g., arithmetic and memory load/store. The cost for each of these operations is then read from a table, e.g., μ-Profiler [140]. This approach is applicable for determining application hot spots and for getting an early idea of the application runtime. It is however not very accurate for modeling non-scalar architectures, such as DSPs and VLIWs. It cannot model complex operations such as multiply-accumulate or *Single Instruction Multiple Data* (SIMD). Additionally, it ignores several of the performance variations introduced by the target compiler, e.g., target specific optimizations and register spilling.

Emulation-based: Emulation-based approaches mimic the effects of the target compiler. This allows to account for target-specific optimizations and for the effects of the compiler backend phases: code selection, scheduling and register allocation. The more details of the target architecture and its compiler are emulated, the more accurate the estimation is. At the same time, the more details are emulated the slower the emulation becomes. This approach, in contrast to table-based, can estimate time for irregular architectures with special instructions.

In this book, the *Totalprof* [88] emulation-based profiler is used. TotalProf [88] is a retargetable source code profiler that mimics the effects of compiler backends and several dynamic effects of modern architectures (e.g., cache misses and branch mis-prediction). Compiler effects are modeled by a so-called *virtual backend*. This backend does not produce target binary code, but a host-based low-level code that mimics the target ISA. The timing reported by the backend scheduler is annotated back to this low-level code. As such, the timing of full portions of code is estimated. The granularity can vary from a set of operations, a loop iteration or an entire function, as opposed to single operations as it is done in the table-based approach.

Simulation-based: This approach uses *Instruction Set Simulators* (ISSs) to estimate time for different PEs. This method does not require compiler emulation since the ISSs directly run target binary code, hence the inaccuracies introduced by Totalprof's virtual backend disappear. In the last years, simulation speed has seen a steep increase, from *interpretive* simulation (e.g., Simple Scalar [34]), to *Just-In-Time Cached Compiled* (e.g., in [203]) and *Dynamic Binary Translation* (e.g., in [133]). These advances made it possible to apply simulation-based approaches within programming flows. This book uses the ISSs from Synopsys PA [250] and generated with Synopsys PD [252]. Despite the increase in simulation speed, target binary simulation is still usually slower than executing a host binary with timing annotations.

Measurement-based: This method estimates performance by using measurements from the target MPSoC. This method is potentially the most accurate, but requires the actual board which may not be available in early design phases or may not be accessible to all software developers. Due to its intrusive nature, measurement-based estimates may deviate from real application timing.

Analytical: Analytical or static performance estimation tries to find theoretical bounds on the *Worst-Case Execution Time* (WCET), without actually executing the code.[1] Using compiler techniques, all possible control paths are analyzed and bounds are computed by using an abstract model of the architecture. This task is particularly difficult in the presence of caches and other architectural features that lead to non-deterministic timing behavior. For such architectures, the WCET might be too

[1] This method is not used in this book, but is listed for the sake of completeness.

pessimistic and thus induce bad decisions (e.g., wrong schedules). A good survey of static techniques can be foundin [289].

2.2.3 Notation

Throughout this book, the following conventions are used:

- Scalars are represented by lower case letters in normal type (a, b, c).
- Sets, tuples and sequences are represented by capital letters in normal type (A, B, C). Sometimes, more than one letter is used so that it is easier to recognize the concept that is referred to, e.g., PE could represent a processing element.
- Families of sets are denoted by capital letters in calligraphic type. The same notation is also sometimes used for sets of tuples, in order to distinguish the set name from its elements' names. For example, the set of all processing elements is denoted $\mathscr{PE}$. The power set of set A is represented by $\wp(A)$.
- Functions are represented by Greek letters (α, β, γ). For the sake of readability, some functions are given full names. This eases reading through the pseudocode of some algorithms. $\mathscr{I}(\alpha)$ denotes the image of a function α.
- Usually the context in which an entity (e.g., variable, set and tuple) is contained is given in superscript. Subscripts are used to represent count or to identify an element within a collection. As example, if SOC denotes a given hardware platform, the tuple PE_i^{SOC} represents the i-th PE of the platform SOC. This convention is sometimes ignored if the context is obvious.

2.3 Multi-Application Problem

This section provides a definition of the multi-application problem which is more precise than the intuitive description given in Chap. 1. Before formally stating the problem in Sect. 2.3.2, several fundamental definitions are given in Sect. 2.3.1.

2.3.1 Definitions

The multi-application problem in Fig. 1.5 involves an *architecture*, a set of *applications* and a set of *constraints*. Its solution is a *runtime configuration* data base that specifies how applications are to be run depending on the execution scenario. These concepts are formalized in the following sections.

2.3.1.1 Architecture

A heterogeneous MPSoC is composed of processing elements of different types, storage elements, peripherals and interconnect. This section defines the elements which are relevant to the multi-application mapping problem and introduces their models.

Definition 1 A **processor type** (PT) stands for a processor architecture that is instantiated, possibly multiple times, in the MPSoC. It is not only distinguished by its *Instruction Set Architecture* (ISA), but also by its timing and power characteristics. It can be associated with the reference name of the processor core, e.g., ARM926EJ-S or TI C64x+.

A processor type is modeled as a triple $PT = (\mathscr{CM}^{PT}, X^{PT}, V^{PT})$. $\mathscr{CM}^{PT}$ is an abstract *cost model* that can be seen as a set of functions that associate a numerical value with an application. Examples of these values are execution time and energy consumption. X^{PT} is a set of attributes that give extra information about the core and its supporting runtime system. The attributes model application independent costs, e.g., penalties for a cache misses or boot-up time. The attribute $x_{\text{cs}}^{PT} \in X^{PT}$ is the amount of time required to perform a context switch. The attribute $x_{\text{tasks}}^{PT} \in X^{PT}$ models the maximum amount of tasks that can share this type of resource. V^{PT} is a set of variables $V^{PT} = \{v_1, \ldots, v_n\}$, where v_i is defined over a domain $D_{v_i}^{PT}$. Processor variables as well as other variables defined in this chapter are used to model free parameters that are to be set by the mapping process. For example, the variable $v_{\text{SP}}^{PT} \in V^{PT}$ represents the scheduling policy used by the processor. The domain of this variable describes the policies supported by the processor type, e.g., priority-based or round-robin. The set of all processor types in the MPSoC is denoted $\mathscr{PT}$.

Definition 2 A **processing element** (PE) is an instance of a given processor type and thus inherits its cost model, attributes and variables. The set of all PEs in the MPSoC is denoted $\mathscr{PE} = \{PE_1, PE_2, \ldots, PE_{n_{PE}}\}$. For convenience, let PE_i^v be the i-th PE of type v, $PE_i^v \in \mathscr{PE}$, $v \in \mathscr{PT}$. Additionally, let $\mathscr{PE}^v$ denote the set of all PEs of type v. Consequently, $\mathscr{PE} = \sqcup_{v \in \mathscr{PT}} \mathscr{PE}^v$, with $\sqcup$ being the disjoint union.

Definition 3 A **communication resource** (CR) refers to hardware modules that can be used to implement communication among application tasks. Shared and local memories as well as hardware communication queues are examples of CRs. A CR is modeled as a pair of attributes that describe hardware properties, $CR = (x_{\text{MEM}}^{CR}, x_{\text{CH}}^{CR})$. The amount of memory of the CR is modeled by the attribute x_{MEM}^{CR}, whereas x_{CH}^{CR} models the amount of logical communication channels that can be mapped to the CR.

Definition 4 A **communication primitive** (CP) represents software APIs that can be used to communicate among application tasks in the target platform. It is modeled as a 4-tuple $CP = (PE_i, PE_j, S \subseteq \mathscr{CR}, \mathscr{CM}^{CP})$ that expresses how a task in PE_i can communicate with a task in PE_j, where $i = j$ is allowed. It further expresses that the actual communication uses a subset S of the resources of the platform. This allows to represent a wider range of software communication means, e.g., *Inter-Processor Communication* (IPC) over more than one resource. It is possible to distinguish communication means implemented differently over the same resource, e.g., lock-less queues, semaphore-protected queues and queues with packaging over a single shared memory. $\mathscr{C}\mathscr{M}^{CP}$ is a cost model of the communication primitive that consists of functions that associate communication volume with a numerical value. An example of such a function, denoted $\zeta^{CP}: \mathbb{N} \to \mathbb{N}$, associates a cost in terms of time with any amount of bytes to be transmitted. The set of all CPs in an MPSoC is denoted by $\mathscr{C}\mathscr{P}$.

Definition 5 A **platform graph** (SOC) is a multigraph $SOC = (\mathscr{P}\mathscr{E}, \mathscr{E})$ that represents the target platform. $\mathscr{E}$ is a multiset over $\mathscr{P}\mathscr{E} \times \mathscr{P}\mathscr{E}$ that contains an element $e_{ij} = (PE_i, PE_j)$ for every $CP \in \mathscr{C}\mathscr{P}$. For convenience, let CP_{ij} denote the communication primitive associated with edge e_{ij}.

2.3.1.2 Applications

Applications are the second main input to the flow in Fig. 1.5. Different application types are modeled differently as discussed later in Sects. 2.4–2.6. This section introduces an abstract model that allows to formulate the multi-application problem in a general way.

Definition 6 An **application** (A) is a triple $A = (\mathscr{M}^A, V^A, \mathscr{K}^A)$. $\mathscr{M}^A$ is an abstract model of the application. Depending on the model, three application *types* can be distinguished: sequential, parallel and SDR. V^A is a set of application variables that are given a value within a domain after the mapping process. $\mathscr{K}^A$ is a set of constraints defined on V^A. Depending on the constraints, three application *classes* can be distinguished: hard real-time, soft real-time and non real-time (or best-effort). Hard real-time applications must fulfill every single deadline, whereas an average analysis suffices for soft real-time applications. The distinction between hard and soft real-time serves to configure the runtime system, so that more critical applications are given preference. The set of all applications in the multi-application problem is denoted by $\mathscr{A}$.

Application variables are used to define constraints and measure the quality of the mapping results. The exact semantics of the variables varies with the application type. During the analysis phases, models are extended by adding variables that help to steer the mapping process. Variables can express more than timing characteristics. They can be used to enforce mapping constraints among other constraints as discussed in Sect. 2.3.1.4.

In this book, hard and soft real-time applications are defined as in the wider context of real-time *systems* [150], where the compiler is expected to *guarantee* at design time that the *system* will react within the specified time, provided a load characterization (or *load hypothesis*). As with traditional C programming, it is up to the user to ensure that both the application specification and the input stimuli lead to a high coverage. Additionally, with no assumptions on the target hardware, the tool flows presented in this book cannot claim strong guarantees due to the various sources of timing unpredictability [262]. Finally, this book is not concerned with other aspects that are typically associated with real-time systems, such as failure models and error recovery.

Application variables allow to model the single-application problem as a general *Constraint Satisfaction Problem* (CSP) [267]. Consider an application $A = (\mathscr{M}^A, V^A, \mathscr{K}^A)$ with application variables $V^A = \{v_1, \ldots, v_n\}$, where each variable v_j is defined over a domain $D^A_{v_j}$. A constraint $K^A_i \in \mathscr{K}^A$ is a pair $(S_i \subseteq V^A, R_i)$, with $S_i = \{v_1, \ldots, v_m\}$ a subset of application variables and R_i an m-ary relation ($m = |S_i|$) on $\times_{v \in S_i} D^A_v$. The result of a mapping process can be seen as an *evaluation* of the application variables $\varepsilon : V^A \to \bigcup_{l=1}^n D^A_{v_l}$, with $\forall v_l \in V^A, \varepsilon(v_l) \in D^A_{v_l}$. That is, it assigns a value to every variable in its respective domain. A result satisfies a constraint $K^A_i = (S_i = \{v_1, \ldots, v_m\}, R_i) \in \mathscr{K}^A$ if the variable evaluation belongs to the region defined by R_i, i.e., $(\varepsilon(v_1), \ldots, \varepsilon(v_m)) \in R_i$. Consider as an example the five-task application model in Fig. 2.2. Suppose that the variable set consists of the time stamps at which each task finishes its execution $V^A = \{t_1, \ldots, t_5\}$, with $D^A_{t_i} = \mathbb{N}$ for every variable. A timing constraint k^t_i for each task is then represented as (V^A, R), with $R = \{(v_1, \ldots, v_5) \in \mathbb{N}^5, v_i \leq k^t_i, i \in \{1, \ldots, 5\}\}$.

The CSP described above is accompanied by a single-objective optimization problem that serves to finally select a solution among all feasible ones. Depending on the application class, a different objective is minimized: the application makespan for best-effort applications and the resource usage for real-time applications.

Application models partition the set $\mathscr{A}$ into a sequential subset ($\mathscr{A}^{\text{seq}}$), a parallel subset ($\mathscr{A}^{\text{kpn}}$) and an SDR subset ($\mathscr{A}^{\text{sdr}}$), so that $\mathscr{A} = \mathscr{A}^{\text{seq}} \sqcup \mathscr{A}^{\text{kpn}} \sqcup \mathscr{A}^{\text{sdr}}$. Similarly, application constraints partition the set into subsets of hard real-time applications ($\mathscr{A}^{\text{hrt}}$), soft real-time applications ($\mathscr{A}^{\text{srt}}$) and non real-time applications ($\mathscr{A}^{\text{nrt}}$).

How applications are expected to interact during runtime is specified in the form of a *multi-app* description as shown in Fig. 1.7. This description can be specified in form of a graph or a set of *use cases*, as follows.

Definition 7 An **application concurrency graph** (ACG) is an edge-weighted graph $ACG = (\mathscr{A}, E^{ACG}, W^{ACG})$. An edge $e_{ij} = (A_i, A_j) \in E^{ACG} \subseteq \mathscr{A} \times \mathscr{A}$ indicates that applications A_i and A_j may run simultaneously. The weights on the edge, $W^{ACG}(e)$, serve to mark that certain combinations of applications are more important than others.

Definition 8 A **use case** (UC) represents a set of applications that can run concurrently. It is modeled as a pair $UC = (S^{UC}, p^{UC})$ where $S^{UC} \subset \mathscr{A}$ is the set of applications and p^{UC} is a weighting factor. As with edge weights in the ACG, this factor is used to mark which use cases are more important for the user. It can be thought as the probability that the applications in S^{UC} actually run simultaneously. For convenience, the notation $A \in UC$ is used to represent that $A \in S^{UC}, UC = (S^{UC}, p^{UC})$.

Note that any fully connected subgraph of the ACG, i.e., any clique, determines a use case. Worst-case multi-application scenarios are determined by the maximal cliques of the graph. Let $UCG = (S^{UC} \subset \mathscr{A}, E^{UC} \subseteq E^{ACG}, W^{UC} \subseteq W^{ACG})$ be a subgraph of the ACG, so that S^{UC} is a clique. In this case, the weight of the use case is defined as

$$p^{UC} = \prod_{e \in E^{UC}} W^{UC}(e) \tag{2.1}$$

Equation 2.1 is motivated by the computation of the joint probability of independent events. Assume that the weights on the ACG edges represent the probability of two applications running concurrently, i.e., $\forall e_{ij} = (A_i, A_j) \in E^{ACG}, W^{ACG}(e_{ij}) = P(A_i \wedge A_j)$. Assuming statistical independence, the probability that all applications in a use case run simultaneously is given by

$$P\left(\bigcap_{A_i \in UC} A_i\right) = \prod_{A_i \in UC} P(A_i) = \left(\prod_{e \in E^{UC}} W^{UC}(e)\right)^{\frac{1}{|S^{UC}|-1}} \tag{2.2}$$

where the term in Eq. 2.1 can be identify on the right-hand side of Eq. 2.2.

Definition 9 An **application concurrency set** (ACS) provides an alternative representation to the ACG, where all use cases are explicitly defined in a set $ACS = \{UC_1, \ldots, UC_n\}$.

A graph representation provides a compact way to represent applications' interaction. However, a graph representation does not offer enough control to represent a given list of multi-application scenarios. Consider the case of three applications, where any combination of two applications is allowed. A graph representation would inevitably include a spurious use case with the three applications in it, since the graph is fully connected itself.

2.3.1.3 Performance Estimation Functions

As mentioned in Sect. 2.2.2, performance estimation plays an important role in the mapping process. Timing information for both source code and for data communication is needed. The timing estimators are modeled as functions within the cost model of processor types and communication primitives. In this book, time is measured in terms of cycles of the platform's main clock. Therefore, the cost models of resources that use a different clock have to be scaled accordingly. In this way, the costs are independent of the specific frequency used in the platform, which may be variable.

Consider an application $A = (\mathscr{M}^A, V^A, \mathscr{K}^A)$ to be executed on an $SOC = (\mathscr{PE}, \mathscr{E})$, with processor types $\mathscr{PT} = \{PT_1, \ldots, PT_n\}, PT_i = (\mathscr{CM}^{PT_i}, X^{PT_i}, V^{PT_i})$ and communication primitives $\mathscr{CP} = \{CP_{ij}, CP_{ij} = (PE_i, PE_j, S, \mathscr{CM}^{CP_{ij}})\}$.

Definition 10 Communication cost estimation is the process of estimating the amount of time needed to transfer data by using a given primitive CP_{ij}. The cost is modeled by a function $\zeta^{CP_{ij}} \in \mathscr{CM}^{CP_{ij}}$, $\zeta^{CP_{ij}}: \mathbb{N} \rightarrow \mathbb{N}$. The function is parameterized by three constants: *offset*, *start* and *stair*. The *offset* is a constant overhead in terms of clock cycles that models time spent in synchronization and initialization. Usually these values do not depend on the amount of bytes to be transferred. The *start* parameter models bytes that can be transferred during the initialization phase. The *stair* parameter is a pair of values that determines how the time increases as a function of the transferred bytes. For example, the pair (4, 1) tells that every 4 bytes the communication cost is increased by 1 cycle. Let the stair be described as (s_1, s_2), the communication cost for b bytes is defined by:

$$\zeta^{CP_{ij}}(b) = \begin{cases} offset & \text{if } b < start \\ offset + s_2 \cdot \lceil (b - start + 1)/s_1 \rceil & \text{otherwise} \end{cases} \tag{2.3}$$

This simple model allows to represent communication primitives based on different underlying physical communication channels, such as dedicated hardware queues and buses.

Definition 11 Source code performance estimation is the process of estimating the amount of time needed for a given computation executing on a given processor type PT_i. This is also modeled by functions $\{\zeta_1^{PT_i}, \ldots, \zeta_l^{PT_i}\} \subseteq \mathscr{CM}^{PT_i}$, where each function is defined as $\zeta_j^{PT_i}: S \subseteq \mathscr{M}^A \rightarrow \mathbb{N}$. As an example, let $S = \{T1, \ldots, T5\}$ for the application in Fig. 2.2a. The execution time of the i-th task for the two different processors can then be expressed as $\zeta^{PT_1}(T_i)$ and $\zeta^{PT_2}(T_i)$.

Compared to communication cost estimation, source code performance estimation is a more intricate process. Any of the techniques described in Sect. 2.2.2 can be used for this purpose. The exact definition of the functions $\zeta^{PT_{i_j}}$ depends on the application type (sequential, parallel or SDR).

2.3.1.4 Constraints

Consider an application $A = (\mathscr{M}^A, V^A, \mathscr{K}^A)$ to be executed on an $SOC = (\mathscr{PE}, \mathscr{E})$. Recall that $K_i^A = (S_i \subseteq V^A, R_i)$, with variable $v \in S_i$ defined over a domain D_v^A. Several constraint types are allowed, defined by the nature of v and D_v^A.

Definition 12 Timing constraints are defined over application variables that relate to timing. Depending on the application model, these variables could represent application makespan, task finishing times, etc. Since time is measured in terms of clock cycles, these variables are defined over the natural numbers, i.e., $D_v^A \subset \mathbb{N}$.

Definition 13 Processing constraints are defined over the platform computing resources. They allow to constrain the mapping process, so that tasks are only mapped to a given subset of the PEs, i.e., $D_v^A \subset \mathscr{PE}$, or a given processor type u, i.e., $D_v^A = \mathscr{PE}^u$.

Definition 14 **Memory constraints** are defined over the platform communication resources. They allow to constrain the whole memory requirements of the application and the individual memory requirements of communication channels. For these constraints, $D_v^A \subset \mathbb{N}$.

2.3.1.5 Runtime Configuration

For an application $A = (\mathscr{M}^A, V^A, \mathscr{K}^A)$ and platform $SOC = (\mathscr{PE}, \mathscr{E})$ a configuration describes the results of the mapping process. More formally, let $S_{\text{proc}} \subset \mathscr{M}^A$ denote the model elements that relate to processing (e.g., tasks in the example in Fig. 2.2a). Similarly, let $S_{\text{comm}} \subset \mathscr{M}^A$ denote the elements that relate to communication (e.g., edges in Fig. 2.2a).

Definition 15 A **runtime configuration** (RC^A) is a triple $RC^A = (\mu_p, \mu_c, \mu_a)$ of functions. The mapping of processing to PEs is specified by $\mu_p\colon S_{\text{proc}} \subset \mathscr{M}^A \to \mathscr{PE}$. The mapping of application communication channels is defined by $\mu_c\colon S_{\text{comm}} \subset \mathscr{M}^A \to \mathscr{E}$. Finally, the function μ_a is a mapping of platform variables and application variables to their corresponding domains, i.e., $\forall PE_i^{PT} \in \mathscr{I}(\mu_p), \forall v \in V^{PT}, \mu_a(v) \in D_v^{PT}$, similarly, $\forall v \in V^A, \mu_a(v) \in D_v^A$. In other words, μ_a fixes the runtime configuration for all processing elements (e.g., scheduling policy) and remaining free variables of the application (e.g., tasks' finishing time and memory consumption). A set of runtime configurations for a given application A is denoted $\mathscr{RC}^A$.

Definition 16 A runtime configuration is **valid** if the following conditions are met:

1. Application constraint satisfaction: As mentioned before, a mapping can be seen as an evaluation of the application variables, $\varepsilon\colon V^A \to \bigcup_{l=1}^n D_{v_l}^A$. The application constraints are satisfied if

$$\forall K_i^A = (S_i = \{v_1, \ldots, v_m\}, R_i) \in \mathscr{K}^A, (\varepsilon(v_1), \ldots, \varepsilon(v_m)) \in R_i \qquad (2.4)$$

2. Platform constraint satisfaction: The assignment must not violate the attributes provided by the platform resources. For example, memory allocation on communication primitives must not exceed the maximum memory available in the CR, nor the amount of logical channels ($x_{\text{MEM}}^{CR}, x_{\text{CH}}^{CR}$ in Definition 3).
3. Logical validity: For every application communication channel $C \in S_{\text{comm}}$, with producer task $\text{src}(C) \in S_{\text{proc}}$ and consumer $\text{dst}(C) \in S_{\text{proc}}$, it must hold:

$$\text{src}(\mu_c(C_i)) = \mu_p(\text{src}(C)) \wedge \text{dst}(\mu_c(C_i)) = \mu_p(\text{dst}(C)) \qquad (2.5)$$

The last condition ensures that tasks that communicate with each other run on processors that can communicate. As mentioned in Sect. 2.3.1.2, from all valid configurations the single application flows would select the one that minimizes the makespan (for best-effort) or the resource usage (for real-time).

Definition 17 A **use case runtime configuration** ($\mathscr{RC}^{UC}$) is a set of runtime configurations. Given a use case $UC = (S^{UC} = \{A_1, \ldots, A_n\}, p^{UC})$, $\mathscr{RC}^{UC} = \{RC^{A_1}, \ldots, RC^{A_n}\}$. A $\mathscr{RC}^{UC}$ is **valid** if each $RC^{A_i} \in \mathscr{RC}^{UC}$ is valid and if they are *jointly* valid, i.e., the validity is not violated by mapping all applications in the UC to the platform.

2.3.2 Problem Statement

With the previous definitions it is now possible to state the multi-application problem.

Definition 18 Multi-application problem: Given an application set $\mathscr{A}$, a graph $ACG = (\mathscr{A}, E^{ACG}, W^{ACG})$ (or a set ACS) and a target platform model $SOC = (\mathscr{PE}, \mathscr{E})$, find a valid use case runtime configuration $\mathscr{RC}_i^{UC}$ for every use case $UC_i \in ACS$.

As can be inferred from this definition, solving the multi-application problem is a challenging task. It is therefore divided into sub-problems that first address single application problems. A logical division arises from the different application models. The sequential problem for $A \in \mathscr{A}^{\text{seq}}$ is stated in Sect. 2.4, the parallel problem for $A \in \mathscr{A}^{\text{kpn}}$ in Sect. 2.5 and the SDR problem for $A \in \mathscr{A}^{\text{sdr}}$ in Sect. 2.6.

2.4 Sequential Code Problem

As mentioned in Sect. 2.1, a possible approach for mapping a sequential application ($A \in \mathscr{A}^{\text{seq}}$) consist in selecting the PE on which the application meets the constraints. An automatic flow following this approach would have two disadvantages. First, it would fail if no PE can run the application within its time constraint. Second, it would not profit from the parallel computing power of the target MPSoC.

The sequential problem is therefore addressed as a parallelism extraction problem from a sequential C program. Once a parallel specification is obtained, it becomes a parallel mapping problem. Before formally stating the problem, the next section gives background knowledge on compiler technology and parallelism extraction.

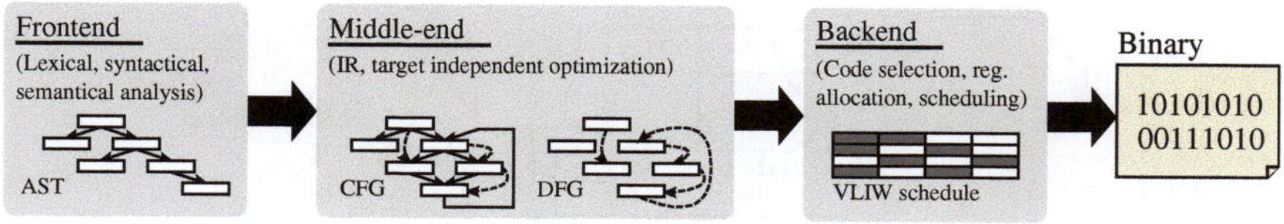

Fig. 2.3 Phases in a traditional uni-processor compilation flow

2.4.1 Compilers and Parallelism Extraction

A compiler is typically divided into three phases: The *frontend*, the *middle-end* and the *backend* [3, 193] (see Fig. 2.3). The frontend checks for the lexical, syntactic and semantic correctness of the application. The main data structure delivered by the frontend is the so-called *Abstract Syntax Tree* (AST). An AST is a structured representation of the application according to the syntax of the underlying programming language. It is used for program analysis and for code selection in the backend phase of the compiler. ASTs provide a language-independent abstraction of the application known as *Intermediate Representation* (IR). The middle-end performs different analyses on the IR that enable target-independent optimizations that mainly aim at improving the performance of the generated code. Two data structures are key for the optimizations in the middle-end, namely the *Control Flow Graph* (CFG) and the *Data Flow Graph* (DFG). Finally, the backend produces target binary code. The backend itself is also divided into phases. Typical backend phases include *code selection*, *register allocation* and *instruction scheduling*. These phases are machine dependent and therefore require a model of the target architecture.

For the purpose of parallelism extraction, the compiler phases perform conceptually the same tasks but at a different, *coarser* level. While a traditional compiler optimizes for *Instruction Level Parallelism* (ILP), a coarser parallelism is sought when targeting a multi-processor platform. Therefore, this section introduces some compiler concepts that are needed to obtain an application model ($\mathscr{M}^A$) that is suitable for parallelism extraction. A more comprehensive treatment of compilers can be found in [3, 193]. This section also introduces basic concepts of parallelism.

2.4.1.1 Intuitive Presentation

Consider the sample application presented in Fig. 2.4a. Suppose that the call to function `f2` does not modify the contents of array `A`, so that the first and the second loops are independent. Notice also that in the second loop, the variable `s` is only updated every four iterations. Based on these observations it is possible to represent the application as the abstract graph shown in Fig. 2.4b, where some details were omitted for the sake of clarity. In the graph, solid lines represent *control flow* induced by the sequential specification, and dashed lines represent *data dependencies*. Control edges are annotated with the amount of times each node follows its predecessor

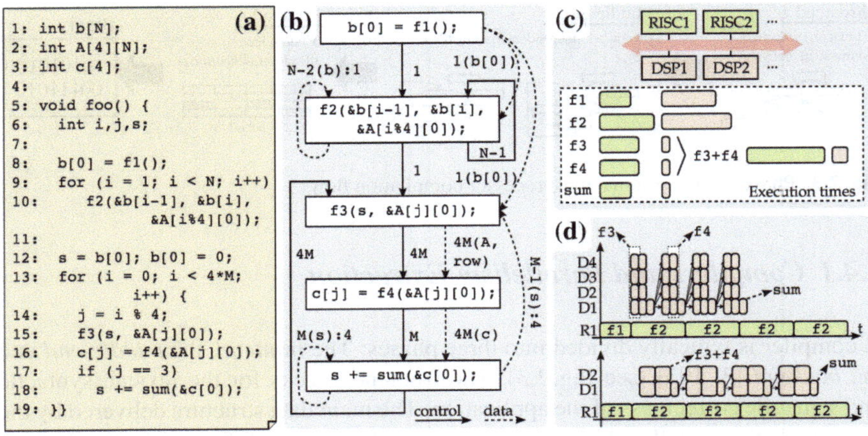

Fig. 2.4 Sequential code problem. **a** Original C code. **b** Control and data flow representation. **c** Timing information. **d** Two possible mappings

in an execution of the application. Besides the edge count, data edges are annotated with the variable that produced the dependency. In the case of loops, an additional *dependency distance* is annotated after the colon. The distance represents the amount of iterations between a write to and a read from a variable. From this graph, it follows that after `f1` returns, both loops can execute in parallel. Furthermore, every four iterations of the functions `f3` and `f4` can be run in parallel as well. This graph now exposes the parallelism that was originally hidden in the application. Now assume a platform with two different processor types and the execution times represented by colored bars in Fig. 2.4c. With this platform information, two possible schedules are shown in Fig. 2.4d. In the schedules it is assumed that `N = 4` and `M = 4`. The upper schedule exploits all the parallelism exposed by the application. However, it is not necessarily better than the second schedule, which produces the same application makespan while only using three PEs.

This example shows the main steps needed to address the sequential problem. The steps will be further discussed in the following sections. (1) Obtain a graph representation of the application that displays all kinds of dependencies (see Sects. 2.4.1.2–2.4.1.4). (2) Discover hidden parallelism in the graph (see Sect. 2.4.1.5). (3) Finally, select a parallel specification that better suits the target platform (see Sect. 2.4.1.6).

2.4.1.2 Control Flow Analysis

The example in Fig. 2.4 intuitively introduced the concept of the control flow of a program. The following definitions formalize the concepts.

Definition 19 An **intermediate representation** (IR^A) of an application is a pair $IR^A = (S^A_{\text{stmt}}, S^A_{\text{f}})$, where $S^A_{\text{stmt}} = \{s^A_1, \ldots, s^A_n\}$ is the set of all IR-statements. IR-

(a)

```
1: int main()
2: {
3:    int s=0, i;
4:    for (i = 0; i < 10; i++)
5:    {
6:      s += 1;
7:    }
8:    return 0;
9: }
```

(b)

```
1:  int main() {
2:     int s_3i, i_4;
3:     int t1, t2, t3, t4, t5;
4:     s_3 = 0; i_4 = 0;
5:     t1 = i_4 < 10;
6:     t5 = !t1;
7:     if (t5) goto LL1;         BB1
8:  LL3:
9:     t4 = s_3 + i_4;
10:    s_3 = t4;                 BB2
11: LL2:
12:    t2 = i_4;
13:    t3 = t2 + 1;
14:    i_4 = t3;
15:    t1 = i_4 < 10;
16:    if (t1) goto LL3;
17: LL1:
18:    return 0; }               BB3
```

Fig. 2.5 Example of three-address code representation. **a** Original C code. **b** 3AC representation

statements are a *Three-Address Code* (3AC) representation of the original source code statements. $S_{\mathrm{f}}^{A} = \{f_1^A, \ldots, f_m^A\}$ is the set of all functions defined in the application. Each function $f \in S_{\mathrm{f}}^{A}$ maps to a subset of statement $S_{\mathrm{stmt}}^{f^A} \subset S_{\mathrm{stmt}}^{A}$.

An example of a 3AC representation of a simple application is shown in Fig. 2.5. The example shows how C constructs are lowered into simpler statements, labels and `goto` statements. The for loop in Lines 4–7 in Fig. 2.5a is implemented in Lines 7–16 in Fig. 2.5b. The shaded regions in Fig. 2.5b are *Basic Blocks* (BBs) which are basic analysis elements in compiler theory.

Definition 20 A **basic block** (BB^A) of an application A is a maximal sequence of consecutive IR-statements $BB^A = (s_1^{BB^A}, \ldots, s_n^{BB^A})$ in which flow of control enters at the beginning and leaves at the end without halt or possibility of branching except at the end [3]. The first statement, $s_1^{BB^A}$, is called the leader of the basic block. The set of all basic blocks in a function f^A is denoted by $\mathscr{BB}^{f^A}$ and in an application A by $\mathscr{BB}^A$.

In the example in Fig. 2.5b, three basic blocks are defined by the statements in Lines 2–7, Lines 9–16 and Line 18, represented by shaded areas. Note that label LL2 does not define a new basic block, since this label is never jumped to. Which basic blocks are executed is controlled by conditions (see for example Line 7 in Fig. 2.5b). This is formally modeled as *control dependencies*.

Definition 21 **Control dependence** (δ^c) is a relation between two statements. A statement s_2^A is control dependent on a statement s_1^A, denoted $s_1^A \, \delta^c \, s_2^A$, if its execution depends on the execution of s_1^A.

In the example in Fig. 2.5b, $s_7 \, \delta^c \, s_9$ and $s_7 \, \delta^c \, s_{18}$, where statements are identified by the lines they appear on. The control dependence relation can be extended to basic blocks and ignored for statements inside a basic block, since they are always executed if the first statement is executed. For example, $BB_1 \, \delta^c \, BB_2$ and $BB_1 \, \delta^c \, BB_3$. Dependencies at the level of basic blocks are easily visualized by means of a CFG.

Definition 22 A **control flow graph** (CFG^{f^A}) of a function $f^A \in S_{\mathrm{f}}^{A}$ is a directed graph $CFG^{f^A} = (\mathscr{BB}^{f^A}, E_c^{f^A})$ in which there is a control edge $e_{ij} =$

$(BB_i^A, BB_j^A) \in E_c^{f^A} = \mathcal{BB}^{f^A} \times \mathcal{BB}^{f^A}$ if $BB_i^A \ \delta^c \ BB_j^A$. For a node $BB^A \in \mathcal{BB}^{f^A}$, $\mathrm{Succ}(BB^A)$ and $\mathrm{Pred}(BB^A)$ are the sets of direct successors and predecessors in the graph, i.e., $\mathrm{Succ}(BB^A) = \{v \in \mathcal{BB}^{f^A}, (BB^A, v) \in E_c^{f^A}\}$ and $\mathrm{Pred}(BB^A) = \{v \in \mathcal{BB}^{f^A}, (v, BB^A) \in E_c^{f^A}\}$.

In order to discover loops and find regions in a CFG, the concepts of *dominance* and *postdominance* are important. Given two nodes $n_1, n_2 \in \mathcal{BB}^{f^A}$, it is said that n_1 dominates n_2, denoted n_1 dom n_2, if every control path that reaches n_2 from the *start* node must go through n_1. Similarly, the node n_2 postdominates n_1, denoted n_2 pdom n_1, if every control path starting from n_1 and ending in the *end* node must go through n_2. The start and end nodes are special nodes that are usually added to the CFG of a function to ease analysis. This is simply done by adding edges from the start node to every node without predecessors, and from every node without successors to the end node.

2.4.1.3 Data Flow Analysis

Data Flow Analysis (DFA) serves to gather information at different program points, e.g., about available defined variables (*reaching definitions*) or about variables that will be used later in the control flow (*liveness analysis*). For the purpose of parallelism extraction, it is crucial to know where data is produced and used. Only with this information it is possible to insert correct communication statements if a portion of the application is offloaded to a different PE. It also serves to know if it is beneficial at all to parallelize a computation. In the example in Fig. 2.4b the backward edge to function `f2` prevents the first loop of the program to be parallelized (see the Gantt Charts in Fig. 2.4c). The following definitions serve to obtain a graph representation of a function, similar to the one in Fig. 2.4b.

Definition 23 A **data dependence** $(\delta^f, \delta^o, \delta^a)$ between two statements s_1^A, s_2^A indicates that s_2^A cannot be executed before s_1^A. There are three different kinds of dependencies:

- *Read After Write* (RAW, also *true/flow dependence*): Statements s_1^A and s_2^A are RAW-dependent, denoted $s_1^A \ \delta^f \ s_2^A$, if s_1^A writes a resource that s_2^A reads thereafter.
- *Write After Write* (WAW, also *output dependence*): Statements s_1^A and s_2^A are WAW-dependent ($s_1^A \ \delta^o \ s_2^A$) if s_2^A modifies a resource that was previously modified by s_1^A.
- *Write After Read* (WAR, also *anti-dependence*): Statements s_1^A and s_2^A are WAR-dependent ($s_1^A \ \delta^a \ s_2^A$) if s_2^A modifies a resource that was previously read by s_1^A.

Definition 24 A **data flow graph** (DFG^{f^A}) of a function $f^A \in S_f^A$ is a directed multigraph $DFG^{f^A} = (S_{\mathrm{stmt}}^{f^A}, E_d^{f^A})$ where $S_{\mathrm{stmt}}^{f^A}$ is the set of statements of the

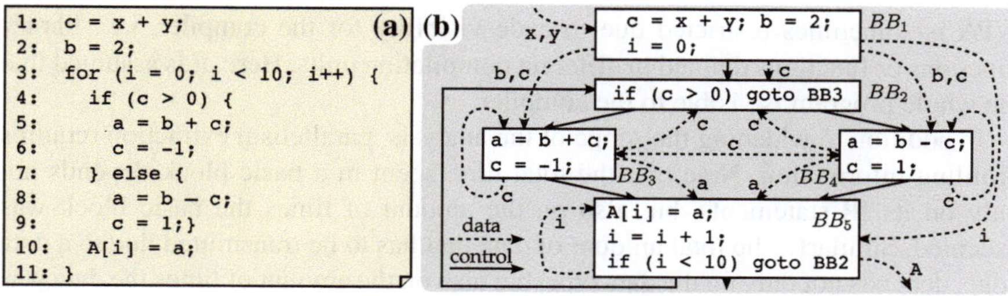

Fig. 2.6 CDFG Example. **a** Original C code. **b** Control and data flow representation

function and $E_d^{f^A}$ is a multiset on $S_{\text{stmt}}^{f^A} \times S_{\text{stmt}}^{f^A}$. There is a data edge $e_{ij} = (s_i^A, s_j^A) \in E_d^{f^A}$ if $s_i^A\, \delta^{f,o,a} s_j^A$. Edges may be labeled by the kind of dependency they represent.

Definition 25 A **control-data flow graph** ($CDFG^{f^A}$) of a function $f^A \in S_{\text{f}}^A$ is an annotated directed multigraph $CDFG^{f^A} = (\mathscr{BB}^{f^A}, E_c^{f^A} \cup E_d^{f^A*}, \text{var}_{\text{size}})$ where $E_d^{f^A*}$ is the set of data flow edges $E_d^{f^A}$ *summarized* at the basic block level. $\text{var}_{\text{size}}: E_d^{f^A*} \rightarrow \mathbb{N}$ is a function that returns the size in bytes of the data type associated with the variable that caused the data flow edge.[2]

For parallelism extraction, every function in a sequential application can be modeled as a CDFG (or similar graph representations). An example of a CDFG is shown in Fig. 2.6b for the code snippet in Fig. 2.6a. For clarity, the data edges in the CDFG are annotated with the variable that originates the dependency instead of the variable size. Notice that the value of variable `c` in Line `4` could have been defined in Line `1`, Line `6` or Line `9`. That is the reason for the three incoming edges to BB_2 labeled with variable `c`. A similar analysis can be performed for the rest of the edges in the figure.

For the C language, the problem of determining all data dependencies statically at compile time is NP-complete and in some cases undecidable. This is due to the use of pointers [111] and indexes to data structures that can only be resolved at runtime. For this reason, dependencies are sometimes tracked at runtime to get less conservative data dependency edges, allowing a more aggressive parallelism extraction. This kind of analysis is known as dynamic DFA and is employed in the MAPS framework.

2.4.1.4 Inter-procedural Analysis and Profiling

Until now, there is a fairly compact representation of functions within an application. However, parallelism extraction requires a wider analysis scope, often regarded as *Whole Program Analysis* (WPA). WPA is usually enabled by means of a *Call Graph* (CG) that expresses how functions are called to make up an application. In practice,

---

[2] In this book, it is assumed that the size of data types does not depend on the processor type the code is compiled to run on. In reality, the actual size depends on the individual target compilers.

WPA is sometimes restricted due to code visibility for the compiler, i.e., library functions or functions defined in different compilation units. Here, it is assumed that the whole program is visible to the compiler.

In addition to widening the scope of the analysis, parallelism extraction requires profiling information. Note that the total time spent in a basic block depends not only on its IR-statements, but also on the amount of times the basic block was executed. Similarly, the total amount of data that has to be transmitted due to a data edge, depends not only on the data type, but also on the amount of times the data was actually produced and used. As an example, consider the information annotated to the edges in the example in Fig. 2.4b. The CG and the profiling concepts are formalized in the following.

Definition 26 A **call graph** (CG^A) of an application $A \in \mathscr{A}^{\text{seq}}$ is a directed multi-graph $CG^A = (S_{\text{f}}^A, E_{\text{cg}}^A, S_{\text{stmt}}^A, \sigma^A)$. E_{cg}^A is a multiset of edges over $S_{\text{f}}^A \times S_{\text{f}}^A$, where $e_{ij} = (f_i^A, f_j^A) \in E_{\text{cg}}^A$ expresses that function f_i^A calls f_j^A. S_{stmt}^A is the set of all statements in the application (recall Definition 19). Finally, the function σ^A associates every edge with a *call site*, i.e., the exact statement that calls the function. It is therefore defined as $\sigma^A\colon E_{\text{cg}}^A \to S_{\text{stmt}}^A$. Naturally, $\sigma^A(e_{ij}) \in S_{\text{stmt}}^{f_i^A}$.

Definition 27 A **sequential application element set** ($\mathscr{S}\mathscr{E}^A$) is a set that contains all *elements* of a sequential application. An application element can be a statement, a basic block, a control edge, a data edge or a function. Let $E_c^A = \bigcup_{f^A \in S_{\text{f}}^A} E_c^{f^A}$ and $E_d^A = \bigcup_{f^A \in S_{\text{f}}^A} E_d^{f^A *}$ be the sets of all control and data edges in the application. The set of all elements is defined as $\mathscr{S}\mathscr{E}^A = S_{\text{stmt}}^A \cup \mathscr{B}\mathscr{B}^A \cup E_c^A \cup E_d^A \cup S_{\text{f}}^A \cup E_{\text{cg}}^A$.

Definition 28 A **sequential profile** (π^A) of an application $A \in \mathscr{A}^{\text{seq}}$ is a function $\pi^A\colon \mathscr{S}\mathscr{E}^A \to \mathbb{N}$ that returns the amount of times a given application element was executed during a profiling run.

Recall the general definition of an application $A = (\mathscr{M}^A, V^A, \mathscr{K}^A)$ in Definition 6. With the definitions presented in this section, it is now possible to formally define the application model used in the sequential code problem.

Definition 29 A **sequential application model** ($\mathscr{M}^A$) for an application $A \in \mathscr{A}^{\text{seq}}$ is a triple $\mathscr{M}^A = (\mathscr{S}\mathscr{E}^A, CG^A, \pi^A)$. Let the notation $e \in \mathscr{M}^A$ refer to $e \in \mathscr{S}\mathscr{E}^A$.

Recall the source code performance estimation functions in Definition 11, $\zeta_j^{PT_i}\colon S \subseteq \mathscr{M}^A \to \mathbb{N}$ for processor type PT_i. It is now possible to refine this general definition into $\zeta_j^{PT_i}\colon S_{\text{stmt}}^A \cup \mathscr{B}\mathscr{B}^A \cup S_{\text{f}}^A \to \mathbb{N}$, where the index j refers to the different means for obtaining the estimate (see Sect. 2.2.2). In particular, the table-based performance estimation can be represented by a function $\zeta_{\text{tb}}^{PT_i}\colon S_{\text{stmt}}^A \to \mathbb{N}$. The *static cost* of a basic block BB^A can be then computed as $\zeta_{\text{tb}}^{PT_i,\text{st}}(BB^A) = \sum_{s^A \in BB^A} \zeta_{\text{tb}}^{PT_i}(s^A)$. By using the profiling information, it is also possible to compute the *dynamic cost* of a basic block BB^A as $\zeta_{\text{tb}}^{PT_i,\text{dy}}(BB^A) = \pi^A(BB^A) \cdot$

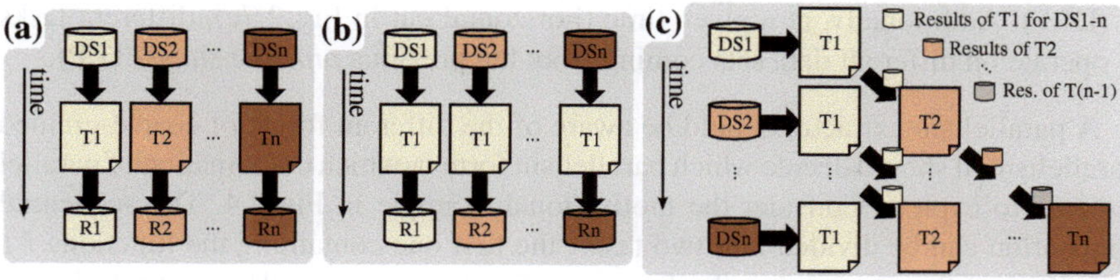

Fig. 2.7 Forms of parallelism. Different tasks (T1, T2,...Tn), their corresponding input data set (DS) and output results (R) are represented with different colors. **a** TLP. **b** DLP. **c** PLP

$\zeta_{\text{tb}}^{PT_i,\text{st}}(BB^A)$. Similarly, the estimation of the total time spent in a function f^A, can be computed as $\zeta_{\text{tb}}^{PT_i,\text{dy}}(f^A) = \sum_{BB^A \in \mathscr{BB}^{f^A}} \zeta_{\text{tb}}^{PT_i,\text{dy}}(BB^A)$. The average cost, per function call, can be also easily computed as $\zeta_{\text{tb}}^{PT_i,\text{dy}}(f^A)/\pi^A(f^A)$. With this simple model, it is possible to obtain the execution times that were assumed given in the example in Fig. 2.4c.

2.4.1.5 Forms of Parallelism

The sequential application model provides the information that is needed to extract parallelism. There are different kinds of parallelism that can be hidden in an application. While a traditional compiler focuses on exploiting *Instruction Level Parallelism* (ILP), parallelism extraction for MPSoCs happens at a coarser level. The most prominent types of coarse parallelism are illustrated in Fig. 2.7 and are described in the following.

- *Task (or functional) Level Parallelism* (TLP): In TLP, a computation is divided into multiple tasks that operate in parallel on different *data sets*, as illustrated in Fig. 2.7a. Tasks may have dependencies to each other, but once a task has its data ready, it can execute in parallel with the already running tasks in the system. The DAG example in Fig. 2.2 exposes TLP. This kind of parallelism is a form of *Multiple Instruction Multiple Data* (MIMD) [87].
- *Data Level Parallelism* (DLP): In DLP, a computation is replicated into several equal tasks that operate in parallel on different data sets, as illustrated in Fig. 2.7b. This kind of parallelism can be seen as a generalization of SIMD [87]. DLP can be found in multimedia applications, where a decoding task performs the same operations on different portions of an image or video (e.g., *macro blocks* in H.264).
- *Pipeline Level Parallelism* (PLP): In PLP, a computation is broken into a sequence of tasks which are repeated for different data sets as shown in Fig. 2.7c. This kind of parallelism can be seen as a generalization of software pipelining [159]. PLP can be seen as a mixture of DLP and TLP. Every diagonal sequence of tasks in Fig. 2.7c perform the same computation (i.e., T1, ... Tn) on different data sets

(DLP). Additionally, at a given time (horizontal cut in Fig. 2.7c), different tasks operate on different data sets coming from the previous *pipeline stage* (TLP).

A parallelism extractor should be aware of the different forms of coarse-grained parallelism. It should decide which parallelism form or which combination of parallel patterns to exploit. Consider the motivational example in Fig. 2.4. The sequential application can be divided into two tasks, the first one containing the functions `f1` and `f2`, and the second one the functions `f3`, `f4` and `sum`. These two tasks are an example of a TLP partition of the original application. The second task can be partitioned as well, with two possibilities shown in Fig. 2.4d, being both an example of DLP.

2.4.1.6 Parallelism Extraction

Given a sequential application model $\mathscr{M}^A = (\mathscr{S}\mathscr{E}^A, CG^A, \pi^A)$ where functions are modeled as graphs (e.g., $CDFG^{f^A}$), parallelism extraction is usually implemented by means of *graph clustering* or *graph partitioning*. A good introduction to the graph clustering terminology and a survey of graph clustering methods can be found in [230]. Two good examples of graph clustering algorithms are *METIS* [158] and *MCL* [69]. Fundamental data clustering techniques are surveyed in [127]. This section only focuses on the definitions needed for the problem statement.

Definition 30 A **clustering of a graph** $G = (V, E)$ is a collection of non-empty sets $\mathscr{C}^G = \{C_1, \ldots, C_k\}$ such that $\bigcup_{i=1}^{k} C_i = V$. An element $C_i \in \mathscr{C}^G$ is called a cluster. Each subset $C_i \in \mathscr{C}^G$ has a corresponding *induced subgraph* $G_i = (C_i, E_i)$, with $E_i = \{e = (u, v) \in E, u, v \in C_i\}$.

Definition 31 A **partition of a graph** $G = (V, E)$ is a clustering $\mathscr{C}^G = \{C_1, \ldots, C_k\}$ where the sets $C_i \in \mathscr{C}^G$ are pairwise disjoint, i.e., $\forall i, j, i \neq j, (C_i \cap C_j = \emptyset)$.

Definition 32 A **hierarchical clustering/partition of a graph** $G = (V, E)$ is a finite ordered list of clusterings/partitions $(\mathscr{C}_1^G, \ldots, \mathscr{C}_l^G)$ where each $\mathscr{C}_{i+1}^G, 1 \leq i < l$ can be obtained by joining elements of $\mathscr{C}_i^G$.

Definition 33 A **partitioned graph** of a graph $G = (V, E)$ and partition $\mathscr{C}^G$ is a graph $G' = (V', E')$. There is a node $v'_i \in V'$ for every set $C_i \in \mathscr{C}^G$, and there is an edge $e'_{ij} = (v'_i, v'_j) \in E'$ if there was an edge $e_{ij} = (v_i, v_j) \in E$, with $v_i \in C_i$ and $v_j \in C_j$. A partitioned graph is also regarded as the induced graph of a partition.

A partitioned graph explicitly exposes TLP within a function. Notice however, that other forms of parallelism are not explicit. In order to expose PLP and DLP, the graph has to include additional information about the nature of the cluster. For example, how many versions of the cluster can be started in parallel in the case of DLP. Similarly, a sequence of clusters may expose PLP, which has to be additionally annotated. These parallelism patterns are added to the graph as annotations.

Definition 34 A **parallel annotation** (PA^C) of a cluster C in a hierarchical, partitioned graph contains information about the different forms of parallelism that can be exploited. It is generally defined as a triple $PA^C = (\{\text{DLP}, \text{PLP}\}, X^{PA^C}, V^{PA^C})$. The first element distinguishes between DLP and PLP. Note that TLP is not annotated, since it is expressed by default. X^{PA^C} is a set of attributes associated with the annotation. V^{PA^C} is a set of variables $\{v_1, \ldots, v_k\}$ associated with the annotation, where a variable v_i can take a value within a domain $D_{v_i}^{PA^C}$.

The definition of a parallel annotation becomes clearer if the two cases (DLP, PLP) are analyzed separately. In the case of DLP, the attribute set is empty, while the variable set contains a single variable that specifies how many copies of the task are to be generated. The domain of this variable is $\{1, 2, \ldots, m\} \subset \mathbb{N}$, where m is determined by the maximum amount of data parallel tasks. This can be bounded by the maximum *trip count* of the loop that originated the task, i.e., the maximum number of iterations of the loop.

In the case of PLP, the attribute set contains a collection of clusters from the previous partition in the hierarchy, $X^{PA^C} = \{X_1\}$. As an example, consider a cluster from a graph partition $\mathscr{C}_{i+1}^G$ that consist of m clusters of the previous partition hierarchy $\{C_1, \ldots, C_m\} \subseteq \mathscr{C}_i^G$. The PLP attribute would be the set $X_1 = \{C_1, \ldots, C_m\}$, which indicates the partitions C_i the pipeline can be composed from. Apart from the attribute, two variables are defined for a PLP annotation $V^{PA^C} = \{v_1, \upsilon_2\}$. The first one determines how many pipeline stages are to be implemented. This variable can take a value that is bounded by the amount of partitions, i.e., $D_{v_1}^{PA^C} = \{1, \ldots, m\} \subset \mathbb{N}$, with $m = |X_1|$. The second variable is a function that determines which partitions are to be mapped to which pipeline stages, i.e., $\upsilon_2 \colon X_1 \to D_{v_1}^{PA^C}$. The domain of this variable is the set of all possible assignments of clusters to pipeline stages, i.e., $D_{\upsilon_2}^{PA^C} = X_1 \times D_{v_1}^{PA^C}$ for a given value of v_1.

Definition 35 A **parallel-annotated graph** is a hierarchical, partitioned graph $G' = (V', E', \mathscr{PA}^{V'})$, where $\mathscr{PA}^{V'}$ is a set of parallel annotations associated with the nodes in V'. Since a data parallel loop can be sometimes implemented as a pipeline, there may be more than one annotation per node.

Definition 36 **Parallel-annotated application model:** ($\mathscr{M}_{\text{par}}^A$) Consider a sequential application model $\mathscr{M}^A = (\mathscr{SE}^A, CG^A, \pi^A)$ (in Definition 29) in which functions are modeled as graphs (e.g., $CDFG^{f^A}$). A parallel-annotated application model contains a parallel-annotated graph for every function in CG^A. It is represented as $\mathscr{M}_{\text{par}}^A = (\mathscr{SE}^A, CG_{\text{par}}^A, \pi^A)$.

The model with parallel annotations ($\mathscr{M}_{\text{par}}^A$) explicitly exposes all kinds of parallelism available in the application. If the application is to be run on a parallel platform, the model has to be refined. That is, the parallelism settings must be fixed. For example, if a cluster exposes DLP, the amount of data parallel tasks must be set.

In the example in Fig. 2.4d two possible DLP configurations are shown, for 2 and 4 tasks to execute `f3` and `f4`. This is captured by a *parallel implementation option*.

Definition 37 A **parallel implementation option** ($\mathscr{P}\mathscr{I}^A$) of a parallel-annotated application model $\mathscr{M}^A_{\text{par}} = (\mathscr{S}\mathscr{E}^A, CG^A_{\text{par}}, \pi^A)$ is the result of (1) selecting a single clustering result from the hierarchical clustering ($\mathscr{C}_i^{CDFG^{f^A}}$ for all $f^A \in S^A_{\text{f}}$), (2) selecting only one parallel annotation for nodes containing more than one and (3) assigning values to the variables of the selected parallel annotations. It is represented as $\mathscr{P}\mathscr{I}^A = (\mathscr{S}\mathscr{E}^A, CG^A_{\text{pi}}, \pi^A)$.

Definition 38 A **suitable parallel implementation** for a target platform model $SOC = (\mathscr{P}\mathscr{E}, \mathscr{E})$ is a parallel implementation that exposes all *relevant* parallelism for a later mapping phase, i.e., it only contains parallel tasks that produce a considerable speedup without wasting resources.

As an example consider the application analysis in Fig. 2.4. Intuitively, it is clear that the parallelism provided by the upper configuration in Fig. 2.4d is not *relevant*, since the whole execution is already lower-bounded by the runtime of the functions `f1` and `f2`. How to define *relevant* parallelism to obtain a *suitable* parallel implementation is not further formalized in this chapter. Initial works on parallelism extraction focused on optimizing a single graph property, e.g., the *ratio cut* [278] or a given similarity measure [48, 230]. However, a single optimization criterion is not general enough to produce good results in such a complex problem. The subjective criteria used to determine a suitable implementation are built in the heuristics described in Chap. 5.

2.4.2 Problem Statement

The definitions from the previous sections now allow to define the sequential problem as:

Definition 39 **Sequential code problem:** Given a sequential application $A \in \mathscr{A}^{\text{seq}}$ specified using the C language and a target platform model $SOC = (\mathscr{P}\mathscr{E}, \mathscr{E})$, find a suitable parallel implementation $\mathscr{P}\mathscr{I}^A = (\mathscr{S}\mathscr{E}^A, CG^A_{\text{pi}}, \pi^A)$.

The problem is solved in three steps. (1) First, the sequential application model $\mathscr{M}^A = (\mathscr{S}\mathscr{E}^A, CG^A, \pi^A)$ is built. (2) Then, it is analyzed to obtain the parallel-annotated model $\mathscr{M}^A_{\text{par}} = (\mathscr{S}\mathscr{E}^A, CG^A_{\text{par}}, \pi^A)$. (3) Finally, from all possible options allowed in $\mathscr{M}^A_{\text{par}}$ the most suitable parallel implementation is selected.

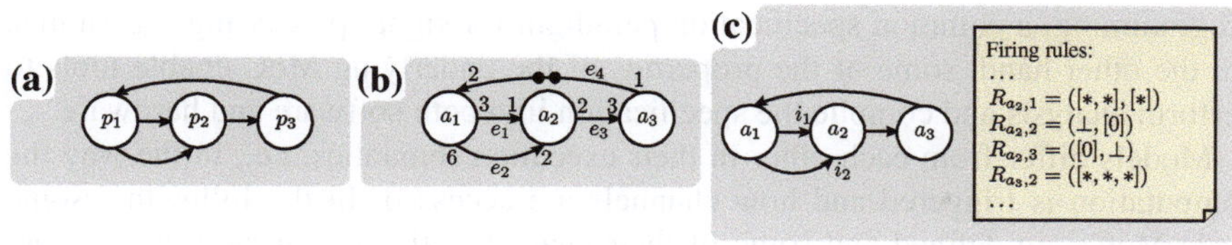

Fig. 2.8 Example of three different concurrent MoCs. **a** KPN model. **b** SDF model. **c** DDF model

2.5 Parallel Code Problem

The parallel code problem deals with finding a valid runtime configuration (see Definition 16) for a parallel application ($A \in \mathscr{A}^{\text{kpn}}$) within the multi-application specification or coming from a parallelism extraction process. There are many different ways of specifying parallel applications. MAPS uses the CPN language which allows to represent applications as a mixture of MoCs. Therefore, this section first provides an overview of process networks and dataflow models in Sect. 2.5.1. Thereafter, the CPN language is briefly introduced in Sect. 2.5.2. This book focuses on mapping techniques for applications that follow the KPN model. New definitions and refinements of the definitions in Sect. 2.3.1 are presented in Sect. 2.5.3. Finally, the parallel code problem is stated in Sect. 2.5.4.

2.5.1 Process Networks and Dataflow Models

Parallel programming models based on concurrent MoCs have been gaining momentum in the embedded domain, especially for describing signal processing applications. The MoC theory originated from theoretical computer science, with the goal of formally describing a computing system. It was initially used to compute bounds on complexity (the "big O" notation —$O(\cdot)$). In the early 80s, MoCs were used to model VLSI circuits. Later in the 90s, they started to be applied for modeling parallel applications. The formalism behind some MoCs makes it possible to analyze application properties (timing, memory consumption). This made such programming models so attractive for embedded software.

A big collection of concurrent MoCs have been proposed for embedded programming. The most prominent ones are *Synchronous Dataflow* (SDF) [166], *Cyclo-Static Dataflow* (CSDF) [28], *Boolean Dataflow* (BDF) [164], *Dynamic Dataflow* (DDF) [33], *Process Networks* (PN) and *Kahn Process Networks* (KPN) [135]. These programming models have in common a directed graph representation of the application, in which nodes represent computation and edges represent logical communication channels. An example of such a graph is shown in Fig. 2.8a.

Apart from explicitly exposing parallelism, MoC-based programming models became attractive for two reasons. On the one hand, they are well suited for graphical programming, a common specification paradigm for signal processing algorithms. On the other hand, some of the properties of the underlying MoC enable tools to perform analysis and compile the specification into both software and hardware.

Models differ from each other in their execution semantics, i.e., in the way the computation is triggered and how channels are accessed. In the following, some basic MoCs are defined and some of their characteristics are introduced. A more comprehensive treatment can be found in [26, 128, 163, 239].

2.5.1.1 Synchronous Dataflow (SDF)

Definition 40 An **SDF graph** is an annotated multigraph $G = (V, E, W)$, where nodes represent computation and edges represent queues of data items or *tokens*. Nodes are denoted $a \in V$ and are called *actors*. W is a set of annotations, $W = \{w_1, \ldots, w_{|E|}\} \subset \mathbb{N}^3$. An annotation is a triple of integers for every edge, $w_e = (p_e, c_e, d_e)$. Given a channel $e = (a_1, a_2) \in E$, p_e represents the number of tokens produced by every execution of actor a_1, c_e represents the number of tokens consumed in every execution of actor a_2 and d_e represents the number of initial tokens present on edge e. Initial tokens are also called *delays* and represent data dependencies over different iterations of the graph. SDF is also regarded as *Multi-Rate Dataflow* (MRDF).

Definition 41 A **Homogeneous SDF** (HSDF) is an SDF graph $G = (V, E, W)$ with unit token production and consumption rates, i.e., $\forall e \in E, w_e = (1, 1, d_e)$. Every SDF can be turned into an HSDF [239]. HSDFs are also regarded as *Single-Rate Dataflow* (SRDF).

The execution of an SDF is controlled by data in its edges. Every actor is executed or *fired* once it has enough tokens in its input channels. The amount of tokens needed for an actor to fire is determined by the second component of the edge annotation (c_e).

It has been shown that it is possible to compute a static schedule for an SDF graph [166]. Two major scheduling approaches can be identified: *blocked* and *non-blocked*. In the former, a schedule for one cycle is computed and repeated without overlapping, whereas in the latter, the executions of different iterations of the graph are allowed to overlap. For computing a blocked schedule, a *complete cycle* in the SDF has to be determined. A complete cycle is a sequence of actor firings that brings the SDF to its initial state. Finding a complete cycle requires that (1) enough initial tokens are provided in the edges and (2) there is a non trivial solution for the system of equations $\Gamma \cdot \mathbf{r} = 0$. Γ is called the *topology matrix* of the SDF and is defined as $[\Gamma_{ij}] = p_{ij} - c_{ij}$, where p_{ij} and c_{ij} are the number of tokens that actor j produces to and consumes from channel i respectively. The vector $\mathbf{r}$ is called *repetition vector*.

An example of an SDF graph is shown in Fig. 2.8b with delays represented as dots on the edges (see edge e_4). According to the SDF execution model, once there are 2 tokens on e_4, a_1 will fire, producing 3 tokens to e_1 and 6 to e_2. For the SDF in the example, $W = \{(3, 1, 0), (6, 2, 0), (2, 3, 0), (1, 2, 2)\}$, its topology matrix is:

$$\Gamma = \begin{pmatrix} 3 & -1 & 0 \\ 6 & -2 & 0 \\ 0 & 2 & -3 \\ -2 & 0 & 1 \end{pmatrix}$$

and a repetition vector is $\mathbf{r} = [1\ 3\ 2]^T$, meaning that after 1 execution of actor a_1, 3 executions of actor a_2 and 2 executions of actor a_3 the SDF graph returns to its original state. By *unfolding* the SDF according to its repetition vector and removing the feedback edges (those with delay tokens) one obtains a DAG [239]. Using this procedure, the problem of mapping an SDF is turned into a DAG problem (see Sect. 2.2.1).

2.5.1.2 Dynamic Dataflow

Definition 42 A **DDF graph** is a directed multigraph $G = (V, E, \mathscr{R})$, with $\mathscr{R} = \{R_{a_1}, \ldots, R_{a_{|V|}}\}$ a family of sets, one set for every node $a \in V$. Edges have the same semantics as in the SDF model. Actors, in turn, have a more complex firing semantics determined by a set of *firing rules* in $\mathscr{R}$. Every actor $a \in V$ has a set of firing rules $R_a \in \mathscr{R}$, $R_a = \{R_{a,1}, \ldots, R_{a,r}\}$. A firing rule for an actor a with p inputs is a p-tuple $R_{a,i} = (c_1, \ldots, c_p)$ of conditions, which describe a sequence of tokens that has to be available at the given input queue. Parks introduced a notation for such conditions in [210]. The condition $[X_1, X_2, \ldots, X_n]$ requires n tokens with values $X_1, X_2, \ldots, X_n$ to be available at the top of the input queue. The conditions $[*]$, $[*, *]$, $[*_{(1)}, \ldots, *_{(m)}]$ require at least 1, 2 and m tokens respectively with arbitrary values to be available at the input. The symbol $\perp$ represents any input sequence, including an empty queue. For an actor a to be in the ready state at least one of its firing rules need to be satisfied.

An example of a DDF graph is shown in Fig. 2.8c. In this example, the actor a_2 has 3 different firing rules. This actor is ready if there are at least two tokens in input i_1 and at least 1 token in input i_2 ($R_{a_2,1}$), or if the next token on input i_2 or i_1 has value 0 ($R_{a_2,2}$, $R_{a_2,3}$). Notice that more than one firing rule can be activated, in this case the dataflow graph is said to be non-determinate.

Note that an SDF can be seen as a simplification of a DDF model,[3] in which an actor with p inputs has only one firing rule of the form $R_{a,1} = (n_1, \ldots, n_p)$ with $n \in \mathbb{N}$. Additionally, the amount of tokens produced by one execution of an actor on every output is also fixed.

[3] Being more closely related to the so-called *Computation Graphs* [139].

2.5.1.3 Kahn Process Networks

Definition 43 A **KPN** is a directed multigraph $G = (V, E)$. Computational nodes in V are called *processes* and edges represent infinite token queues or First-In First-Out (FIFO) buffers. Processes do not feature the firing semantics of actors in dataflow graphs. Instead, they are allowed to be in one of two states: ready or blocked. The blocked state can only be reached by reading from *only one* empty input channel, commonly denoted as *blocking reads* semantics. A KPN is said to be determinate, i.e., the history of tokens produced on the communication channels does not depend on the scheduling [135]. Formally, a process is modeled as a functional mapping from input *streams* to output *streams*, where streams are possibly infinite sequences of data elements [135, 210].

An example of a KPN is shown in Fig. 2.8a. Note that the graph does not describe how processes access channels, i.e., how processes consume and produce tokens to the channels like SDF graphs do. Each process has its own control flow and may produce and consume data with arbitrary patterns that may be controlled by incoming data. This makes KPNs flexible enough to represent a wider range of applications, but at the same time, makes it difficult to analyze them. In fact, for general KPNs it is not possible to compute a static schedule, so that they are scheduled dynamically.

There are two major ways of dynamically scheduling a KPN: *data* and *demand* driven. In data-driven scheduling, every process in the KPN is moved to the ready state once sufficient data is available at its inputs. A dynamic scheduler then decides which process gets executed on which processor at runtime. A demand-driven scheduler first schedules processes with no output channels. This processes execute until a read blocks in one of the input channels. The scheduler then triggers only the processes from which data has been requested (*demanded*). This process continues recursively.

Data-driven scheduling exploits the available parallelism in the KPN, but can lead to unbounded accumulation of tokens in the channels. This is the case for processes with no input channels which would be always in the ready sate. Unbounded token accumulation can be avoided by setting an upper limit on the FIFO channels and implementing *blocking writes* semantics, i.e., a process would not only block on a read from an empty channel, but also on a write to a full channel. The process of determining the sizes of the FIFO queues in a KPN is known as *buffer sizing*. Introducing blocking writes to the KPN model can lead to *artificial deadlocks*, i.e., deadlocks that would not occur in the case of unbounded channels. Demand-driven scheduling, in turn, produces only the tokens that are needed. As a consequence, there is no unbounded accumulation of tokens in the channels. However, due to the way processes are activated, demand-driven produces a sequential execution of the network and introduces a high context switching overhead. In practice, hybrid data and demand-driven schedulers are used to schedule KPNs.

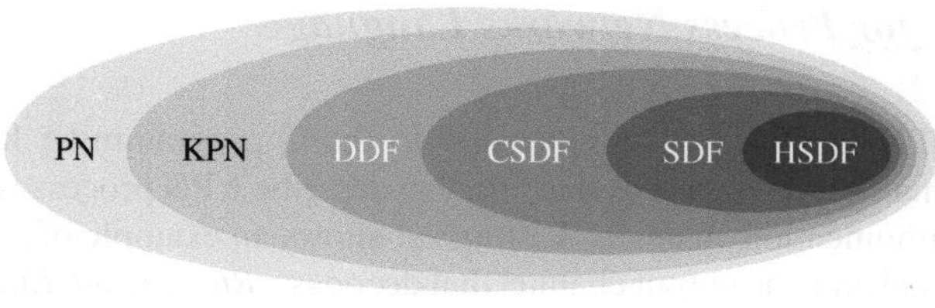

Fig. 2.9 Graphical representation of the expressiveness of different MoCs. A graph in an inner model can also be expressed by a graph in an outer model

2.5.1.4 Comparison of MoCs

When selecting an underlying MoC for an application specification, an implicit trade-off between *expressiveness* and *analyzability* is made. Static models (e.g., SDF) are more amenable to design-time analysis. For example, the questions of *termination* and *boundedness*, i.e., whether an application runs forever with bounded memory consumption, are decidable for SDFs but undecidable for BDFs, DDFs and KPNs [210]. Static models are however not general enough to represent applications with data-dependent communication patterns. With dynamic models (e.g., DDF or KPN) this kind of behavior can be modeled. As a consequence of this higher expressiveness, it is more difficult to reason at design time about possible runtime configurations (mapping, scheduling and buffer sizes).

Besides analyzability and expressiveness, the complexity of specifying an application using a MoC varies. This can be measured by the amount of information a programmer has to provide in the specification. In the examples in Fig. 2.8 it is clear that the KPN MoC displays the lowest specification effort and DDF the highest.

The expressiveness of some MoCs is shown graphically in Fig. 2.9. In addition to the models introduced in the previous section, the figure includes CSDF and PN. CSDF is an extension to SDF that allows to model several, predefined, static, cyclic behaviors [28]. PNs are KPNs without blocking reads semantics and are therefore more general. The graphical representation in Fig. 2.9 shows that an application represented in an inner model can also be represented in an outer model. Every HSDF is also an SDF, every SDF is also a CSDF and so forth. It is arguable whether or not DDF applications are a subset of KPN applications, since they can be non-determinate as discussed in Sect. 2.5.1.2. The figure refers to determinate DDFs.

The work presented in this book deals with parallel applications represented as KPNs. The motivation for addressing KPNs is twofold. (1) A simpler specification increases the acceptability of the language. At the same time, the software productivity is improved by a simpler language. (2) A more expressive model allows to represent modern, dynamic applications, thereby making the solutions applicable to a wider range of problems.

2.5.2 The C for Process Networks Language

The MAPS CPN language is an extension to the C programming language that consists of a small set of keywords that allow to describe KPN processes, SDF actors and FIFO communication channels. Listing 2.1 shows an example of a process with one input channel and one output channel that decodes a *Run-Length Encoded* (RLE) sequence of integers (see Lines 1–9). Channels are declared with the __PNchannel keyword and can be of any type, including user-defined structures (see Line 10). The actual definition of the process (in Line 1) is preceded by __PNkpn and followed by the declaration of input and output channels. The keywords __PNin and __PNout within a process are used to mark portions of code in which channels are accessed. Within these scopes, every access to the channel variable will read/write the same position in the FIFO buffer. In the example, the accesses to channel A in Line 4 will read values of different positions. Instead, all accesses to channel A in Line 8 within an execution of the for loop will return the same value. Given an input stream $\{2, 1, 3, 5, \ldots\}$, the output of the process would be $\{1, 1, 5, 5, 5 \ldots\}$. Finally, the code in Lines 11–12 shows how to instantiate processes and connect them with channels.

```
1  __PNkpn rle_dec __PNin(int A) __PNout(int B) {
2   int cnt, i;
3   while (1) {
4     __PNin(A) { cnt = A; }
5
6     __PNin(A) {
7      for (i = 0; i < cnt; ++i)
8       __PNout(B) { B = A; }
9  }}}
10 __PNchannel int src, dec;
11 __PNprocess src = Source __PNout(src); // Defined elsewhere
12 __PNprocess rle_dec = rle_dec __PNin(src) __PNout(dec);
```

Listing 2.1 Sample code for RLE decoding

2.5.3 KPN Mapping and Scheduling

As mentioned above, computing a mapping of a KPN application is a challenging task. The semantics of the KPN graph representation are much more involved than those of the initial DAG example in Fig. 2.2. Note that the DAG problem is equivalent to the problem of mapping a single iteration of an acyclic HSDF. For both problems, only the timing information is relevant for the mapping process (if the communication cost is ignored). In a KPN, instead, more information about the processes is required. A process cannot be considered a *black-box* like in the case of HSDF, but a model of its internal behavior is needed (e.g., its internal CFG).

Consider the sample KPN application shown in Fig. 2.10a together with the CFGs of its processes. Read and write accesses to channels are represented by the functions `read(...)` and `write(...)`. The channel that is accessed is represented by variables c1–c4 (for channels C_1–C_4). The actual data containers are omitted from the function signature for the sake of clarity. Actual processing is represented by

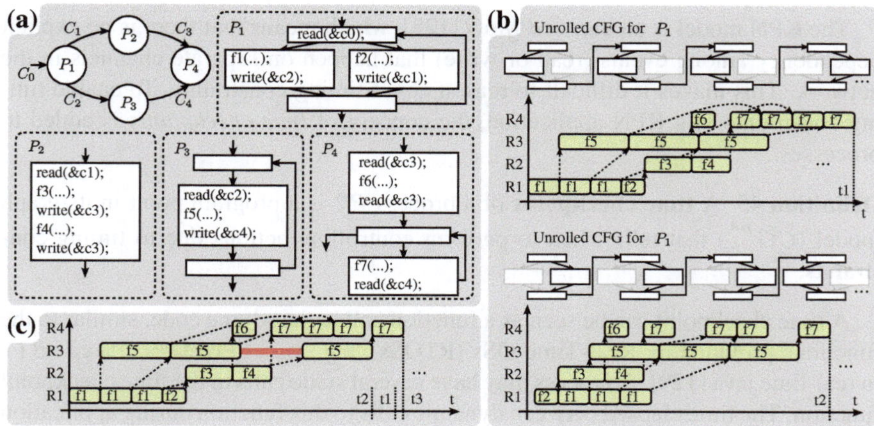

Fig. 2.10 KPN application mapping. **a** KPN application and CFGs for each process. **b** Possible schedules for hypothetical runs of P_1. **c** Example of the effect of buffer sizing for channel $C4$

functions `f1()`–`f7()`. Figure 2.10b shows two possible schedules for two different hypothetical control paths for process P_1 (on top of each Gantt Chart). For simplicity, the example supposes a single processor type, negligible communication costs and a one-to-one mapping of processes to PEs (P_1 mapped to PE R1, P_2 to R2 and so on).

In a real implementation, the theoretically infinite FIFO queues have to be bounded. As a result, processes feature blocking writes semantics in addition to the blocking read semantics, as discussed in Sect. 2.5.1.3. The effect of buffer sizing for this example is shown in Fig. 2.10c for the upper CFG of process P_1 from Fig. 2.10b. All buffer sizes are assumed to be set to one. As a result, process P_3 blocks when writing for the second time to channel C_4 since process P_4 has not yet read the first token. This introduces a blocking time in the third PE (see the red, thinner bar). The Gantt Chart in Fig. 2.10c contains the makespan for the three configurations for comparison purposes. Also note, that this buffer sizing scheme would have no effect in the lower configuration in Fig. 2.10b.

The example in Fig. 2.10 demonstrates that even with a simplified setup, the control paths followed by a process in a KPN greatly influence the application makespan. It also shows that buffer sizing can influence the makespan as well. The remainder of this section formalizes the KPN model for the parallel code problem and refines some of the definitions presented in Sect. 2.3.1.

Note that a process is itself a sequential problem, thus the sequential application model in Definition 29 can be used to describe it.

Definition 44 A **process** (P^A) in a KPN application A is a triple $P^A = (\mathscr{SE}^{P^A}, CG^{P^A}, \pi^{P^A})$. As in Definition 29, $\mathscr{SE}^{P^A}$ is the set of all sequential elements of the process (e.g., basic blocks and functions), CG^{P^A} is the call graph of the process and π^{P^A} is the profiling function $\pi^{P^A}: \mathscr{SE}^{P^A} \rightarrow \mathbb{N}$. The set of all processes of an application is denoted $\mathscr{P}^A$.

The KPN model is an *untimed* MoC [128], which means that there is no explicit dependency among events (read or write) that happen on separate channels in the network. This makes it difficult to reason about timing constraints. To enable timing constraints in a KPN application, the concept of *time checkpoints* is added to processes.

Definition 45 A **time checkpoint** of a process P^A is a program point in its graph model (CG^{P^A}) that tells when to perform controlling actions due to timing constraints.

A time checkpoint can be seen as a function call in the source code, similar to the functions provided by Real-Time OSs (RTOSs), e.g., `waitForNextPeriod()` in real-time java [129]. A process may have several static calls to the time checkpoint function. The time elapsed between dynamic calls to this function during application execution is used to check timing constraints (Definition 12). With time checkpoints enabling a real-time application specification, it is now possible to refine the definition of the KPN graph used in this book.

Definition 46 A **KPN application model** (KPN^A) is an annotated KPN graph $KPN^A = (\mathscr{P}^A, \mathscr{C}^A, \text{var}_{\text{size}})$. $\mathscr{P}^A$ is the set of processes (see Definition 44), some of which may be extended with time checkpoints. $\mathscr{C}^A$ is a multiset of FIFO channels over $\mathscr{P}^A \times \mathscr{P}^A$. Similarly to Definition 25 for a CDFG, $\text{var}_{\text{size}}: \mathscr{C}^A \rightarrow \mathbb{N}$ is a function that returns the size in bytes of the data token associated with a FIFO channel.

Definition 47 A **parallel application element set** ($\mathscr{PAE}^A$) is a set that contains all the *KPN elements* of a parallel application. A KPN element can be a process or a channel, i.e., $\mathscr{PAE}^A = \mathscr{P}^A \cup \mathscr{C}^A$.

Recall the general definition of an application $A = (\mathscr{M}^A, V^A, \mathscr{K}^A)$ in Definition 6. With the definitions presented in this section, it is now possible to formalize the application model used in the parallel code problem.

Definition 48 A **parallel application model** ($\mathscr{M}^A$) for an application $A \in \mathscr{A}^{\text{kpn}}$ is a pair $\mathscr{M}^A = (\mathscr{PAE}^A, KPN^A = (\mathscr{P}^A, \mathscr{C}^A, \text{var}_{\text{size}}))$. A particular set of application variables in V^A determine the amount of memory allocated for each buffer. The variable for the size of buffer $C^A \in \mathscr{C}^A$ is denoted $b^A_{C^A}$, with $\forall C^A \in \mathscr{C}^A, b^A_{C^A} \in V^A$. Let the set of all channel size variables be $V^A_{\text{size}} \subset V^A$. Let also the notation $e \in \mathscr{M}^A$ refer to $e \in \mathscr{PAE}^A$.

2.5.4 Problem Statement

With the new definitions the parallel problem can be stated as:

Definition 49 Parallel code problem: Given a parallel application $A \in \mathscr{A}^{\text{kpn}}$, $A = (\mathscr{M}^A, V^A, \mathscr{K}^A)$ with its behavior specified using the CPN language with underlying

KPN model $\mathscr{M}^A = (\mathscr{PAE}^A, KPN^A = (\mathscr{P}^A, \mathscr{C}^A, \text{var}_{\text{size}}))$ and target platform $SOC = (\mathscr{PE}, \mathscr{E})$, find an optimal valid runtime configuration $RC^A = (\mu_p, \mu_c, \mu_a)$ (see Definition 16).

Recall the general Definition 15, $RC^A = (\mu_p, \mu_c, \mu_a)$ for a $SOC = (\mathscr{PE}, \mathscr{E})$ with communication primitives $\mathscr{CP}$. For a KPN application, this reduces to $\mu_p: \mathscr{P}^A \subset \mathscr{M}^A \rightarrow \mathscr{PE}$ and $\mu_c: \mathscr{C}^A \subset \mathscr{M}^A \rightarrow \mathscr{CP}$. The function μ_a represents the mapping of platform and application variables, as mentioned in Definition 15. The result of the buffer sizing process can be therefore seen as part of the function μ_a. Let $\beta: V^A_{\text{size}} \rightarrow \mathbb{N}$ represent the result of the buffer sizing process, i.e., $\beta(b^A_{C^A})$ is the amount of tokens allocated to channel C^A. For convenience, the alternative representation $\beta: \mathscr{C}^A \rightarrow \mathbb{N}$ is also used.

2.6 Software Defined Radio Problem

The SDR problem deals with finding a valid runtime configuration for an SDR application ($A \in \mathscr{A}^{\text{sdr}}$) within the multi-application description. SDR applications are *waveforms* or transceiver specifications of a given radio standard. They typically feature more stringent constraints than the applications addressed by the parallel code problem. As a consequence, hardware acceleration and specialized library routines are commonly used in their implementations, as discussed in Sect. 1.1.1.

This book addresses SDR applications that describe the two bottom layers of wireless communication standards, namely the *physical* (PHY) and the *Medium Access Control* (MAC) layers [126]. They are modeled as KPN applications using the CPN language. Note however, that the pure software CPN specification does not directly support hardware acceleration. From a C specification of a process it is impossible to infer whether or not a process might be run by a hardware accelerator or might be already implemented as an assembly routine in one of the platform's PEs. SDR applications are therefore treated in a slightly different way than plain CPN applications, as will be discussed in this section.

The SDR problem can be seen as an extension of the parallel code problem, with two new main features. (1) Processes can be *marked* as special computational kernels, with a high-level algorithmic description that allows to seek for support in the target MPSoC. (2) The model of the MPSoC is extended to express which computational kernels are supported by hardware acceleration or by specialized library functions. These extensions were introduced and implemented within the context of the *Nucleus Project*, which is introduced in Sect. 2.6.1. Thereafter, the extensions are formalized in Sect. 2.6.2 and the SDR problem is finally stated in Sect. 2.6.3.

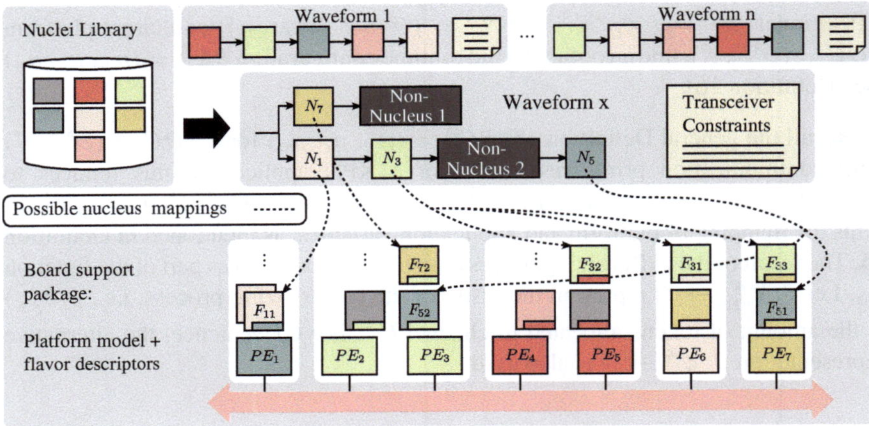

Fig. 2.11 The Nucleus waveform development concept (adapted from [222]). Different waveforms can be composed by using the nuclei in the library. Processor types are represented by different colors

2.6.1 The Nucleus Project

The Nucleus Project is a large scale project of the UMIC Excellence Cluster [271] that involves several chairs from different universities in a common effort to develop novel wireless communication algorithms (e.g., in [175, 299]), novel architectures (e.g., in [31, 292]) and novel tools and methodologies. The solution to the SDR problem presented in this book forms part of the Nucleus methodology.

The main concept behind the SDR tool flow was introduced in [222] and is illustrated in Fig. 2.11. It consists of a component-based methodology in which waveforms are composed out of library blocks called *nuclei*, which are platform-independent algorithmic entities that represent demanding computational kernels common to different wireless communication standards. Examples of such algorithms include, among others, matrix-vector operations, matrix-matrix operations, matrix factorizations, the Fast Fourier Transform (FFT), search algorithms on lists and trees. Nuclei are characterized by variables that serve to parameterize their underlying algorithm, e.g., size of data types and vector lengths. A nucleus can be implemented on different platforms in several ways. Each of these implementations is called a *flavor*. The specification of a flavor contains platform-dependent implementation details, e.g., the PE on which the flavor may run, its communication interface and its data representation. As an example, consider the platform-dependent linear algebra libraries used in HPC, e.g., the *Basic Linear Algebra Subprograms* (BLAS) [68] and the *Linear Algebra PACKage* (LAPACK) [6].

The SDR application mapping is computed by using the algorithmic information of the waveform contained in the nucleus specifications and the available flavors contained in the *Board Support Package* (BSP). This SDR mapping process, sketched by dashed arrows in Fig. 2.11, is different from the mapping process considered so far. Instead of mapping processes to PEs, nuclei are mapped to flavors, where a single PE may contain different flavors for a single nucleus. In Fig. 2.11, nucleus N_1 has two implementations on PE_1 (F_{11} and F_{12}) and nucleus N_3 has three implementations on different PEs (F_{31} on PE_6, F_{32} on $PE_{4,5}$ and F_{33} on PE_7). As shown in the figure, some of the blocks in the application may not correspond to a nucleus (*non-nuclei* blocks). These blocks are treated by a traditional compilation approach.

In addition to determining which flavor to use, the SDR mapping process must consider several new constraints. For example, the hardware interfaces of two flavors that communicate with each other must be configured so that they match. The configuration may include adapting the data sizes and the synchronization approach.

The nucleus methodology, sketched in Fig. 2.11, is not far from that of commercial visual programming languages, e.g., Simulink [184] or LabView [200]. The novelty of the approach lies on a characterization of the flavors with both algorithmic and hardware knowledge, that allows to obtain good application performance while using a high-level transceiver description language. In this way, the final implementation does not suffer from the performance gap observed between C implementations of computationally intensive kernels and their corresponding optimized versions (as assembly routines for parallel architectures or as HW accelerators). Not seldom, this gap extends over orders of magnitude (see for example [143]), which makes traditional compilation approaches ill-suited.

2.6.2 Extensions to the Parallel Problem

This section formally defines the concepts that were intuitively presented in the previous section (nuclei and flavors) and adapt some of the previous definitions to match the new problem elements. Platform-independent computational kernels in the application are modeled as nuclei, defined as follows.

Definition 50 A **nucleus** (N^A) of an SDR application A is modeled as a 5-tuple $N^A = (P^{N^A}, V^{N^A}, \mathscr{K}^{N^A}, IN^{N^A}, OUT^{N^A})$. P^{N^A} is a process in the sense of Definition 44. This functional specification is used in case there is no suitable flavor in the target platform. V^{N^A} is a set of variables $V^{N^A} = \{v_1, \ldots, v_m\}$ associated with the underlying algorithm. Each variable $v \in V^{N^A}$ is defined over a domain $D_v^{N^A}$. $\mathscr{K}^{N^A}$ is a set of nucleus specific constraints imposed by the programmer. These constraints are defined on the nucleus variables, i.e., $K_i^{N^A} \in \mathscr{K}^{N^A}$, $K_i^{N^A} = (S_i \subseteq V^{N^A}, R_i)$. Finally, IN^{N^A} and OUT^{N^A} are sets of input and output ports. The set of all nuclei defined in an application is denoted $\mathscr{N}^A$, and the set of all nuclei defined in the library $\mathscr{N}$.

Nuclei variables are used to model parameterizable algorithms. As an example, an FFT algorithm could be parameterized by the number of points (v_1) and the number of bits used for the data representation (v_2). Once a nucleus is instantiated in a waveform, the programmer may fix the value of a variable (e.g., $v_1 = 1024$ for a 1024-point FFT) or may restrict the domain of a variable (e.g., $v_2 \in \{8, 16\} \subset D_{v_2}^{N^A}$). The setting of variables is modeled by the nucleus-dependent constraint set $\mathcal{K}^{N^A}$. The sets of input and output ports were omitted in the process definition. For a nucleus, they are explicitly modeled (IN^{N^A} and OUT^{N^A}). This enables hardware interface considerations during the mapping process, since hardware interfaces are generally less flexible than software interfaces.

A flavor is a platform-dependent implementation of a nucleus. More formally,

Definition 51 A **flavor** (F^{SOC}) in a target platform SOC is modeled as a 7-tuple $F^{SOC} = (N, V^{F^{SOC}}, \mathcal{K}^{F^{SOC}}, IN^{F^{SOC}}, OUT^{F^{SOC}}, \mathcal{CM}^{F^{SOC}}, \mathcal{PE}^{F^{SOC}})$. The first component is the nucleus $N \in \mathcal{N}$ the flavor implements. $V^{F^{SOC}}$ is a set of flavor variables which contains all variables defined in the nucleus and additional implementation-dependent variables. Each variable $v \in V^{F^{SOC}}$ is defined over a domain $D_v^{F^{SOC}}$ on which constraints ($\mathcal{K}^{F^{SOC}}$) are defined. $IN^{F^{SOC}}$ and $OUT^{F^{SOC}}$ are the sets of input and output ports of the implementation. For every port in the associated nucleus, there has to be a port in the flavor. $\mathcal{CM}^{F^{SOC}}$ is the cost model of the flavor, represented as a set of functions on the flavor variables. Finally, $\mathcal{PE}^{F^{SOC}}$ denotes the set of PEs in the platform that contain this flavor. Naturally, $\mathcal{PE}^{F^{SOC}} \subseteq \mathcal{PE}$, with $SOC = (\mathcal{PE}, \mathcal{E})$. The set of all flavors in the target platform is denoted $\mathcal{F}^{SOC}$.

Implementation-dependent flavor variables allow to model additional algorithmic parameterization and hardware features. Additional algorithmic parameterization may include the parallelization degree or the memory stride. Hardware-related variables describe the way the flavor is interfaced with the rest of the system. This may include sizes of internal buffers, address ranges visible to the ports of the hardware accelerator, and the synchronization strategy used by the flavor.

The cost model of a flavor serves to compute different metrics of the implementation, such as area (in case of reconfigurable architectures), power and timing (latency and throughput). In this book the analysis is restricted to latency, i.e., the cost model replaces the software performance estimation functions from Definition 11. The latency cost function is defined as $\zeta^{F^{SOC}} : \times_{v \in V^{F^{SOC}}} D_v^{F^{SOC}} \to \mathbb{N}$. As an example, consider the implementation of an FFT, with variables that model the number of points (v_1), the number of bits used for the data representation (v_2) and the parallelism degree (v_3). Its latency is then modeled by an arbitrary function, e.g., $\zeta^{F^{SOC}}(v_1, v_2, v_3) = c \cdot v_1 \cdot v_2 / v_3$, with $c \in \mathbb{R}$.

Recall the parallel application model in Definition 48, $\mathcal{M}^A = (\mathcal{PAE}^A, KPN^A = (\mathcal{P}^A, \mathcal{C}^A, \text{var}_{\text{size}}))$. By adding nuclei to the parallel specification, it is now possible to formalize the application model used in the SDR problem.

Definition 52 An **SDR application model** ($\mathcal{M}^A$) for an application $A \in \mathcal{A}^{\text{sdr}}$ is a pair $\mathcal{M}^A = (\mathcal{PAE}^A, KPN^A = (\mathcal{P}^A \cup \mathcal{N}^A, \mathcal{C}^A, \text{var}_{\text{size}}))$. The underlying KPN specification has two different types of nodes, and consequently, $\mathscr{C}^A = (\mathscr{P}^A \cup \mathscr{N}^A) \times (\mathscr{P}^A \cup \mathscr{N}^A)$. The set of parallel elements is also extended from Definition 48 and is now defined as $\mathscr{PAE}^A = \mathscr{P}^A \cup \mathscr{N}^A \cup \mathscr{C}^A$.

In order to map the SDR problem to the parallel code problem, a mapping from nuclei to flavors has to be performed. This includes an assignment of flavor variables to their respective domains. Only with fixed variable values it is possible to lower the specification into an executable description.

The following definitions hold for an SDR application $A \in \mathscr{A}^{\text{sdr}}$ with underlying model $KPN^A = (\mathscr{P}^A \cup \mathscr{N}^A, \mathscr{C}^A, \text{var}_{\text{size}})$ on a platform with flavor set $\mathscr{F}^{SOC}$.

Definition 53 A **nucleus mapping** (NC^A) is a pair of functions $NC^A = (\mu_n, \mu_f)$. μ_n is a partial function on $\mathscr{N}^A$, that assigns a flavor to a subset of the application's nuclei $\mu_n \colon S \subseteq \mathscr{N}^A \to \mathscr{F}^{SOC}$. It is a partial function since some nuclei may have no optimized implementation in the target platform. For such nuclei, its functional specification ($P_i^{N^A}$) is used. μ_f is a mapping of flavor variables to their corresponding domains, i.e., $\forall F^{SOC} \in \mathscr{I}(\mu_n), \forall v \in F^{SOC}, \mu_f(v) \in D_v^{F^{SOC}}$.

Not every mapping would result in an implementable specification. Flavors that communicate with each other must have a *matching* configuration. As an example, consider two flavors F_1 and F_2 that communicate over a single channel. Suppose that the variables v and v' determine the locations in system memory that both flavors can use to communicate, with domains $D_v^{F_1}$ and $D_{v'}^{F_2}$. It is clear that the nucleus mapping results for these variables must be equal and contained in both variable domains, i.e., $\mu_f(v) = \mu_f(v') \in D_v^{F_1} \cap D_{v'}^{F_2}$. The same reasoning extends to more complex interfaces exposed by hardware accelerators in the target platform. More formally,

Definition 54 An **interface matching** ($\equiv_{\text{IF}}$) of two connected flavors is a relation between the flavor variables that describe the ports that connect the flavors. Let F_i and F_j be two connected flavors after a nucleus mapping $NC^A = (\mu_n, \mu_f)$. Let the flavors be connected due to a channel $c \in \mathscr{C}^A$ and let $V_c^{F_i} \subset V^{F_i}$ and $V_c^{F_j} \subset V^{F_j}$ be the set of variables that define the interfacing configuration over connection c for each of the flavors. The port interfaces match, denoted $V_c^{F_i} \equiv_{\text{IF}} V_c^{F_j}$, if $\forall v \in V_c^{F_i}, \mu_f(v) = \mu_f(v') \in D_v^{F_i} \cap D_{v'}^{F_j}$, where $v' \in V_c^{F_j}$ is the variable that corresponds to $v \in V_c^{F_i}$.

A nucleus mapping returns an implementable specification only if all flavors can actually communicate. Formally,

Definition 55 A **matching nucleus mapping** is a nucleus mapping $NC^A = (\mu_n, \mu_f)$ in which all interconnected flavor ports have matching interfaces, i.e., $\forall c_{ij} \in \{(n_i, n_j) \in (\mathscr{N}^A \times \mathscr{N}^A) \cap \mathscr{C}^A, \mu_n(n_i) = F_i, \mu_n(n_j) = F_j\}, V_{c_{ij}}^{F_i} \equiv_{\text{IF}} V_{c_{ij}}^{F_j}$.

By selecting flavors for some nuclei in the SDR application and fixing their configuration parameters so that they match, the SDR application model is lowered to an equivalent plain KPN model.

Definition 56 An **SDR implementation model** ($\mathscr{S}\mathscr{I}^A$) for an application $A \in \mathscr{A}^{\text{sdr}}$ is an implementable KPN model $\mathscr{S}\mathscr{I}^A = KPN^A = (\mathscr{P}^A, \mathscr{C}^A, \text{var}_{\text{size}})$ (see Definition 46) resulting from a matching nucleus mapping $NC^A = (\mu_n, \mu_f)$.

2.6.3 Problem Statement

With the definitions from the previous sections, it is now possible to define the SDR problem, as the problem of selecting the best matching flavors for an abstract waveform description so that the final implementation meets the constraints.

Definition 57 SDR problem: Given an SDR application $A = (\mathscr{M}^A, V^A, \mathscr{K}^A)$, $A \in \mathscr{A}^{\text{sdr}}$, with SDR application model $\mathscr{M}^A = (\mathscr{P}\mathscr{A}\mathscr{E}^A, KPN^A = (\mathscr{P}^A \cup \mathscr{N}^A, \mathscr{C}^A, \text{var}_{\text{size}}))$ and target platform $SOC = (\mathscr{P}\mathscr{E}, \mathscr{E})$ with flavors $\mathscr{F}^{SOC}$, find a matching nucleus mapping $NC^A = (\mu_n, \mu_f)$ so that there is a valid runtime configuration $RC^A = (\mu_p, \mu_c, \mu_a)$ for the resulting SDR implementation model $\mathscr{S}\mathscr{I}^A = KPN^A = (\mathscr{P}^A, \mathscr{C}^A, \text{var}_{\text{size}})$.

Note that after finding a matching nucleus mapping, the problem of finding an optimal valid runtime configuration reduces to the parallel problem in Definition 49. A simple solution approach could therefore be to iteratively build SDR implementation models and seek for a valid runtime configuration until one is found.

2.7 Synopsis

This chapter introduced several concepts relevant to this book, (1) a general overview of the mapping and scheduling terminology in Sect. 2.2.1, (2) a brief survey of performance estimation methods in Sect. 2.2.2, (3) basic compiler technology and parallelism extraction concepts required for the sequential tool flow in Sect. 2.4.1, (4) background knowledge on process networks needed for the parallel tool flow in Sect. 2.5.1 and (5) the context and basic terminology of the SDR tool flow in Sect. 2.6.1. Additionally, the chapter introduced a formal framework that allowed to represent the multi-application problem and its sub-problems as a general constrained optimization problem. The abstract constructs of the different problem statements, mainly variables and constraints, are further refined when presenting the corresponding tool flows in Chaps. 5–8.

This chapter included intuitive, simplified examples to illustrate the basic setup and the goal of the different problems, (1) the fundamental DAG mapping and scheduling problem in Fig. 2.2, (2) the parallelism extraction problem in Fig. 2.4, (3) the KPN mapping problem in Fig. 2.10 and (4) a motivational example for the multi-application problem in Sect. 2.1.

Before delving into the details of the solutions to the problems stated here, the next chapter presents an overview of solutions to similar problems.

Chapter 3
Related Work

Plenty of researchers have identified the issues described in Chap. 1. As a result, one can find a great amount of works with solutions to the problems that gave rise to the software productivity gap. These solutions solve problems that are similar to the problems stated in the previous chapter, with slight variations in the problem setup, e.g., a different input language, a different constraint formulation (or no constraints) or a different platform model (sometimes homogeneous).

This chapter surveys the related work along the topics introduced in Sect. 1.4. Although not directly in the scope of this work, techniques from the desktop computing and HPC domains are also briefly surveyed. This helps to understand the origin, the background and, more importantly, the differentiators of approaches in the embedded domain. Before discussing programming flows, Sect. 3.1 provides a survey of runtime management, a fundamental enabler for parallel execution. Techniques for parallelism extraction and synthesis of parallel specifications are presented in Sects. 3.2–3.3. Section 3.4 discusses approaches in the SDR domain, and Sect. 3.5 addresses the relatively new area of multi-application analysis.

3.1 Runtime Management

Several authors have identified the need to improve the runtime system on MPSoCs. This section discusses other approaches for runtime acceleration and compares them with the OSIP custom processor.

3.1.1 Survey of Runtime Managers

A runtime manager performs several functions in a system, including task scheduling, mapping and synchronization, admission control and resource management.

J. Castrillón Mazo and R. Leupers, *Programming Heterogeneous MPSoCs*, DOI: 10.1007/978-3-319-00675-8_3,

Approaches can be broadly divided into software and hardware solutions. The former offers high flexibility at the cost of a high runtime overhead. The latter reduces the overhead and thus is more efficient. On the downside, hardware solutions are usually less flexible and more difficult to program and integrate in a platform. Besides, they suffer from scalability issues, i.e., the area and energy consumption grow with the problem size.

In the embedded domain, software runtime managers typically follow the *master-slave* paradigm, in which a controller processor is used to orchestrate the execution of tasks on a handful of number crunchers, e.g., VLIWs and hardware accelerators. This approach is the most common in commercial platforms, such as the TI OMAP [62], the Atmel D940 [12], the Cell Broadband Engine [214] and the Samsung S3C6410/S5PC100 processors [38]. Also in academia, authors reserve a processor in the system to perform runtime management, often including resource management (see for example [155]).

Hardware support for task management can be tracked back to the 60s [174] where it enjoyed limited success. Nowadays, with a different technology landscape, several new solutions have been proposed. Hardware assisted OSes for single processor systems have been presented in [149, 194, 198]. Nakano et al. [198] implemented several system calls of the *μITRON* OS into a hardware block called *sTRON*, with speedups ranging from 10 to 50 x with respect to the original software implementation. Kohout et al. [149] introduced a unit called *Real-time Task Manager* (RTM), resulting in a 90 % reduction of the overhead introduced by the RTOS. Murtaza et al. [194] presented a full implementation of an RTOS in silicon, namely the *Silicon RTOS* (sRTOS). In their solution, they remove the scheduling overhead by adding hardware contexts for one-cycle context switches.

With the advent of multi-processor platforms, the idea of offloading OS tasks to a hardware block garnered much more attention, both in the desktop [106, 156] and in the embedded domains [169, 178, 195, 209, 231]. In the desktop domain, acceleration was motivated by the difficulty to scale general-purpose OSes (see for example [61]) and by the lack of support for asymmetrical processing [14]. Hankins et al. [106] introduced the *Multiple Instruction Stream Processor* (MISP) which allows user threads (*shreds*) to be scheduled at user level with low overhead and without OS intervention. This is enabled by the so-called *Sequencer*, a new architectural resource that accelerates MIMD execution. In this approach, however, scheduling is still performed in software. Kumar et al. [156] proposed *Carbon*, an architecture with hardware task queues that enable efficient dynamic scheduling of fine-grained tasks. Both MISP and Carbon technologies are limited in their use since they need specialized processing elements with either a new or an extended ISA.

The motivations in the embedded domain for hardware acceleration were similar to those in the desktop domain. However, with tighter real-time and energy constraints as well as finer task granularities, approaches in the embedded domain are more radical, with more functionalities offloaded to hardware. Park et al. [209] introduced the so-called *Hardware OS Kernel* (HOSK). HOSK is a coprocessor that performs scheduling in a homogeneous cluster of simplified RISC processors with a low context switch overhead (1 % for 1 kcycle tasks). With no software support nor programming model, programming a HOSK-based MPSoC may prove difficult. Additionally it is not clear how to integrate HOSK into a traditional component based design with off-the-shelf processing cores.

Seidel [231] introduced the *CoreManager*, a hardware block that performs scheduling, synchronization and data management on heterogeneous MPSoCs. Later, in [176], Limberg et al. presented extensions for real-time processing. They also devised a programming environment for applications described in an SDF-like MoC. The CoreManager is aware of task dependencies and uses this information to move data and trigger tasks at runtime. The authors reported a low scheduling overhead of 60 cycles, at the cost of a hardware complexity that grows quadratically with the problem size. This technology was integrated into the *Tomahawk* [177] chip for multimedia and baseband processing, capable of the recent Long Term Evolution (LTE) standard. In recent publications, the functionality of the CoreManager was implemented for the ARM926 programmable processor [10]. The complexity and lack of flexibility of the hardware implementation motivated this migration to software.

Lippett [178] presented the *SystemWeaver* hardware block for heterogeneous MPSoC mapping and scheduling. SystemWeaver features a better scalability but suffers from a higher design complexity. Finally, hardware extensions for embedded linux have been proposed, both for scheduling [195] and for security [84].

Industry has also seen the need for efficient task management. Paulin e tal. proposed the *MultiFlex* architecture, which contained a HW accelerator for task mapping and communication [213]. This accelerator appears to have evolved into the hardware synchronizer in the CEA and STMicroelectronics P2012 platform [242] as discussed in [212]. Recently, TI included hardware support for scheduling in the new version of their Keystone architecture [29].

3.1.2 This Book in Perspective

Runtime management approaches for multi-processor systems are summarized in Table 3.1. The table shows how hardware and software represent a tradeoff between efficiency and flexibility. The fourth column (*Restriction*) refers to constraints that the approach impose in the whole hardware platform. The first two approaches require a special ISA, while the next two pure hardware solutions (HOSK and CoreManager) require dedicated interfaces and PEs. The last three approaches, instead, support standard hardware interfaces and can therefore be interfaced with off-the-shelf cores.

The last row refers to the runtime manager used in this book. As will be seen in Chap. 4, OSIP is an ASIP that provides special instructions for task mapping, scheduling and synchronization. As it is common in an ASIP approach, OSIP represents a tradeoff between pure software and pure hardware solutions. Therefore, it attains a performance close to that of hardware implementations while retaining the flexibility of software.

Table 3.1 Comparison of different approaches for runtime management

	Domain	Implementation	Restriction	Efficiency	Flexibility	Scalability	Heterogeneity
Hankins [106] (MISP)	HPC	HW/SW	ISA	✓	✓	✓	✓
Kumar [156] (Carbon)	HPC	HW	ISA	✓✓	✗	✓	✗
Park [209] (HOSK)	Embedded	HW	Dedicated	✓✓	✗	✗	✗
Seidel [231], Limberg [176] (CoreManager)	Embedded	HW	Dedicated	✓✓✓	✗	✗	✓
Arnold [10] (CoreManager)	Embedded	SW	–	✓	✓	✓	✓
Lippett [178] (SystemWeaver)	Embedded	HW	–	✓✓	✗	✓	✓
OSIP	Embedded	HW/SW	–	✓✓	✓	✓	✓

3.2 Parallelism Extraction from Sequential Code

Automatic parallelism extraction from sequential specifications represents one of the most wanted compiler features. It has been so even since the single core era, during which a great deal of effort was invested in the extraction of fine-grained ILP. An overview of automatic parallelization techniques can be found in [17, 85, 144, 159, 171]. Several of this research results materialized in frameworks such as the *Stanford University Intermediate Framework* (SUIF) compiler framework [290] and the Open64 compiler [102]. Today, in the Multicore era, extracting parallelism automatically has become even more important for two reasons. On the one hand, humans are used to think and therefore program in a sequential manner. On the other hand, years of sequential programming have left industry with billions of lines of sequential code, which cannot be simply thrown away.

Initial works on coarse-grained parallelism extraction were based on traditional *static* compiler analysis [97, 98, 105, 229]. Researchers soon realized that static techniques, i.e., solely based on code analysis, were not enough to uncover parallelism (see, for example, the discussion in [153]). As a result, several works based on dynamic analysis or speculative execution began to appear around the mid 2000s, both in the embedded and the desktop/HPC domains. Most of the works focus on exploiting TLP within loops (PLP), with less emphasis on coarse-grained DLP.

3.2.1 General-Purpose Domain

Works in the desktop computing and HPC domains are characterized by the assumption of a homogeneous target platform and the sole goal of reducing the application makespan. Ottoni et al. [207] proposed an automatic pipeline extraction with the so-called *Decoupled Software Pipelining* (DSWP) algorithm. The algorithm performs clustering on the application graph and uses a pipeline balancing strategy to improve throughput. This extraction is done at the granularity of instructions, thus is expensive in the absence of specialized hardware. Thies et al. [264] presented an approach to exploit pipelined parallelism from C applications with user annotations that serve to define a functional pipeline. The compiler then uses dynamic analysis to identify the data flowing between pipeline stages. Bridges et al. [32] described an extension to DSWP that supports *Thread Level Speculation* (TLS) together with extensions to the C language that allow to express, for example, that the order in which two functions are called is irrelevant. The approach followed by Rul et al. [227] is similar to the one presented in this book. They employ a combination of static and dynamic data flow analyses to obtain a graph representation of the application in which parallelism patterns are searched for. A different approached is followed by Tournavitis et al. [266], who applied machine learning techniques to decide whether or not to parallelize a loop and which OpenMP [260] scheduling policy to use. This approach applies for OpenMP-enabled platforms, which are uncommon in the embedded domain.

3.2.2 Embedded Domain

As opposed to the software-driven nature of HPC approaches, early works in the embedded domain appear to be inspired by *High-Level Synthesis* (HLS) techniques (see for example [59]). This is characterized by resource constraints considerations and a target timing to be met. As an example, Karkowski et al. [138] described a methodology to map a loop to a pipeline of ASIPs. Instead of supposing existing hardware, this approach could be used to synthesize an application specific SoC.

After initial attempts, the results of parallelizing compilers were unsatisfactory. While in the desktop domain, programmers are satisfied with a *good* application speedup, this is usually not the case in the embedded domain. Embedded programmers commonly require a near to optimal parallelization, i.e., one that achieves the desired execution time with the least energy consumption. For that reason, embedded software companies are used to allocate more time to application development. Embedded application experts, as opposed to general-purpose programmers, are accustomed to manually tweak code for irregular architectures (DSP and ASIPs). With this background in mind, some works focused on helping the designer to improve their code rather than directly performing automatic parallelism extraction. Researchers at the *Interuniversity MicroElectronics Centre* (IMEC) designed an analysis tool called *CleanC* [121]. The tool identifies several code patterns that hide parallelism in C code and leaves it up to the programmer to fix them. In this way, the refactored code is more amenable for parallelism extraction by other programmers or by automatic parallelization tools. Chandraiah et al. [51] went a step forward by actually including code transformations for parallel execution on heterogeneous MPSoCs. They do this in an interactive framework where transformations are applied under the designer's full control. A similar approach is followed by the authors of the *MPSoC Parallelization Assist* (MPA) [13, 189]. In MPA, the user provides the application code and a high-level parallelization specification called *ParSpec*. By using profiling traces and a high-level simulator, the user can iterate quickly and modify the parallel specification. MPA inserts communication and synchronization primitives automatically, except for shared variables which have to be explicitly identified by the programmer. Instead of having a separate parallel specification, Reid et al. [224] devised extensions to the C language to embed this information in the C source code itself.

Instead of circumventing the shortcomings of parallelizing compilers with expert application knowledge, some authors opted for restricting the input language, e.g., by prohibiting C `break` statements. They can then safely apply parallelization schemes automatically without caring about the intricacies of the language. With this approach it is possible to derive parallel specifications that constrain themselves to a given MoC, like KPN. That is the case of automatic derivation of PNs from MATLAB code, see Harris et al. [108], and C code, see Verdoolaege et al. [275]. These works support only a subset of their respective input languages, namely so-called *(Static) Affine Nested Loop Programs* (SANLPs), i.e., programs that consist of a single loop whose bounds are affine expressions of the iterators. Several relaxations for dynamic loop bounds and while loops have been presented in [196, 197, 241]. Recently, Geuns et al. [94] published an approach that supports NLPs with unknown upper loop bounds. Additionally, communication is implemented with circular buffers with overlapping windows, which are more powerful than FIFO buffers. Even with these extensions, the authors were able to prove deadlock-free execution.

The last three examples show that by restricting the input language it is possible to automatically derive parallel implementations. These implementations are generally formally founded and therefore display properties which are well received in the embedded domain, e.g., determinism or deadlock-free execution. These solutions, however, do not directly address the problem of generic legacy code. Besides, there is an effort involved in rewriting an application to be compliant with the restricted sequential specification. This includes refactoring the code into an NLP that contains function calls with a sensible granularity for the parallelization to make sense. It is arguable that this effort is comparable to that of rewriting the application using a simple abstract parallel programming model.

Two last works from Weng et al. and Cordes et al. are worth highlighting, since they tackle automatic parallelism extraction from almost arbitrary C code.

Weng et al. [279] presented a partitioning algorithm for applications in the network processing domain. They employ an approximation of the ratio cut algorithm [278] which is adapted to produce balanced partitions. In this way they obtain a relatively balanced functional pipeline. The pipeline stages are thereafter mapped to the target platform by using random search. Weng's approach is similar to the MAPS partitioning approach in [48]. It is however restricted to the regular workloads and processor arrays that are common in network processing. Besides, their assembly level profiling would make it difficult to integrate it into an interactive framework.

The work by Cordes et al. [58] complements the MPA tool by adding automatic parallelism extraction. They do so by using a hierarchical task graph, similar to [97], and applying *Integer Linear Programming* (ILP) to produce a partition. Their ILP formulation takes into account constraints which are common in embedded systems. The authors later applied genetic algorithms for multi-objective optimization [57].

Table 3.2 Comparison of parallelism extraction approaches

	Domain	Method	Parallelism	Input	Heterogeneity
Ottoni [207]	HPC	Clustering	PLP	C	✗
Bridges [32]	HPC	Clustering	PLP	C+ext	✗
Tournavitis [266]	HPC	Machine learning	PLP	C	✗
Rul [227]	HPC	Pattern	PLP	C	✗
Karkowski [138]	Embedded	Clustering	PLP	C	✗
Verdoolaege [275]	Embedded	–	PN	C (SANLP)	–
Geuns [94]	Embedded	–	PLP	C (NLP)	–
Weng [279]	Embedded	Clustering	PLP	C	✗
Cordes [57, 58]	Embedded	ILP, GA	TLP	C	✗
VectorFabrics [274]	HPC	Pattern	D,T,PLP	C	✓
Compaan [56]	Embedded	–	PN	C (SANLP)	✗
MAPS	Embedded	Clustering, Pattern	D,T,PLP	C	✓

3.2.3 Commercial Solutions

Some of the techniques in the previous sections have made their way into commercial products. The most prominent examples are VectorFabrics [274], Compaan [56] and CriticalBlue [60]. The former provides sequential code partitioning tools that are similar to the ones presented in this book. VectorFabrics *Pareon* tool explicitly searches for patterns of parallel execution in the application. It suggests several patterns to the user with an estimated speedup and lets him decide which to use. VectorFabrics does not provide code generation facilities, but instead, exports so-called *recipes*. The recipes are detailed steps that the user can follow to parallelize the application. Support for heterogeneous embedded systems is under development in VectorFabrics. Compaan, in turn, has its origin in the previously cited work of Harris [108]. Its *Hotspot Parallelizer* can automatically translate a C program into a parallel version. As in [108], Compaan can only analyze a subset of the C language, and produces a process network version of the application. They provide a Pthreads backend for generic Multicores and a backend for *Field Programmable Gate Arrays* (FPGAs). CriticalBlue's Prism tool provides code analysis based on simulation traces as well. Internally, Prism emulates the parallel execution of an application given a parallelization strategy. This allows to explore different strategies and perform *what if* analysis. Prism supports a wider range of platforms than VectorFabrics but provides less help for actually generating code, which is entirely left to the programmer.

3.2.4 This Book in Perspective

Sections 3.2.1–3.2.3 presented works targetting manual, semi-automatic and automatic parallelism extraction. In order to better place this book into context, approaches for automatic parallelism extraction are listed with their main features in Table 3.2. The *method* (third column in the table) roughly describes how partitions are obtained.

The entry *pattern* refers to algorithms that explicitly search for parallelism patterns in the application IR. As can be seen from the table, most approaches target parallelism in loops (PLP) and very few are explicitly meant for heterogeneous platforms. If a feature does not apply to an approach, the cell entry is marked empty ('–'). For example, in the approaches by Geuns [94] and Verdoolaege [275], there is no partitioning method, since the tasks are determined by the code lines within the nested loop. Similarly, from their publications it is not possible to assess if the approach can target heterogeneous platforms or not.

The last row in Table 3.2 stands for MAPS parallelization tools presented in this book, which use both graph clustering and pattern-based algorithms. Since the input language is not constrained, the analysis phase and the algorithms are designed so that they can be used interactively. This allows to use the programmer's a priori knowledge to improve the initial sequential specification. Additionally, The parallelization does not restrict itself to loops and is aware of the entire call hierarchy. The parallelization framework supports heterogeneous platforms, with different means of sequential performance estimation as discussed in Sect. 2.2.2.

3.3 Synthesis of Parallel Specifications

The alternative to sequential programming is parallel programming, in which an application is expressed as a set of parallel tasks. A myriad of parallel programming languages have been proposed along the years (see [286]). This section starts by providing background on the main parallel languages used in the general-purpose domain. The flexibility inherent to these languages makes them error-prone and therefore less applicable to the embedded domain. Nevertheless, they are often used in code which is automatically generated from more abstract parallel programming models or even from automatic parallelization tools. The section then focuses on abstract parallel programming models used in the embedded domain. These models are usually based on a MoC that allows a more formal approach to code generation in the presence of constraints, commonly regarded as *software synthesis*.

3.3.1 General-Purpose Parallel Programming Models

There are manifold general-purpose parallel programming models. Among the most used, one finds models that are built on top of sequential languages like C by providing libraries (e.g., Pthreads [240], MPI [238]) or language extensions (e.g., OpenMP [260]). Models are typically classified by the way tasks or processes interact, with the most common ways being shared memory (e.g., Pthreads or OpenMP) and distributed memory or message passing (e.g., MPI).

Pthreads and MPI are the most traditional programming models. In the former, the programmer is responsible for synchronizing the threads to prevent data corruption in shared memory regions. This is known to be an error-prone task [165]. In the latter, the programmer does not have to care for synchronization but for explicitly distributing the data among the processors. This is a time consuming task that requires a very good understanding of the application. Although MPI has its roots in HPC for regular computations, it is also used in desktop computing and even in embedded systems. In both MPI and Pthreads programming models, the user has to specify the task mapping explicitly. The work of the parallel compiler is therefore simplified. In contrast to MPI and Pthreads, OpenMP eases programming by allowing the programmer to implicitly define tasks, task mapping and even synchronization. It is the job of the compiler and the runtime system to insert communication and synchronization barriers as well as to map tasks to processors. However, if the parallelization directive entered by the user is not correct, wrong code will be generated. Tracking down these wrong assumptions is a cumbersome task.

From the late 2000s, programming models that were initially dedicated for graphic processing in GPUs started to be used for general-purpose applications (see GPGPU [208]). In this area, two prominent programming models are worth mentioning, namely the Open Computing Language (OpenCL) [145] and Nvidia's *Compute Unified Device Architecture* (CUDA) [204]. OpenCL strives at portability and may therefore be outperformed by CUDA, which has a stronger compiler support. Both programming models are targeted for the large arrays of processing elements that characterize GPUs, usually with an identical flow of control. For this reason, they are not in the focus of this book.

General-purpose parallel languages have not gained much acceptance in deeply embedded systems. They are neither safe nor *thin*, i.e., their implementations typically incur in a high runtime overhead. The safeness issue has also motivated research in new programming models in the desktop and HPC communities, e.g., *Transactional Memory* (TM) [2] and *Thread Level Speculation* (TLS) [152]. However, these models require hardware support in order to attain good performance. The area and power overhead added by this extra hardware may become a showstopper for a later adoption in the embedded domain.

3.3.2 Embedded Parallel Programming

In contrast to the *application-centric* or programmatic approach followed in the general-purpose domain, programming models in the embedded domain are either, *hardware-centric* or *formalism-centric* [11]. Hardware-centric approaches strive for efficiency and require expert programmers. They expose architectural features in the language, e.g., memory and register banks. SoC companies typically offer hardware-centric programming models. In academia, in turn, formalism-centric models are more widespread, with the intention of producing *"correct by construction"* code. These models have clear underlying semantics which enables better analysis within programming flows (see Sect. 2.5.1). The works discussed in this section are all formalism-centric.

Initial research on formal parallel programming was based on theoretical models of concurrent systems, e.g., *process calculi*. A prominent example of such models are Hoare's *Communicating Sequential Processes* (CSP) [112, 113]. These works mainly focused on *rendezvous* or synchronous communication, in which a process or task that produces a data item would wait until the item has been read by the consumer. Later, more interaction models were analyzed, described and formalized, making it possible to reason about concurrent processes with asynchronous communication,e.g., in [168]. These early works laid the foundations of two of the main frameworks for analysis and synthesis of MoCs, namely Ptolemy [76] and Metropolis [15]. The former focuses on system modeling aspects and on establishing a framework for mixing and analyzing different MoCs, whereas the latter defines a methodology for mapping computation and communication onto a platform.

In the field of digital signal processing, graphical programming models became quite popular, since they provide algorithm designers with a natural way of specifying an application. Commercial examples include MATLAB Simulink [184], National Instruments LabVIEW [200] as well as Synopsys *Signal Processing Worksystem* (SPW) [253] and System Studio [254]. Dataflow semantics (introduced in Sect. 2.5.1) are a common underpinning of most graphical programming models. Several publications addressed the problems of scheduling, buffer sizing and deadlock-free execution for different types of dataflows. Pioneering works on SDF graphs, BDF and CSDF were published by Lee et al. in [166], in [164] and Bilsen et al. in [28] respectively. Synthesis of SDF graphs for DSPs can be found in [23, 239]. A mature tool set for generating and analyzing SDF, CSDF and so-called *Scenario-Aware Dataflow* (SADF) can be found in the SDF^3 project [78].

As mentioned in Sect. 2.5.1.4, static models are restricted in the kinds of applications they can express. Therefore, with embedded software becoming more complex, the analyzability-expressiveness tradeoff moved towards more expressive models. Two main lines of recent related work relevant to this book can be distinguished; (1) works that extend static dataflow with dynamic behavior, and (2) works that directly address PN applications. They are discussed in the following.

3.3.2.1 From Static to Dynamic Models

Several authors have proposed extensions to static models that retain some of the original formal properties. Bhattacharyya et al. [24] presented extensions towards dynamic, multidimensional (MDSDF) and windowed dataflow models. They derived balance equations, similar to those of SDF, for multi-dimensional and windowed (C)SDFs.

For more general models, synthesizing a static schedule becomes more difficult. In order to allow synthesis of DDF graphs, authors normally describe or automatically find clusters in the graph that feature static behavior. Plishker et al. [217] proposed a scheduling algorithm for a special kind of DDF, called *Core Functional Dataflow* (CFDF) [216]. The application is specified by using the functional *Dataflow Interchange Format* (DIF) [216], which allows to mix several kinds of dataflow actors, e.g.,

SDF, CSDF and BDF. This format is complex and is therefore not likely to become widely accepted in industry. In their approach, they decompose the graph and its actor modes into subgraphs that can be scheduled statically, leveraging existing techniques. The switch between static portions of the application happens dynamically at runtime. With this strategy, the authors obtain a reduced makespan and buffer sizes as compared with a dynamic Round-Robin (RR) scheduler. Results were obtained on host machines [217, 233] and not in target DSP systems. Falk et al. [82] proposed a clustering algorithm with a similar goal. After clustering, they synthesize the QSS as a Finite State Machine (FSM) that can be later used to perform dynamic scheduling at runtime. In [83], Falk et al. improved the clustering technique by using a rule-based approach. *Rules* allow to describe the clusters implicitly, without having to enumerate them. They present a considerable improvement in the application execution time and throughput. The algorithm is tested on an FPGA, where a MicroBlaze processor is used to execute every actor, and communication links are synthesized directly in hardware [82]. Therefore, neither actor nor communication mapping are addressed.

Clustering approaches can sometimes reduce the parallelism available in the initial specification and may introduce deadlocks. Wiggers et al. [281–283] follow a different approach for scheduling DDF graphs. Instead of searching for static clusters within a DDF graph, they restrict the dynamic nature of the input graph from the specification itself. They propose several incremental extensions to the SDF model, with the more expressive being *Variable-rate Phased Dataflow* (VPDF) [283]. A VPDF model allows to represent actors with different execution phases, each with a variable number of tokens. The analysis is embedded in the *Omphale* tool [27], which uses the sequential *Omphale Input Language* (OIL) for application specification. By using a sequential language, the user is relieved from the duty of writing directly the data flow graph. For their different models, they devised a strategy that determines the buffer capacities required to meet a throughput constraint. Their strategy starts by computing lower and upper linear bounds on a static schedule. These bounds are used to determine the starting time of the first firing of the actors through a constrained linear optimization. Thereafter, the buffer sizes are computed. The implementation is scheduled via *Time Division Multiplexing* (TDM) on a cycle-true simulator of a two-ARM7 platform.

3.3.2.2 Dynamic Models

Other authors directly address dynamic models without making any assumption on the application communication patterns. The rather theoretical works on KPN of Parks [210], Basten et al. [18] as well as Geilen and Basten [91] focused on devising a dynamic scheduling policy and buffer resizing strategy to ensure a non terminating execution on bounded memory. The initial resizing strategy of Parks consisted in computing arbitrary buffer sizes at design time. At runtime, once a deadlock is reached, the sizes of all the buffers are incremented. Parks proved that if the program can run infinitely and in bounded memory, it will do so when using his technique. Parks then refined his strategy by increasing the size of each full channel at a time, instead of increasing it for all channels. In this way, a similar schedule is obtainable with less memory consumption. Basten et al. [18] improved Parks technique with a method to determine exactly which channel buffer has to be increased. These works considered neither process mapping nor application constraints. A later work by Cheung et al. [52] analyzed the tradeoff between buffer sizes and application makespan. Instead of attempting to find the buffer sizes that ensure a non terminating execution, they modify the buffer sizes in order to achieve a desired application runtime.

Nikolov [202] presented the *Daedalus* framework for whole system synthesis from a PN specification. This work, however, addresses a more restricted PN model, namely *Polyhedral Process Networks* (PPN). PPNs can be automatically derived from SANLPs (see Sect. 3.2.2). For these special process networks, it is possible to compute a static schedule. Daedalus accepts mappings generated by the *Sesame* tool [79, 215] which uses evolutionary algorithms to compute an application mapping. Additionally, Daedalus provides means for design space exploration and hardware synthesis. Merging and splitting processes in a PPN application for improving throughput have been presented in [186, 187]. Recently, authors proved that it is possible to derive a CSDF model for PPNs (see [16]).

Haubelt and Keinert et al. in [109, 141] presented the *SystemCoDesigner* framework for full SoC synthesis. SystemCoDesigner is a library-based approach, based on SystemC [118], that allows to represent various MoCs. It includes means for high-level performance estimation to accelerate the optimization process, which uses evolutionary algorithms. SystemCoDesigner and Daedalus differ from the approach in this book, since they synthesize hardware rather than attempting to map to an existing platform.

Edwards et al. [74] proposed the so-called *Software/Hardware Integration Medium* (SHIM) for KPN applications with rendezvous communication. In their original publication they proposed a dynamic scheduler and addressed neither mapping nor timing constraints. In later publications they present backends for homogeneous (using Pthreads) and heterogeneous architectures, e.g., for the Cell Broadband Engine [214], in [273].

Thiele et al. proposed the *Distributed Operation Layer* (DOL) in [263] for mapping PN applications onto heterogeneous platforms. In order to compute a mapping, they use evolutionary algorithms as well. The performance estimation of the parallel execution can be obtained by using a simulator or by using the *Modular Performance Analysis* (MPA)[1] framework [50, 104, 116]. MPA implements a formal compositional performance analysis, based on so-called *service curves* from *real-time calculus*. By using these methods, bounds on metrics such as throughput and platform utilization can be computed. However, it does not support general KPNs and, due to the formal analysis, might report pessimistic results. Besides, it is not clear if the formal analysis can be applied to arbitrary architectures and schedulers. A trace-based performance estimation approach, similar to the one used in this book, is also presented in [115].

---

[1] Not to confuse with IMEC's MPA (MPSoC Parallelization Assist)

Table 3.3 Comparison of approaches to synthesis of DDF and PN specifications

	MoC (Language)	Scheduling	Mapping	Evaluation	Constraint	Hetero-geneity
Plishker [217]	DDF (func. DIF)	QSS	–	Host	–	✗
Falk [83]	DDF (–)	QSS	–	Host, FPGA	–	✗
Wiggers [281]	DDF (OIL)	Dynamic	–	Simulation	Time	✗
Parks, Geilen, Basten [18, 91, 210]	KPN (C++)	Dynamic	–	Host	–	✗
Nikolov [202] (Daedalus)	PPN (SANLPs, C+coord.)	Static	Evolutionary algorithms	Simulation	Time, Resources	✗
Edwards [74] (SHIM)	Rendezvous-KPN (Tiny-SHIM)	Dynamic	–	Host/Cell BE	–	✓
Keinert [141] (SystemCoDesigner)	SDF, CSDF, ... (SystemC)	Static, Dynamic	Evolutionary algorithms	Simulation	Time, Resources	✗
Thiele [263] (DOL)	PN (C+coord.)	Static, Dynamic	Evolutionary algorithms	Simulation, MPA	Time	✓
MAPS	SDF, KPN (CPN)	Static, Dynamic	Heuristics	Simulation, Trace replay	Time, Resources	✓

In a later publication, Geilen et al. [92] introduced an analysis framework based on timed actor interfaces. This framework unifies (C)SDF models, automata and service curves. The tracing mechanism and the time-checkpoints proposed for KPN applications in this book fit in the more general considerations formalized in [92].

3.3.3 This Book in Perspective

Synthesis approaches for dynamic dataflows and process networks are summarized in Table 3.3. The second column in the table contains the MoC treated by each of the authors. When possible, the input language is also included. Three types of application specification can be identified: domain-specific languages (OIL, DIF, Tiny-SHIM [74]), C/C++ extensions (MAPS CPN) and pure C along with a so-called *coordination language* [93], like C and XML in the DOL framework [263]. Coordination languages are used to describe process interaction, while process behavior is specified in C. Using a coordination language raises the problem of consistency between two separate descriptions of the same application. Specifications that build on top of the C language are generally preferred because they provide a smooth transition from sequential codes.

The second and third columns in Table 3.3 describe how applications are mapped to the target platform. Most authors use evolutionary algorithms to compute a mapping. In this book, instead, heuristics are used that apply to different applications. Most of the heuristics are fast, allowing better user interaction. Additionally, heuristics can be more robust to inaccurate performance estimates, common in early software development phases. For instance, heuristic decisions based on the KPN topology would hold even with wrong time estimates. The fourth column describes how the synthesis results are evaluated. For fast exploration cycles, having an alternative to cycle-true simulations is a must. For this purpose, DOL uses MPA and MAPS a so-called *trace replay* mechanism. The next column displays the kinds of constraints that are supported by the different flows: time constraints (e.g., latency and throughput) and resource constraints. Finally, the last column shows that only a few frameworks truly support heterogeneous systems.

3.4 Software Defined Radio

The idea to move radio functionality from hardware to software was initially motivated by the military in the late 90s. This would allow a single platform to implement several radio standards. Today, there is a big community performing research in technologies for SDR, which will enable, among others, the deployment of cognitive radios.

3.4.1 SDR Approaches

Besides advances in the Radio Frequency (RF) frontend, the success of SDR greatly depends on new programming methodologies. An ideal SDR methodology should allow designers to describe a radio standard by means of a high-level language, enabling faster waveform development cycles and easier waveform migration across platforms. Three main solutions can be identified: *Component-Based Software Engineering* (CBSE) approaches, *library-based* approaches and *language-based* approaches.

Today CBSE techniques (e.g., Fractal [30]) have found large acceptance and are widely used in many application domains. CBSE methodologies provide a well-structured and well-defined software development process on top of programming languages such as C/C++ or Java. However, employing such technologies directly is not an option in embedded systems targeting SDR solutions. They rely on infrastructures that are neither well suited for real-time computing (e.g., Java Virtual Machine), nor designed for platforms with heterogeneous processing elements and limited memory size. In the domain of wireless communications only few frameworks exist that target the specific needs of SDR systems, among them, Prismtech's Spectra CX tools [218], Zeligsoft [298] and CRC Scari++ [36]. These frameworks are based on the *Software Communication Architecture* (SCA) [134], a standard from the *Joint Tactical Radio System* (JTRS) [132] which is widely accepted in the SDR community. Most of the SCA implementations are based on the *Common Object Request Broker Architecture* (CORBA) [259]. Despite recent speed improvements, current CORBA implementations are not lightweight enough for baseband processing applications. Due to this limitation, the aforementioned frameworks only apply to the MAC and above layers. These frameworks treat PHY-layer applications as a single component.

Several companies have coupled CBSE flows (e.g., Simulink) with SDR development boards, e.g., Lyrtech [80] and Coherent Logix [55]. The latter provides a complete flow from Simulink to a 100-core MPSoC [211]. In order to circumvent the problem of compiling directly from Simulink, the authors allow the user to write optimized C code for the platform and embed it into Simulink blocks. This mixture of CBSE and embedded optimized library functions is similar to this book's approach. It is however, not generalized for several platforms into a suitable methodology. Additionally, these development environments are rather intended for prototyping or for a very low volume market.

Library-based approaches expose a high-level API that hides implementation details to the programmer. The API contains highly optimized functions, often programmed in assembly, for the most common algorithmic kernels supported in the target platform. This approach is commonly followed by commercial platform vendors such as TI. Libraries typically achieve a better performance compared to standard software implementations, especially when targeting irregular architectures, e.g., DSPs or ASIPs. However, they are, by definition, platform dependent and therefore not portable. Given the high variety of embedded processors and co-processor interfaces, it is unlikely that the community will agree on a single SDR API.

Language-based approaches provide more flexibility than library-based ones by adding language constructs to represent common operations. These operations can be then translated to target-specific APIs. Püschel et al. [219] presented *Spiral*, a tool that generates optimized target code provided an abstract mathematical representation of the algorithm and a description of the platform. Such tools can be used to accelerate the process of library implementation, as the SDR case study in [276] showed. The authors claim that the abstract representation in form of the so-called *Operator Language* helps to overcome compiler limitations. The Spiral technology has been recently used in the context of SDR [276]. However, developers are used to program in C and it is unlikely that a radically new specification language will be broadly adopted. In more conservative approaches, authors propose extensions to the C language [1, 191, 206]. They provide more information to the compiler, thereby improving performance but reducing portability. Moorby et al. [191] presented a C dialect for vector operations and memory management. Code written with this dialect is portable across different vector processors. Some language-based approaches tend to become processor-specific and therefore share the drawbacks of library-based approaches. In general, new languages can achieve good performance, but at the cost of high investments in compiler technology, instead of reusing the compilers provided by the processor vendors.

Authors have also used dataflow programming languages to describe radio applications. Shen et al. [234] proposed a design flow for SDR based on the DIF framework. They reuse the existing design processes, e.g., compilation and synthesis flows. Their flow includes functional verification of the waveform and final implementation on a target platform. Portability is achieved by defining function prototypes on header files, which is not as general and flexible as the approach presented in this book.

Table 3.4 Comparison of different approaches to SDR

	Specification	Portability	Efficiency	Layer
Spectra CX [218], Zeligsoft [298] Scari++ [36]	SCA/CORBA	✓✓✓	✓	MAC
Lyrtech [180], Coherent Logix [55]	Simulink	✗	✓✓✓	PHY
Optimized libraries (Spiral [276], C-Dialect[191])	machine-code, low-level C	✓	✓✓	PHY
Shen [234]	dataflow (C-based)	✓✓	✓✓	PHY
MAPS	KPN (CPN)	✓✓✓	✓✓	PHY

Finally, Yehia et al. [296] presented a work on so-called *Compound Circuits* that shares several similarities to the approach described in Chap. 7. They decompose algorithms into primitive HW components and then compose them in order to implement more complex algorithms. The final compound circuit is usually smaller than the sum of its parts, which makes it cost effective while retaining flexibility. This hardware-oriented methodology complements the software-oriented methodology presented in this book. A platform generated with the compound circuits methodology could be abstracted into a description that can be used within the MAPS SDR flow.

3.4.2 This Book in Perspective

The works presented in the previous section show two major driving forces in SDR: (1) A desire for a CBSE programming methodology to achieve cross-platform portability and programming efficiency, mainly pushed by industry and the military, materialized in the SCA, and (2) A quest for providing the full platform performance for algorithmic kernels, either via APIs or specific languages. The main characteristics of the existing approaches are listed in the first four rows of Table 3.4. Abstract programming models achieve better portability at the cost of efficiency.

The last row of the table stands for the SDR flow presented in this book. As discussed in Sect. 2.6.1, the nucleus methodology allows to combine the benefits of CBSE and library-based approaches. Optimized implementations (i.e., flavors) could be generated by an expert, e.g., using Spiral. Flavors could themselves be portable, if developed with a C extension for SDR, e.g., Open Vector Radio [191].

3.5 Multiple Applications

Section 1.1 discussed how hardware and software reuse are fundamental for the continuation of industry roadmaps. Today, embedded devices often contain a separate subsystem for every critical application, e.g., 3G baseband, bluetooth, encryption and multimedia. This common practice eases the verification effort, since every subsystem can be analyzed in isolation. However, it acts against hardware reuse. Approaches like SDR enable both hardware and software *static* reuse, i.e., a given waveform can run on different platforms and a given platform can be used to run several waveforms. In the past five years, research has been conducted to enable true *dynamic* reuse, i.e., several applications running simultaneously on the same platform.

3.5.1 Handling Multiple Applications

Systems that run multiple applications are well known from the general-purpose domain, where OSes have been around for decades. In the embedded domain, the set of applications is often known at design time, so that more design effort can be invested before deploying the software. Already for some time, embedded devices are capable of handling multiple applications running simultaneously. However, in the presence of time-critical applications, guaranteeing that all of them meet their constraints remains an open problem. Today, designers analyze each use case by hand to find a set of valid runtime configurations for each application. It is evident that a manual approach cannot keep pace with the increase in system complexity. For this reason, several (semi-) automatic solutions can be found in the literature. Three kinds of approaches can be distinguished, namely *exact*, *virtualization-based* and *composable*.

3.5.1.1 Exact Approaches

Exact solutions perform an exhaustive search of the design space. The complexity of this approach grows exponentially with the number of applications, so they rapidly become infeasible. Yet, for low-complexity systems, efficient optimization engines have been used to find optimal solutions. Benini et al. [21] presented a technique for multiple applications modeled as an acyclic task graph. They apply a logic-based Benders decomposition approach using constraint programming to solve the mapping problem for each use case. Computed system configurations are stored for later use in runtime mapping decisions. The authors analyze the tradeoff between the migration cost of switching to a new scenario and the efficiency of that scenario. Their hardware platform template consists of several homogeneous processors with private memories and a common shared memory.

3.5.1.2 Virtualization-Based Approaches

In virtualization-based approaches, each application runs on a separate virtual machine. Virtual machines guarantee complete isolation of applications, which is useful for *mixed-criticality* applications. Lightweight virtualization has been proposed for embedded systems, e.g., in [136, 268]. It is however difficult to virtualize all platform components, e.g., I/O, to ensure true isolation.

3.5.1.3 Composable Approaches

In the third type of multi-application analysis, authors focus on composable systems. *Composability* is the degree to which the mapping analysis of an application on the system can be performed in isolation, with as little information from other applications as possible. A composable system guarantees that when multiple applications run simultaneously, the same properties observed in isolation still hold, e.g., deadlock-free, throughput, latency or memory requirements. Some authors propose systems for *pure composability* [4, 20, 107, 192], while others opt for a *relaxed* version [154, 155, 245], as explained in the following.

Pure composability is accomplished when the functional and temporal behavior of an application is the same, irrespective of the presence of other applications in the system. This is achieved by assigning each application exclusive access to platform resources through temporal windows. Even when not used, resources of a given application remain invisible for other applications. This might lead to an over-dimensioned system.

Hansson et al. [4, 107] presented the *Composable and predictable Multi-Processor System on Chip* (CoMPSoC) platform template, an architectural concept for pure composability originated at NXP labs and now maintained by Goossens and Molnos [77]. Their platform uses fine-grained virtualization to allow different applications, potentially written using different MoCs, to be developed and verified in isolation. To achieve this, both the hardware and the software infrastructure are built with the concept of resource reservation to avoid interference among applications. In their approach, the interconnect, the memory controller and the memories are predictable and composable by means of admission control and budget enforcement based on TDM. On this hardware platform, the authors are able to provide guarantees for CSDF applications. They model the effects of TDM and static order schedulers directly in the dataflow graph. Moreira et al. [192], for example, presented an algorithm for multiple real-time jobs modeled as HSDFs. The algorithm guarantees a requested minimum throughput and maximum latency, while minimizing the usage of processing resources. Their approach is applicable to MPSoC platforms designed following the rules presented by Bekooij et al. in [20]. Until now, these approaches have not been extended to more expressive dataflow models. Besides, in CoMPSoC, the problem of determining time budgets for different applications remains open.

In *relaxed composability*, resources are not reserved. As a consequence, resource contention may introduce deviations in the execution properties measured for each application in isolation. Authors define a *composability function* that is used to compute the total requirements of the full system and to perform admission control.

Kumar et al. [155] analyzed the impact of arbitration schemes, such as First-Come First-Served (FCFS) and Round-Robin With Skipping (RRWS), on multiple SDF

Table 3.5 Comparison of different approaches for multi-application support

	Approach	Real-time	MoC	HW Properties	Heterogeneity
Benini [21]	Exact	Hard	Task DAG	Local/Shared Mem.	✗
Moreira [192]	Pure composability	Hard	HSDF	Predictable	✗
Hansson [107] (CoMPSoC)	Pure composability	Hard	CSDF	Predictable	✗
Kumar [155]	Relaxed compos.	Soft	SDF	Tiled with arbiter	✓
Shabbir [232] (CA-MPSoC)	Synthesis	Hard	SDF	Comm. assist	✓
Bamakhrama [16] (Daedalus^{RT})	Synthesis	Hard	acyclic CSDF	Tiled	✗
Thiele [261] (DAL)	–	✗	KPN	Tiled	✓
MAPS	Relaxed compos.	Soft	KPN	–	✓

applications. The authors concluded that static analysis for such scenarios suffers from scalability problems, or provides unreasonable performance estimates, leading to a waste of resources or poor performance. Therefore, they proposed a resource manager that takes over admission control and resource budget enforcement. For admission control, a composability function tests that the resource utilization stays below 100 %. With their approach, they cannot provide hard guarantees, but can achieve a higher platform utilization.

3.5.1.4 Other Approaches

Besides analyzing a set of applications for execution on a fixed platform, authors have also addressed it as a hardware synthesis problem. Shabbir et al. [232] presented a design flow in which a so-called *Communication Assist based MPSoC* (CA-MPSoC) is synthesized for a given set of applications. Hard and soft real-time applications are modeled as SDFs and are scheduled in a non preemptive fashion. Very recently, Bamakhrama et al. [16] presented the *Daedalus^{RT}* framework, an extension to the Daedalus framework discussed in Sect. 3.3.2. Daedalus^{RT} performs schedulability analysis of multiple hard real-time applications. They can generate a preemptive schedule for acyclic CSDF graphs using multiprocessor scheduling techniques [64].

None of the multi-application approaches discussed so far addresses KPN applications. Recently, in the context of the *EUropean REference TILed architecture Experiment* (EURETILE) [81], Thiele et al. presented the *Distributed Application Layer* (DAL) [261]. DAL, an extension to the DOL framework discussed in Sect. 3.3.2, is a software development framework targeting multiple KPN applications for many-tile architectures. The DAL framework includes several considerations for fault-tolerant systems. Derin et al. [66] also proposed a middleware for fault tolerance in NoC-based MPSoCs for KPN applications.

3.5.2 This Book in Perspective

Approaches for multi-application analysis are summarized in Table 3.5. As shown in the table, hard real-time applications can only be handled by pure composability approaches. Until now, such approaches support restricted programming models and hardware, e.g., predictable interconnect. Relaxed approaches can deal with more expressive models, but provide no hard guarantees.

The multi-application flow presented in this book (last row in Table 3.5) follows a pragmatic approach. It makes no assumptions on the hardware platform and supports a wider range of applications via the KPN MoC. However, it cannot provide hard guarantees and is therefore not directly applicable to hard real-time applications. The intention of the multi-application flow is to quickly filter out infeasible runtime configurations and present to the programmer only a small set with the most promising ones. These configurations can then be tested via simulation by the programmer. Such semi-automatic flow accelerates the initial fully manual approach and hence increases productivity.

Chapter 4
MPSoC Runtime Management

This book presents methodologies and algorithms that aim at obtaining efficient application execution on a target MPSoC. Note that the efficiency of a parallel execution is not only determined by the application itself and the quality of the programming flow, but also by the efficiency of the MPSoC's runtime system. This is especially true for applications with fine-grained tasks, like those in the domains of interest of this book. For such small tasks, heavy software stacks for multi-tasking introduce an overhead that restricts the optimization space for the tool flows proposed in this book. In order to alleviate this problem, this chapter presents an ASIP for accelerating runtime management, named OSIP. Along this book, OSIP is used in target platforms to provide low-overhead synchronization and dynamic task scheduling.

This chapter is organized as follows. Section 4.1 presents the OSIP architecture. The hardware and software architecture of OSIP-based MPSoCs is described in Sect. 4.2. Section 4.3 presents results that help to assess the benefits introduced by OSIP in a system-wide context. The chapter ends with a summary in Sect. 4.4.

4.1 OSIP Solution

Providing architectural support for MPSoC runtime systems has been the focus of several academic and industrial projects, as mentioned in Sect. 3.1.1. Pure hardware solutions found little acceptance due to flexibility and scalability issues, which motivated the design of OSIP, a custom processor that retains the efficiency of hardware, while adding the flexibility of software. This section details the design decisions that led to the final OSIP architecture and provides an overview of the solution. As in every ASIP flow, the design starts by analyzing the application that will ultimately run on the custom processor. In the case of OSIP, the application corresponds to a generic mapping and scheduling algorithm, discussed in Sect. 4.1.1. The architecture and the firmware are then described in Sects. 4.1.2–4.1.4.

J. Castrillón Mazo and R. Leupers, *Programming Heterogeneous MPSoCs*, DOI: 10.1007/978-3-319-00675-8_4,

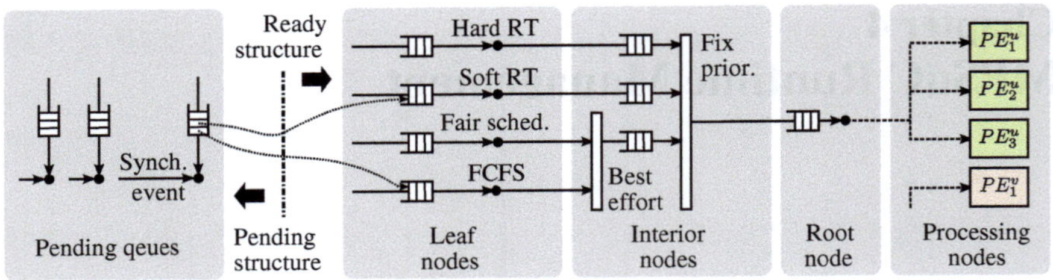

Fig. 4.1 Hierarchical mapping and scheduling

4.1.1 Mapping and Scheduling Approach

There are many mapping and scheduling algorithms in the literature. Therefore, an ASIP for runtime acceleration must not only support a single algorithm, but several of them in a general structure that is open for extension. At the same time, it must account for application classes (e.g., real-time and best-effort) and different processor types. For these reasons, OSIP was designed for a general hierarchical mapping and scheduling approach [100].

The principle of hierarchical mapping is illustrated in Fig. 4.1. Leaf nodes represent task queues that are scheduled according to local, configurable scheduling policies. Interior nodes provide more layers of schedulers, which can be used to differentiate among application classes. Tasks arriving to the root node are mapped to a processor within a *processing class* according to a configurable mapping policy. Processing classes are user-defined and can be determined by *physical* architectural features (e.g., RISC, DSP, VLIW or HW accelerator) or by *logical* features (e.g., cluster of PEs dedicated for an application class). There is no restriction on how to define processing classes. They could even overlap, meaning that a given task can be executed in different processing classes. This implies that logical classes are binary compatible or that there are two different implementations of the same task, e.g., a software and a hardware implementation. Mapping policies can take different criteria into account such as load balancing or power control modes. The algorithm also includes pending queues for task synchronization, shown on the left-hand side of Fig. 4.1. They can be associated with user-defined synchronization events or with system level events (e.g., I/O or timer interrupts). The former provides means for implementing a dataflow-like application execution and the latter allows for a task to become ready upon completion of peripheral processing (e.g., serial communication).

An example with four leaf nodes for hard real-time, soft real-time and two different kinds of best-effort applications is shown in Fig. 4.1. The top-level scheduler implements a fixed-priority policy that schedules soft real-time tasks only in the absence of hard real-time tasks and best-effort tasks in the absence of real time tasks. This separation of application classes makes it easier to implement solutions

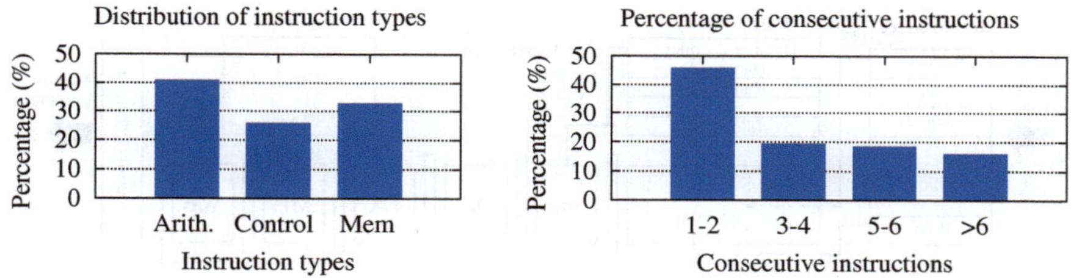

Fig. 4.2 Profiling results for the OSIP application

to the multi-application problem out of the runtime configurations of individual applications.

This hierarchical scheduling approach was implemented first using the C language in order to identify potential bottlenecks that may be suitable for architectural support. Descriptors for tasks, queues and other nodes in Fig. 4.1 were implemented with a common abstract data type (*OSIP-DT*). The queues are cyclic doubly-linked lists of OSIP-DT elements. Four basic scheduling policies are supported, which can be instantiated in any of the decision nodes. The policies are: priority-based, round-robin, first-come first-served (see Sect. 2.2.1) and weighted-fair queue. The latter is an implementation of a fair scheduling strategy [65]. For mapping, the round-robin and load balancing policies are supported.

Two things are required to profile the generic hierarchical algorithm. First, a structure has to be configured by defining the number of entry queues, the number of layers, the size of the queues and the number of processing elements. Second, scheduling events from sample applications have to be recorded to stimulate the scheduling algorithm. Different scheduling events would ripple through the hierarchical structure in different ways, affecting the application profile. The scheduler structure and the scheduling events are generated according to a video decoder and a set of artificial benchmarks (see details in Sect. 4.3). The sequence of application-dependent scheduling events are recorded and used to stimulate the hierarchical algorithm.

The application profiling was carried out on the general-purpose LTRISC processor that comes with the Synopsys PD starter kit [252]. The LTRISC is a simple load-store architecture with a 5-stage pipeline. The resulting instruction-level profile is shown in Fig. 4.2. It shows how the application behavior is almost evenly distributed into arithmetic, control and memory access operations (left-hand side of Fig. 4.2). The profile includes cycles spent in `nops` and pipeline stalls/flushes due to control and data hazards. These cycles are added to the profile of the operation type that caused the hazard (*Arith.*, *Control* and *Mem*). A peculiarity of this profile, is that arithmetic operations are typically separated by memory accesses or branch instructions. This is expressed by the plot on the right-hand side of Fig. 4.2, The plot shows the distribution of the length of sequences of consecutive arithmetic instructions, as observed during the profiling runs. During application execution, 46 % of the sequences contain less than three instructions (first bar in the plot). This control-

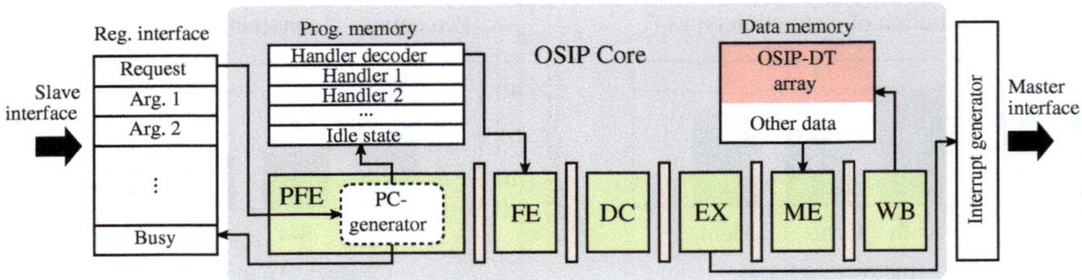

Fig. 4.3 OSIP Architecture

dominated behavior is due to constantly traversing the hierarchy of schedulers and iterating over task queues. The memory-dominated execution is a consequence of repetitive indexing of the abstract node descriptors (OSIP-DT).

4.1.2 OSIP Architecture

A typical ASIP design consists in finding repetitive patterns of consecutive instructions that might be later implemented as a custom instruction to improve the application runtime. Unfortunately, the instruction-level profile of the OSIP application does not display such dominant patterns, rendering the ASIP design more challenging. As a consequence, the architecture optimization focused on providing efficient memory access and reducing control overhead in combination with arithmetic operations.

An overview of the final OSIP architecture is given in Fig. 4.3. In addition to the actual core, two interfaces are provided: a slave register interface and a master interrupt interface. Through the register interface, OSIP can be integrated as a standard peripheral on any memory-mapped communication architecture. The interrupt interface is composed of a set of interrupt ports that serve to trigger task execution on the processors. The core itself is a load-store architecture with a 6-stage pipeline, consisting of a *prefetch* stage (PFE), a *fetch* stage (FE), a *decode* stage (DC), an *execute* stage (EX), a *memory* stage (ME) and a *write-back* stage (WB). The last five stages correspond to the original stages of the baseline LTRISC architecture. The additional PFE stage serves to prefetch the *Program Counter* (PC) upon arrival of a system event, e.g., task creation or task synchronization.

When non-operational, OSIP stays in the idle state, potentially using a low power mode. Once a request arrives through the register interface, the PC generator in PFE stage determines the handler and sets a busy flag that protects OSIP from additional requests. In order to avoid requests loss, a software client must not issue requests while OSIP is in the busy state. Depending on the request, the *handler decoder* on the top of the program memory in Fig. 4.3 selects a handler routine, e.g., task creation or task synchronization. During request processing, interrupt control signals can be generated at the execute stage in order to control system-wide task execution. Finally, after executing the handler, OSIP releases the busy flag, enters the idle state and waits for upcoming requests.The main new instructions added to the basic RISC architecture are:

Compare and Branch: Typically conditional branches are implemented with two instructions, one that operates on the registers and one that modifies the PC. In OSIP, these two steps are merged in a single instruction, with two possible ssyntaxes:

$$enh\_b \quad cond \quad R[idx_1], R[idx_2], BRANCH\_ADDR$$
$$enh\_b \quad cond \quad R[idx_1], \; imm4, \; BRANCH\_ADDR$$

The second operand is either a register or a 4-bit immediate. The latter form is used to check the status or type of an OSIP-DT against some constants defined by the algorithms, e.g., scheduling policy type and mapper policy type.

Memory Access: Accelerating memory access is one of the key techniques to improve OSIP performance. Single access is accelerated by performing index computation to fields in an OSIP-DT in hardware. To allow efficient address generation, all OSIP-DTs are allocated consecutively in a static array at the top of the memory (see Fig. 4.3). This memory region is invisible to the OSIP C compiler and can only be accessed by special instructions. The access granularity was set to a half-byte, so that more information can be encoded in the available 8 words. The syntax of the load/store instructions is:

$$sp\_load/sp\_store \quad R[value], R[idx], W, W\_MASK$$

where $R[idx]$ holds the index to be accessed. W and W_MASK are a word offset and a bit mask within an OSIP-DT. $R[value]$ contains the value to be loaded/stored.

Update and Continue: While traversing the hierarchy, it is common to update the information in the OSIP-DT of a node and then move to the next level of the hierarchy, e.g., loading its parent node. This is supported by a custom instruction with syntax:

$$update \quad R[idx_2], R[idx_1], W, W\_MASK, K$$

This instruction increments by K the field determined by (W, W_MASK) of the OSIP-DT indexed by $R[idx_1]$. At the same time, the node at the next hierarchical level is pre-fetched into $R[idx_2]$. An example of this instruction is given in Listing 4.1, where a counter (located at halfword 0 in word 4 of an OSIP-DT) is incremented by one within a cyclic list (see Line `3`). The termination condition is checked in Line `2`, by comparing for equality against the head of the list. This sample code can be used, for example, to update the priority of a list of tasks, or to update a field that counts the number of tasks below a node (`R[head]`) upon task insertion.

```
_start:
    enh_b eq R[it], R[head], _end
    update   R[it], R[it],   w=4,  hw=0, 1
    b _start
_end:
```

Listing 4.1 Update a list

Compare Nodes: A basic node-comparison instruction is provided, with syntax:

$$cmp\_node \quad R[result], R[rule], R[idx_1], R[idx_2]$$

This instruction directly compares two in-memory OSIP-DTs with indexes $R[idx_1]$ and $R[idx_2]$ according to a comparison rule given in $R[rule]$. This allows to take the comparison operator from memory, according to the configuration of the scheduling algorithm. The result of the comparison is stored in register $R[result]$.

Compare Nodes and Continue: Often, the best OSIP-DT within a list has to be found according to a given rule. A typical implementation would use an iterator and a variable that contains the current best element. To accelerate this, an enhanced version of the `cmp_node` instruction is provided, with syntax:

$$cmp\_node\_e \quad R[result], R[rule], R[curr\_best], R[it]$$

In addition to the comparison, this instruction automatically updates the index of the current best element $R[curr\_best]$. An example for finding the *best* candidate in a list is shown in Listing 4.2. The code in Line 2 checks if the list has been entirely traversed. In Line 3, the iterator is compared against the current best descriptor. At the same time, the best descriptor is updated, in case the candidate results to be better. The code in Line 4 retrieves the index to the next candidate in the list. In this example, the index to the next element is located at the seventh word of the descriptor (index 6).

```
1 _start:
2     enh_b eq     R[it],    R[head], _end
3     cmp_node_e   R[rslt], R[rule], R[best], R[it]
4     sp_load      R[it],    R[it],   w=6,     hw=0
5     b _start
6 _end:
```

Listing 4.2 Find best candidate

4.1.3 Implementation Results

OSIP was developed using the *Language for Instruction Set Architectures* (LISA) of Synopsys PD [252]. From the LISA description, a *Register Transfer Level* (RTL) description and the software tool-chain (compiler, assembler, linker and simulator) were automatically generated. The RTL description was synthesized with Synopsys Design Compiler [249] for the Faraday 90 nm standard cell library under typical conditions (supply voltage 1.0 V, temperature 25 °C). Table 4.1 shows the synthesis

Table 4.1 OSIP synthesis results

	Area (kgates)	Frequency (MHz)	Power (mW/MHz)
OSIP	41	613	0.067
ARM926EJ-S[a]	322 (160) – approx.[b]	470 (250)	0.20 (0.11)
CoreManager[176][c]	492	200	1.4 (0.7)[d]
HOSK[209]	45,9[e]	–	–

[a] Values correspond to speed optimized and area optimized in parenthesis.
[b] Area reported by ARM in 90nm TSMC technology without caches: 1.01 (0.5) mm^2 [9]. Approximation to gate count made by considering a Faraday NAND gate with an area of 3.136 μm^2.
[c] Approximate values reported by the authors in a personal communication for a 130nm technology.
[d] Value in parenthesis corresponds to the architecture with clock-gating enabled.
[e] Value reported for up to 128 threads. Technology unknown.

results in terms of area, frequency and power estimation. The latter was obtained using Synopsys PrimeTime PX [251], while running the benchmarks presented in Sect. 4.3. The table includes similar numbers for the ARM926EJ-S processor and for two hardware solutions from Sect. 3.1. Note that these values cannot be used for a direct comparison due to inherent differences among the approaches.

4.1.4 OSIP Firmware

In contrast to hardwired solutions, OSIP is programmable and hence the user has the freedom to define new scheduling algorithms. In order to leverage the performance of OSIP, the custom instructions must be used, which requires a deep understanding of OSIP's architecture. To alleviate this problem, a firmware is provided that contains template code and a library of low-level functions that wrap specialized instructions by *Compiler Known Functions* (CKFs). The template code includes sample handler routines that can be used as a basis to implement custom, more complex handlers. Besides, it also provides implementations of several basic scheduling and mapping algorithms.

The prototypes of three firmware functions are shown in Listing 4.3. The first function hides dedicated instructions for memory access (accessing hidden memory region and linking elements). The second function finds a location for the new task inside the list described by `SchedDesc` with a given insertion policy. This function increments the number of elements in the hierarchy (see Listing 4.1). The last function pops a task from a given list with a given policy. The policy determines how the *winner* is selected (see `R[rule]` in Listing 4.2).

```
CreateSchedLayer(OSIPDT ParentLayer, OSIPDT *pNewScheduler);
InsertTask(OSIPDT SchedDesc, OSIPDT *pTaskDesc,
           POLICY InsertPol);
PopTask(OSIPDT SchedDesc, OSIPDT *pTaskDesc,
        POLICY SchedPolicy);
```

Listing 4.3 Sample firmware functions

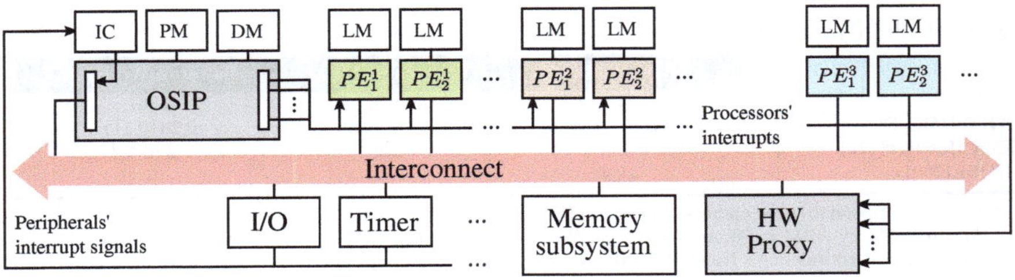

Fig. 4.4 Generic OSIP-based platform

4.2 Platform Integration

This section describes the hardware and software integration of OSIP in heterogeneous MPSoCs and introduces the ESL models used in the test platforms of this book. The integration from the hardware and software perspectives is addressed in Sects. 4.2.1–4.2.2. The ESL models are the matter of Sect. 4.2.3.

4.2.1 Hardware Integration

Figure 4.4 shows a generic view of the hardware of an OSIP-enabled MPSoC. The OSIP processor is shown in the upper left corner of the figure, with its program memory (PM), its private data memory (DM) and an interrupt controller (IC). Three conditions are needed for a correct hardware integration, namely (1) OSIP register interface must be reachable through the interconnect, (2) processors' interrupt signals must be all generated by OSIP and (3) all peripheral interrupt signals must be rerouted through OSIP. The peripheral interrupts are decoded by OSIP's interrupt controller (IC), which creates task synchronization events. An additional hardware block is required if the platform contains processing elements without interrupt support (see PE_1^3 and PE_2^3 in Fig. 4.4). This block is called *HW Proxy* and is shown in the bottom right corner of the figure. The HW Proxy allows to interface hardware accelerators and simple processors by capturing the interrupt signals from OSIP, processing OSIP's requests and retaining the information in internal queues. The processing elements can then perform polling on the proxy's interface and extract tasks from its queues.

4.2.2 Software Integration

The processors in an OSIP-based platform interact with OSIP by sending low-level commands to its register interface. These commands serve five main different purposes: (1) Modify the scheduling hierarchy by creating scheduler descriptors and configuring their policies, (2) create tasks and push them into queues, either in the ready or in the pending state, (3) create pending queues, (4) ask for tasks from OSIP upon receiving an interrupt signal, and (5) update information about the running task, e.g., update the priority according to a priority inheritance protocol.

There are a total of 50 different commands with a binary encoding designed to reduce the traffic in the interconnect. These commands are too detailed to be directly used by the programmer. For this reason, a set of high-level multi-tasking APIs and corresponding data structures were designed. This lightweight software stack includes typical task management functions, such as `CreateTask`, `SuspendTask` and `DeleteTask`, as well as standard synchronization APIs, such as `Wait` and `Signal`.

Apart from runtime support for multi-tasking, low-level routines for boot-up and interrupt handling are required for software integration. The architecture-agnostic part of this low-level software is written in C. Different versions of the architecture-dependent part are provided for the processing elements used in this book.

4.2.3 System Level Modeling

This section discusses ESL modeling issues for the virtual platforms used in this book. OSIP's *client* processing elements and other system components are described in Sect. 4.2.3.1, whereas the models of OSIP itself are treated in Sect. 4.2.3.2.

4.2.3.1 Platform Models

LTRISC: As mentioned before, the LTRISC is a simple 5-stage pipeline processor distributed with Synopsys PD. Being the baseline architecture of OSIP, it serves to measure the impact of the new instructions described in Sect. 4.1.2. Additionally, the LTRISC is used in simple virtual platforms for case studies that are concerned with functional correctness. Due to the LTRISC non-optimized architecture, it is not used in case studies in which the actual application execution time matters. As system model, the virtual platforms use a cycle-accurate simulator from the *Processor Support Package* (PSP) generated with Synopsys PD.

LTVLIW: This is 4-slot Harvard VLIW architecture derived from the VLIW sample model that comes with Synopsys PD, extended with a multiplier and a bus interface. As with the LTRISC, a cycle-accurate PSP is used in the virtual platforms. Since the LTVLIW model has no interrupt, it is interfaced with the HW Proxy. For multi-tasking, an implementation of Protothreads [71] on top of OSIP APIs was developed.

IRISC: IRISC stands for the *ICE RISC core* developed at the ICE chair. It is a RISC processor with a load-store Harvard architecture featuring a fully inter-locked 5-stage pipeline. Compared to the LTRISC, it has an optimized general-purpose instruction set. A cycle-accurate simulator of this processor is used in this book. In contrast to the LTVLIW and the LTRISC, the IRISC has interrupt support. This eases software integration into OSIP-based MPSoCs. The OSIP software stack, including boot-up code, interrupt handling and task switching was ported to this processor.

ARM926EJ-S: This is an instruction-accurate model included in the libraries of Synopsys PA [250]. It is used to benchmark OSIP and as host processor for several of the virtual platforms in this book. The OSIP software stack is also available for this architecture.

AMBA AHB Bus: The default system interconnect used in the virtual platforms of this book is the *Advanced High-performance Bus* (BUS) of the *Advanced Microcontroller Bus Architecture* (AMBA) protocol standard [8]. The model in Synopsys PA is used for simulation.

HW Proxy: This model is only included in OSIP-based platforms that include processing elements with no interrupt support, e.g., the LTVLIW. Its behavior is modeled using an untimed SystemC Transaction Level Modeling (TLM) approach.

Memories: The memory architecture varies across the virtual platforms used in this book. The internal memories of some processors are modeled by the processor simulator itself. External memories are modeled using SystemC TLM 2.0.

4.2.3.2 OSIP Models

As with other LISA processors, an automatically generated, cycle-accurate model of OSIP is available for simulation. However, since the general structure of OSIP's application is clearly defined, it is possible to model the timing behavior in a more abstract way using timing annotations. By doing so, the simulation speed is increased by several orders of magnitude while retaining an acceptable simulation accuracy. Due to the diversity of low-level commands and the open configuration of the hierarchy, it is not possible to characterize OSIP's timing by a single latency and throughput equation. Instead, the model uses the formalisms of *Tagged Signal Models* (TSMs) [167] and time-annotated Communication Extended Finite State Machines (tCEFSMs). Such a model has been successfully applied to model virtual processors in [147].

In the TSM formalism, *processes* (computing nodes) communicate through *signals*. Signals, in turn, are represented by a set of *events* that are pairs containing a *value* and a *tag* (usually time). In an OSIP-based system an event is equivalent to a change on the hardware interface of OSIP at a given time (access to the register interface, peripheral events). A signal is a set of events that trigger a given functional behavior in OSIP. For example, a signal for creating a new task is composed of a chronologically ordered set of events: acquire OSIP spin-lock, receive task's information and trigger task creation. OSIP is represented by a tCEFSM $F_{\text{osip}} = (Z, z_0, I, f, O, U)$, where:

- Z: The set of explicit states. OSIP might be in one of two states: *idle* or *busy*.
- z_0: The initial state (*idle*).
- I: The set of input signals, which contains all possible low-level commands and incoming interrupt signals from peripherals.

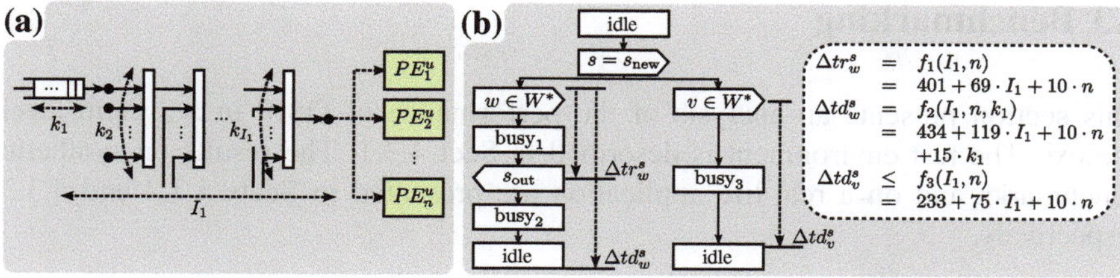

Fig. 4.5 Example of OSIP tCEFSM model

- O: The set of output signals, which contains interrupt signals to every core in the platform.
- $U = \{u_1, u_2, \dots\}$: The set of implicit states, which model the internals of OSIP, e.g., size of internal queues, state of the register interface.
- $f : Z^* \times I \to Z^* \times O$: The state transition function, where $Z^* = Z \times W^*$ and $W^* = W(u_1) \times W(u_2) \times \dots$, with $W(u_i)$ the set of all possible values that the i-th variable can have. The transition function is modeled by a functional C++ model of OSIP.

The execution of the firmware and the individual instructions of OSIP were statically analyzed in order to derive timing equations and bounds. These timing relations are defined on the input signals and the values of the implicit state variables (W^*). There are two kinds of timing equations for a value $w \in W^*$ and input signal $s \in I$:

- $\Delta td_w^s = f_d^s(w)$: Represents the *time delay* during which OSIP stays in the busy state.
- $\Delta tr_w^s = f_r^s(w)$: Represents the *response time* at which OSIP produces an output.

Consider the example in Fig. 4.5. A general hierarchy is shown in Fig. 4.5a, with several parameters that define it ($I_1, k_1, k_2, \dots$) and form part of the implicit state variables. Figure 4.5b presents a sample transition diagram with the annotated timing equations for the incoming signal s_{new}, which models task creation. The diagram shows two possible paths. The one on the left is followed if the new task leads to an interrupt signal generation. The interrupt is generated after a response time Δtr_w^s modeled by the first equation on the right-hand side of the figure. After the interrupt is generated, OSIP remains busy for additional $\Delta td_w^s - \Delta tr_w^s$ cycles.

The second path in the diagram corresponds to the case in which the task is added to the structure without generating an output. The processing time along this path varies with parameters that are not modeled by state variables. For this reason, only an upper bound is provided for this path (see the third equation on the right-hand side of the figure). More complex scenarios are modeled in a similar way.

Besides accelerating system simulation, the abstract model helps MPSoC and firmware designers in several ways. For example, typical interrupt latencies (Δtr_w^s) under different load scenarios can be derived easily from the model. Furthermore, the difference between Δtd_w^s and Δtr_w^s defines design constraints for low-level subroutines on the MPSoC cores, e.g., time constraint for a context switch.

4.3 Benchmarking

This section presents an analysis of the performance of OSIP in a system-level context. The test environment is described in Sect. 4.3.1. The results on synthetic benchmarks and on a real-life application are presented in Sects. 4.3.2 and 4.3.3 respectively.

4.3.1 Experimental Setup

The performance of OSIP was tested on a virtual platform with a variable number of ARM926EJ-S processors. The processors are connected to an AHB bus and have access to a shared memory. The ARM processor was selected because it has the highest performance from the list in Sect. 4.2.3.1. A higher processing speed helps to better stress the platform, and OSIP in particular. Note that a homogeneous setup was chosen, since the nature of the client processors does not affect OSIP timing. Heterogeneous OSIP-based platforms are used later in this book. The clock of the virtual platform is set to 200 MHz (including the OSIP clock). In this section, timing results are given in terms of cycles.

To isolate the effect of OSIP in the system, an ideal interconnect, caches and coherence protocols were assumed. In this way, the measured overhead comes from OSIP itself and other factors are left aside. These factors have to be solved separately in order to optimize an MPSoC and are not in the focus of this book.

The purpose of the next two sections is to compare the performance of OSIP against other programmable approaches and to assess the quality of the formal model. OSIP is therefore compared against the following models:

1. *LT-OSIP*: This model corresponds to the OSIP functionality running on the cycle-accurate LTRISC simulator. This allows to observe the improvement of OSIP with respect to its baseline architecture. For this purpose, the OSIP firmware was ported to LTRISC. The LTRISC processor was extended with master and slave interfaces.
2. *ARM-OSIP*: Similarly, this model corresponds to the OSIP functionality running on the ARM926EJ-S simulator. This allows to compare the performance of OSIP against a widespread commercial architecture. Besides, having an ARM to steer application execution within a heterogeneous MPSoC is a common practice. This comparison serves to assess the gain obtained if OSIP were used instead.
3. *ABS-OSIP*: This is a SystemC TLM model of OSIP that includes the formal model presented in Sect. 4.2.3.2. Comparing OSIP against ABS-OSIP serves to assess the quality of the formal model.
4. *UT-OSIP*: This is an untimed TLM model that corresponds to the ABS-OSIP model without timing annotations. Comparing against UT-OSIP allows to judge how far OSIP is from an ideal runtime manager that processes requests in *zero-time*.

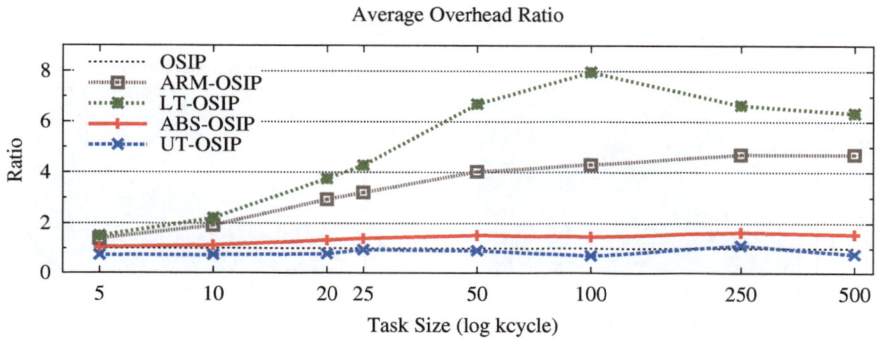

Fig. 4.6 OSIP's synthetic benchmark results

4.3.2 Synthetic Benchmarking

A synthetic multi-tasking application was manually created which allows to measure the scheduling overhead for tasks of different sizes and different load situations. The following parameters are used in this benchmark:

- Number of ARM processors: $n \in \{2, 4, 8, 12, 15\}$.
- Average application load (length of the task queues): $m \in \{2, 4, 8, 12, 16\}$.
- Task size (in kcycle): $t \in \{5, 10, 20, 25, 50, 100, 250, 500\}$.

The results of a total of 1000 simulation runs[1] are summarized in Fig. 4.6. The figure shows the average *overhead* for different tasks sizes relative to the overhead introduced by OSIP. The overhead is computed by measuring the percentage of the time that each processor spends outside the application code. This includes the time spent inside OSIP API calls, the time spinning on the OSIP interface and the time the processor is in the idle state. The overhead is averaged over all combinations of n and m. OSIP introduces 3.4 and 4.9 times less overhead than ARM-OSIP and LT-OSIP respectively. Figure 4.6 also shows that the abstract model closely matches OSIP's performance. The average overhead of ABS-OSIP is only 1.3 higher than that of OSIP. The timing estimation error is due to conservative assumptions in the abstract model. Finally, note that UT-OSIP introduces an overhead that is comparable to that of OSIP (only 1.18 times smaller). The UT-OSIP curve measures the overhead introduced by the multi-tasking software layer which does not disappear even if OSIP processed requests in zero-time.

The overhead shown in Fig. 4.6 translates into a task-size limit below which a system becomes impractical. For a 15-processor configuration, OSIP hits this limit at 25 kcycle, whereas ARM-OSIP and LT-OSIP hits it at 100 and 250 kcycle respectively. For an 8-processor configuration the observed limits were 10 kcycle for OSIP,

[1] The total of 1000 simulations results from all possible parameter combinations: 5 processor configurations, 5 load configurations and 8 task-size configurations.

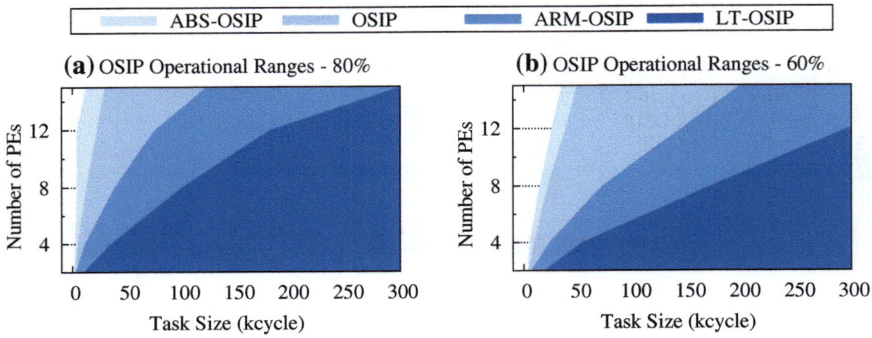

Fig. 4.7 OSIP operational ranges. **a** Areas of operation under 80 % OSIP utilization. **b** Areas of operation under 60 % OSIP utilization

25 kcycle for ARM-OSIP and 50 kcycle for LT-OSIP. These limits are determined by the load of the OSIP implementation—the more loaded the OSIP processor is, the more delay is introduced in the system. By fixing the maximum allowed load of OSIP, it is possible to identify *operational ranges* for the different implementations of OSIP. These ranges are shown in Fig. 4.7 for two typical situations. Figure 4.7a shows the operational ranges for a maximum permissible load of 80 %, while in Fig. 4.7b the load is restricted to 60 % or less. In the figure, every operational point (task size, processor count) to the right of a shaded area is allowed. For example, all implementations can run tasks of 250 kcycle on a 8-core configuration. If the task size is reduced to 50 kcycle, LT-OSIP ceases to *operate* in both 80 and 60 % cases and ARM-OSIP remains operational only in the lax 80 % case. In summary, the figure shows that OSIP truly extends the range of operation as compared to the ARM926EJ-S processor and the LTRISC baseline architecture.

Finally, across all the different operational points it is possible to measure the average scheduling latency. In the case of OSIP, this latency was of 930 cycles, measured from the moment the request arrives at OSIP's interface to the moment the task starts executing in the target processor. Embedded OSes with multi-processor support feature delays of an order of magnitude higher. For example, the latency in RTEMS, RTLinux and VxWorks have been reported to be 21, 20 and 120 kcycle respectively [243].

4.3.3 H.264 Application

Synthetic benchmarking serves to observe general characteristics of the OSIP solution. However, it says little about the impact of these characteristics on real-life applications. For this reason, this section analyzes the performance of an H.264 video decoder with the different implementations of OSIP. To quantify the application performance, the average frame-rate is used, measured in frames-per-second

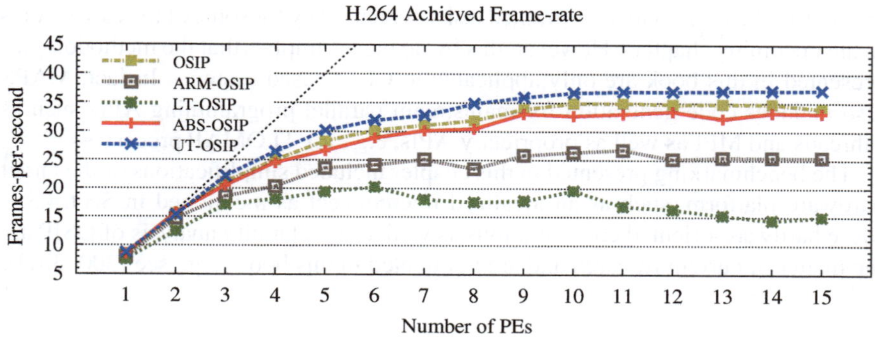

Fig. 4.8 OSIP's results for H.264 decoding

(fps). The parallel implementation of the H.264 decoder follows the 2D-wave concept [35]. Due to this parallelization approach and the video format, the application has a theoretical maximum speedup of 8x which represents the maximum amount of *macro blocks* that can be processed in parallel. The true maximum depends on the video sequence. *Dynamic* sequences typically feature more dependencies among macro blocks, which reduces the maximum parallelism.

The benchmarking results of the H.264 application are shown in Fig. 4.8. These results are in accordance with the trends observed in the synthetic benchmarks. UT-OSIP, the version that reported the lowest average overhead, is also the version that produces the best application performance. Similarly, the worst application performance is achieved with LT-OSIP, which also reported the highest average overhead in Fig. 4.6. As the figure shows, ARM-OSIP and LT-OSIP saturate at around 25 and 20 fps, while OSIP achieves a maximum of 34.9 fps. In order to achieve a framerate of 25 fps, 7 cores are needed in an ARM-OSIP-based platform while 4 processors are enough when using OSIP. Also in this study, the abstract model closely follows the performance of OSIP.

The straight thin line in Fig. 4.8 shows the maximum speedup, computed by multiplying the rate on a single processor of around 8 fps by the number of processing elements. The deviation observed from the theoretical maximum is due to the overhead of the multi-tasking APIs and the restricted parallelism of the application. The maximum speedup achieved by OSIP was of 4.36x, by ARM-OSIP of 3.12x and by LT-OSIP of 2.5x. Note that the curves in Fig. 4.8 start to saturate at around 5 and 6 processors. This saturation point possibly indicates a true maximum speedup of 5–6x as opposed to the theoretical maximum of 8x.

4.4 Synopsis

This chapter presented the OSIP processor and its associated lightweight runtime system. It also introduced the simulation models used in the virtual platforms throughout this book. As the benchmarking showed, an OSIP-based MPSoC can efficiently execute applications with fine-grained tasks, which is key for some of the case studies in the upcoming chapters. However, this by no means implies that the methodologies presented in this book are only applicable to OSIP-based systems. In fact, MAPS also includes code generation for mainstream parallel programming APIs such as Pthreads and MPI as well as proprietary APIs, e.g., for TI OMAP processors.

The benchmarking presented in this chapter included simplifications of the target hardware platform, such as an idealized interconnect as mentioned in Sect. 4.3.1. More hardware-oriented considerations as well as a scalability analysis of OSIP are the matter of current research and are not treated in this book (e.g., see [300, 301]).

Chapter 5
Sequential Code Flow

Chapter 1 alluded to the problem of sequential code in current embedded systems, showing that methodologies and tools are needed to help migrate legacy code to new parallel platforms. As discussed in Sect. 3.2, state-of-the-art solutions differ from each other in the problem setup, e.g., input language, programming restrictions, parallel output and target platform characteristics. This chapter describes a solution to the sequential problem with the setup presented in Sect. 2.4.

The chapter is organized as follows. Section 5.1 describes the overall tool flow to obtain a parallel specification from a sequential C application. Sections 5.2–5.4 provide details about the three phases of this flow: analysis, parallelism extraction and code generation. This is followed by a case study in Sect. 5.5. Thereafter, Sect. 5.6 lists the deficits of the current implementation. The chapter ends with a summary in Sect. 5.7.

5.1 Tool Flow Overview

An overview of the sequential flow is shown in Fig. 5.1. The inputs to the flow are the application code itself and the architecture model. The main goal of this flow is to obtain a *suitable parallel implementation* that can be then transformed into a parallel program, e.g., using CPN or Pthreads. Although this chapter's focus is on methodologies and algorithms for parallelism extraction, a semi-automatic code generator is also included in the flow.

The flow in Fig. 5.1 is divided into three phases. (1) The analysis phase, in Fig. 5.1a, produces a graph representation of the application that includes profiling information, i.e., the sequential application model from Definition 2.29. This phase accounts for the first step of the sequential problem in Sect. 2.4.2. (2) The parallelism extraction phase, in Fig. 5.1b, accounts for the remaining two steps of the sequential problem statement. (3) Lastly, the final backend phase in Fig. 5.1c, is in charge of exporting the parallel implementation in different formats. The tool flow uses the open

J. Castrillón Mazo and R. Leupers, *Programming Heterogeneous MPSoCs*, DOI: 10.1007/978-3-319-00675-8_5,

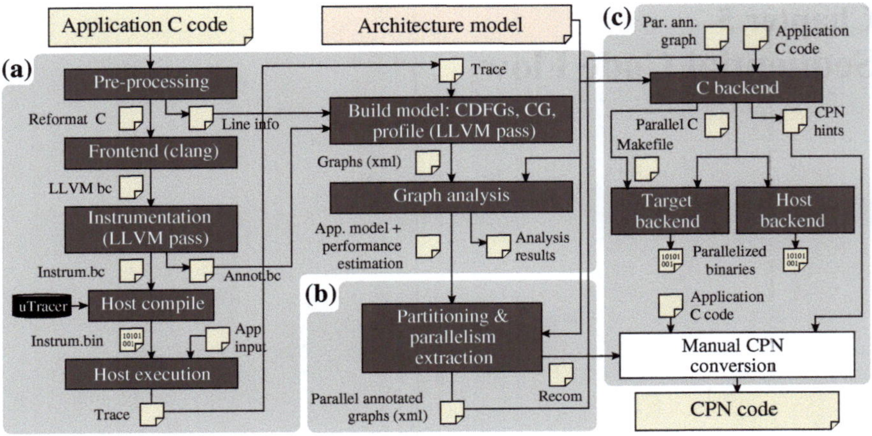

Fig. 5.1 Sequential tool flow. **a** Analysis phase: Tracing, graph model construction and analysis. **b** Extraction phase. **c** Backend phase

source projects LLVM [161] and Clang [160]. These projects are briefly introduced in Sect. 5.1.1. Thereafter, Sect. 5.1.2 gives an overview of the tool flow components.

5.1.1 Clang and LLVM

Previously, the MAPS sequential flow was implemented on top of the LANCE compiler framework [172]. To improve the robustness of the flow and extend its capabilities to the ANSI C99 standard, the compiler framework was migrated to LLVM. This migration also makes it possible to reuse the static code analysis facilities that continue to be improved within the active LLVM community. This includes state-of-the-art points-to analysis [162] and array analysis for polyhedral optimizations [265].

LLVM is an open source compiler project that includes a collection of subprojects. It is currently used in industry by several companies, such as Adobe, Apple, Cray, Nvidia, National Instruments and SUN Microsystems. Among LLVM subprojects, the sequential flow uses the LLVM Core libraries, which include the LLVM IR and the Clang C/C++ frontend. Two main LLVM *passes* are implemented in the sequential flow: a code instrumentor and an application model builder (see Fig. 5.1a). The LLVM IR is a 3AC executable representation of the application (see Sect. 2.4.1.2). Its original version included only 31 RISC-like instructions for a virtual load-store architecture. The IR uses virtual registers in *Static Single Assignment* (SSA) form [161], which simplifies data dependency analysis. Three more features of LLVM are important within the sequential flow. First, it provides a unified interface to index arrays through the instruction `getelementptr`, simplifying array analysis. Second, it has a type system that is both language and target independent, which is beneficial in the context of heterogeneous MPSoCs. Finally, its rich debugging information enables good interaction with the user. This makes it possible to report critical data dependencies to the programmer by using the original variable names.

Clang was selected as the frontend of the sequential flow since it is the native LLVM frontend. Besides, its *Rewriter* class is used for code generation (see Sect. 5.4). With rich source code information in the internal AST, the code generator retains most of the original C code.

5.1.2 Tool Flow Components

The sequential flow shown in Fig. 5.1 is composed of several smaller components. The brief description provided in this section intends to give an overall understanding of the *functionality* of each component. The techniques used to achieve this functionality are discussed later in this chapter.

5.1.2.1 Analysis Phase

The purpose of the analysis phase in Fig. 5.1a is to obtain a model of the sequential application as in Definition 2.29. This includes graph representations of the functions that make up the application, e.g., CDFGs as in Definition 2.26, a call graph as in Definition 2.26, together with profiling information, as in Definition 2.28. This is achieved through the following steps.

Pre-Processing: This component changes the format of the C code (*Reformat C* in the figure). It ensures that every line of code contains no more than one C statement. This makes it easy to back-annotate analysis results to the source code on a per-line basis. It also exports a file that allows later tools to relate statements to lines of code (*Line info* in the figure).

Frontend: The Clang C/C++ frontend is used to produce LLVM *bytecode* (*bc* in the figure), which is an on-disk binary representation of the LLVM IR.

Instrumentation: This component inserts *bookkeeping* functions that record the sequence of basic blocks that are executed and the way memory is referenced. Basic block instrumentation enables obtaining a sequential profile, i.e., the values of π^A for elements in $(S^A_{\text{stmt}} \cup \mathscr{B}\mathscr{B}^A \cup S^A_{\text{f}} \cup E^A_c \cup E^A_{cg}) \subset \mathscr{I}\mathscr{E}^A$ (see Definition 2.28). Memory instrumentation enables dynamic DFA, which is needed to construct the CDFGs. Furthermore, it provides the profiling information for the data edges, i.e., for elements in $E^A_d \subset \mathscr{I}\mathscr{E}^A$. Two files are produced after instrumentation: the instrumented and the annotated bytecode (*Instrum.bc* and *Annot.bc* in the figure). The latter contains the LLVM IR annotated with unique identifiers, used to associate profiling information with original application elements.

Host Compile and Execution: The instrumented version is compiled for the host and linked to the implementation of the bookkeeping functions (*uTracer* library in the figure). After host execution, the instrumented binary produces a trace file that records all the information tracked during execution.

Building a Model: This component is in charge of actually constructing the application model (see Definition 2.29) and annotating the profiling information. It builds the functions' static CFGs, DFGs and the application CG. It then annotates the execution counts extracted from the trace file. It also uses the memory tracing information in the trace to enrich the data flow information in the graphs. The unique identifiers that appear in the trace file are used to retrieve the IR elements. The line information is used to associate values collected for the IR with source code lines.

Graph Analysis: This component performs analysis on the application model and adds target-dependent information. It identifies hot spots and collect information at the level of loops and functions. The architecture model is used for performance estimation. The results of the analysis are also exported as files (*Analysis results* in the figure). These results can be visualized in the MAPS IDE.

5.1.2.2 Parallelism Extraction Phase

The second phase in Fig. 5.1b encompasses graph clustering and parallelism pattern detection algorithms (recall Sect. 2.4.1.6). It produces parallel annotated graphs, as in Definition 2.35, for the functions in the application. If instructed to do so, it also uses the information in the architecture model to produce a suitable parallel implementation. Additionally, this phase also produces recommendations to the programmer (*Recom* in the figure), which help identifying variables that hinder parallelism.

5.1.2.3 Backend Phase

The purpose of the last phase in Fig. 5.1c is to produce a parallel implementation, based on the output of the previous phase. The components of this phase are:

C Backend: This backend produces a parallel implementation on top of the MPI and Pthreads APIs (*parallel* C in the figure). It also produces compiler scripts that use the actual target tool-chains (*Makefile* in the figure).

Target and Host Backends: The MPI-Pthreads implementation of the application can be compiled either for the target or for the host. The host backend makes it possible to comfortably debug the functionality of the parallel implementation on the host. The target backend serves to check the actual timing behavior of the parallel implementation.

Manual CPN Conversion: This is the only part of the flow that is not automated. It represents a manual process in which the C application is turned into a CPN specification with the help of the CPN hints and the *recommendations*. In contrast to the *fixed* and general purpose MPI-Pthreads implementation, a CPN version makes it possible to use the complete software *synthesis* flow presented in Chap. 6, with more elaborate mapping of computation to processing elements and communication to communication APIs.

Algorithm 5.1 Control Flow Instrumentation

procedure CFINST($IR = (S_{stmt}, S_f)$)
    for $f \in S_f$ **do**
        $s_1 \leftarrow$ FIRST(S^f_{stmt}), $s_n \leftarrow$ LAST(S^f_{stmt})
        if $f =$ `main` **then**
            INSERTBEFORE(IR, `_MT_Init()`, s_1), INSERTAFTER(IR, `_MT_Exit()`, s_n)
        end if
        INSERTBEFORE(IR, `_FT_EnterFunction`(f), s_1)
        INSERTAFTER(IR, `_FT_ExitFunction`(f), s_n)
        for $BB \in \mathcal{BB}^f$ **do**
            INSERTBEFORE(IR, `_BBT_EnterBB`(BB), $s_{lead} \in BB$) // s_{lead}: *BB leader*
        end for
    end for
    for $s \in S_{stmt}$ **do**
        if $s =$ `call` $f \in S_f$ **then** INSERTBEFORE(IR, `_FT_StCallingIRStm`(s), s)
        end if
    end for
    return IR // *Instrumented IR*
end procedure

5.2 Application Analysis

This section details the main components of the analysis phase in Fig. 5.1, namely *tracing* (instrumentation and execution) as well as model construction and analysis. Tracing is discussed in Sect. 5.2.1 and model construction in Sect. 5.2.2. Section 5.2.3 revisits the problem of performance estimation in the context of the sequential flow. Finally, Sect. 5.2.4 describes the results exported by the graph analysis component.

5.2.1 Application Tracing

The tracing process takes a sequential application (and its input) and produces a trace file. This process is similar to the original MAPS tracing process in [47]. It is therefore only briefly reviewed. Application tracing starts by instrumenting the application IR. This is done by *control flow* and *memory* instrumentation passes.

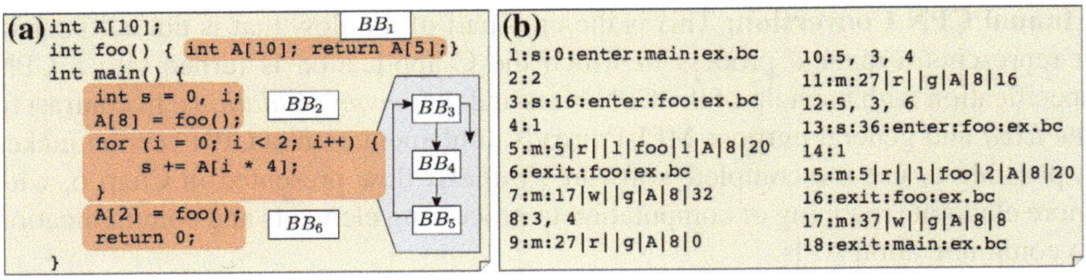

Fig. 5.2 Trace file example. **a** Application code (`ex.c`). **b** Trace file

Algorithm 5.1 shows the pseudocode of the control flow instrumentation pass. It receives as input the application IR (see Definition 2.19) and returns an instrumented version of it. The functions FIRST and LAST in Line 3 return the first and the last statements from the set of function statements (by construction, every function has a single entry and exit point in the flow). The functions INSERTBEFORE and INSERTAFTER insert call statements to the IR before and after a given statement. For every function in the IR, the instrumentor inserts a call to `_FT_EnterFunction` before its first statement and a call to `_FT_ExitFunction` after its last statement (Lines 7–8). In the case of the `main` function, the functions `_MT_Init` and `_MT_Exit` are added as well. The entrance to every basic block is also instrumented by adding a function call to `_BBT_EnterBB` before every basic block leader (Line 10). Finally, every function call is instrumented by adding a call to `_FT_StCallingIRStm` before every `call` statement.

Algorithm 5.2 Memory Instrumentation

1: **procedure** DDFAINST($IR = (S_{\mathrm{stmt}}, S_{\mathrm{f}})$)
2:     INITGLOBALVARS(IR)
3:     **for** $f \in S_{\mathrm{f}}$ **do** INITLOCALVARS(f)
4:     **end for**
5:     **for** $s \in S_{\mathrm{stmt}}$ **do**
6:         **if** $s =$ `load` $\vee$ `store` **then**
7:             $i \leftarrow$ GETACCESSINFO(s), INSERTBEFORE(IR, `_DDFA_TraceMem`(i), s)
8:         **end if**
9:     **end for**
10:     **return** IR
11: **end procedure**

The pseudocode for the memory instrumentation pass is shown in Algorithm 5.2. The function INITGLOBALVARS inserts instrumentation calls for every global variable defined in the application. Similarly, the function INITLOCALVARS instruments the local variables of every function. The code in Lines 5–9 instruments all memory access instructions (`load` or `store`). Since LLVM does not model registers at the bytecode level, accesses to local variables are also implemented by pointer dereferencing. For this reason, the function GETACCESSINFO first analyzes the type of access to distinguish among ordinary local variables, arrays and true pointer access. For example, array accesses are characterized by a call to the `getelementptr`

Table 5.1 Instrumentation functions

	Arguments	Description
_MT_Init	–	Initialize tracing, e.g., open trace file for writing
_MT_Exit	–	Finalize tracing, e.g., close trace file
_FT_EnterFunction	Function name	Track call stack, record function enter in trace file including the ID of the calling statement
_FT_ExitFunction	Function name	Update call stack, record function exit in trace file
_BBT_EnterBB	Basic block ID	Store basic block ID to the trace file
_FT_StCallingIRStm	Statement ID	Store calling statement ID in a variable, used at function enter
_DDFA_TraceMem	Access info (statement ID, read/write, address, baseType)	Record memory access information

function mentioned before. For an array access, the offset that is accessed is passed to the instrumentation function `_DDFA_TraceMem`.

The main instrumentation functions are listed in Table 5.1. These functions are all implemented in the *uTracer* runtime library, which is linked to the binary to obtain the instrumented host executable (see Fig. 5.1a). After running this executable, the trace file is finally produced. Figure 5.2 shows an example of a trace file for a simple application. For the sake of clarity, instead of using the IR (`ex.bc`), the control flow is annotated to the original C code in Fig. 5.2a, with the basic block identifiers generated during tracing. The trace file in Fig. 5.2b first shows that the `main` function started executing after being called from *dummy* statement s_0 (`s:0` in Line `1`). Thereafter, BB_2 was entered, during which function `foo` was called from statement s_{16} (in Lines `2–3`). Within `foo`, BB_1 is executed (Line `4`), which contains an access to array `A`. The access information in Line `5` specifies that: (1) the statement causing the access is s_5, (2) it is a read access (`'r'`), (3) the array is local (`'l'`), (4) it is in the stack of function `foo`, (5) when `foo` is called from call site `1`, (6) the array name is `A`, (7) its base type is `8` which stands for `int` and (8) the offset of the access is `20`, i.e., the fifth element of the array. Compare this information with the read accesses to the *global* array `A` within the for loop in Line `9` and Line `11` of the trace file. Notice also, that when `foo` is called for the second time, the call site information is the only thing that changes (see Line `15`). The call site information added to the memory instrumentation enables *context-sensitive* DFA.

5.2.2 Building the Application Model

After obtaining the execution trace, the actual model construction starts. A detailed flow of this process is shown in Fig. 5.3. As inputs it takes the trace file, the line information generated during pre-processing and the annotated bytecode of the appli-

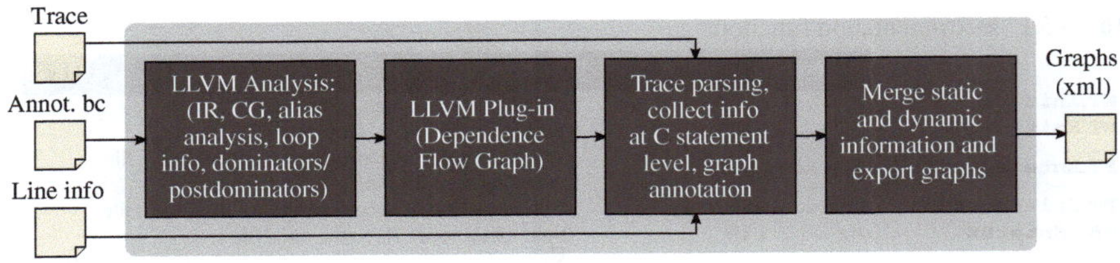

Fig. 5.3 Flow for building an application sequential model

cation. Recall that this annotated version includes unique identifiers assigned to statements and basic blocks during instrumentation. This allows to map the identifiers in the trace with actual IR elements.

The first component of the flow in Fig. 5.3 performs standard static LLVM analysis, which includes, among others, control and data flow analysis as well as call graph generation. The second component is a plugin added to LLVM that changes the CDFG representation to a *Dependence Flow Graph* (DFG) representation.[1] The DFGs of all the functions in the IR are built using the technique described in [131]. In addition to normal CDFG nodes, DFGs have *switch* and *merge* nodes. These nodes are used to add control information to data dependencies. As a consequence, data dependency information (*def-use* chains) is more explicit in a DFG, which eases the later clustering and code generation process. Additionally, the DFGs generated by the second component include loop entry and exit nodes. These nodes are used to collect dependency information and to annotate the results of loop analysis for partitioning.

The third component in Fig. 5.3 is in charge of parsing the execution trace and producing the sequential profile ($\pi^A : \mathcal{SE}^A \rightarrow \mathbb{N}$ in Definition 2.28). Building the control flow profile, i.e., for elements in $(S^A_{\text{stmt}} \cup \mathcal{BB}^A \cup E^A_c) \subset \mathcal{SE}^A$, is a straightforward process that consists in counting appearances of basic block identifiers in the trace. The same holds for function profiling, i.e., for elements in $S^A_{\text{f}} \subset \mathcal{SE}^A$. Additionally, the calling statements exported in the trace allows to distinguish different call sites and add profiling information to the edges in the CG, i.e., for elements in $E^A_{\text{cg}} \subset \mathcal{SE}^A$.

The memory information in the trace is used to extend the static data flow analysis with dynamic information. In other words, it is used to complete the set of data edges E^A_d and define the values $\pi^A(e) \forall e \in E^A_d \subset \mathcal{SE}^A$. The way the dynamic data flow information is collected from the trace is described in [47].

[1] If not said otherwise, in the rest of this book, DFG stands for a function's Dependence Flow Graph and not for a Data Flow Graph (see Definition 2.24).

```
1 int foo() {
2   int a;
3   int *pa = &a;
4   a = 0;              // Definition observed by static DFA
5   *pa = 1;            // Definition observed by dynamic DFA
6   a += 42;            // Use variable a
7   return a;
8 }
```

Listing 5.1 Conflicting static and dynamic DFA

The last component in the flow makes sure that the final DFGs are consistent by merging static and dynamic information and solving potential *conflicts*. Conflicts appear when a dependency is not recognized by the static analysis. New edges coming from the dynamic analysis may invalidate initial static edges. Consider as an example the code in Listing 5.1. If static analysis happens to miss the definition in Line 5 of variable a, a data flow edge would be created from the statement in Line 4 to the statement in Line 6. The dynamic information would add an edge between the statements in Lines 5–6 which invalidates the previous static edge. Such conflicts appear since the static analysis is not carried out in a conservative fashion. A conservative static flow analysis would add many *may-dependencies* that would clutter the DFGs hampering parallelism extraction.

5.2.3 Sequential Performance Estimation Revisited

The different methods of performance estimation used in this book were presented in Sect. 2.2.2. In the sequential flow, fine-grained performance estimation is needed at the level of basic blocks or even statements. For this reason, some of the methods become impractical. An annotation-based approach, for example, would require the programmer to specify the execution time of every basic block for each processor type. Instrumentation at the basic block level would introduce too much overhead for simulation and measurement based approaches. This could be circumvented by only executing the portions of code that are proposed for partitioning. Notice however that this would require the partitioning process to include several execution rounds (on the simulator or on the target architecture). The turn around time of these two approaches would be therefore prohibitively large.

For the aforementioned reasons, the graph analysis component of the sequential flow (see Fig. 5.1) only uses table and emulation based estimation approaches. The table-based approach works as discussed in Sect. 2.4.1.4, with the IR operations defined by the LLVM basic instructions.

The table-based approach was extended in order to cope with non-scalar architectures, e.g., VLIW. The extension consists in multiplying the result of the estimation for a basic block by a factor. Consider a basic block BB with DFG DFG^{BB} to be executed on an architecture with k parallel functional units. Let t_{ASAP} be the time reported by an *As Soon As Possible* (ASAP) scheduler over the DFG. Recall the table-based sequential execution, as defined in Sect. 2.4.1.4, $\hat{t}_{\text{seq}} = \sum_{s \in BB} \zeta_{\text{tb}}^{PT}(s)$. The cost estimation for the target architecture is then given by

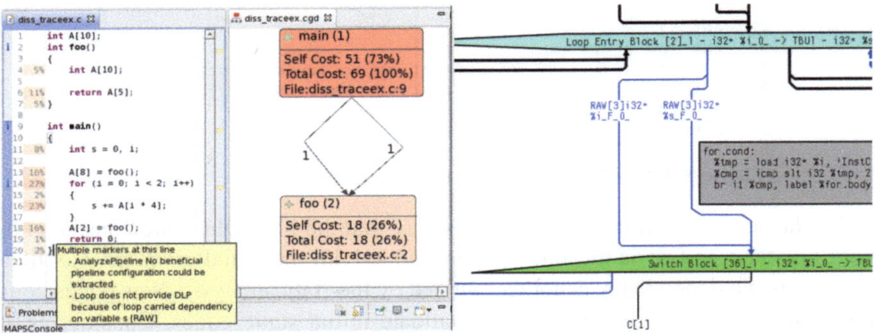

Fig. 5.4 Sample analysis results in the MAPS IDE

$$\zeta_{\text{tb}}^{PT,\text{st}}(BB) = \max\left(\frac{1}{k}, \frac{t_{\text{ASAP}}}{\hat{t}_{\text{seq}}}\right) \cdot \hat{t}_{\text{seq}} \tag{5.1}$$

5.2.4 Analysis Results

The last component of the analysis phase traverses the application graphs and exports execution statistics. This includes a *line profile*, a list of hot spots and annotated graphs for visualization. The line profile shows which lines in the source code are executed the most. The list of hot spots helps the programmer to identify where to focus the analysis. In addition to the execution statistics, the graph analysis component collects information at the level of loops and performs an early analysis of potential parallelism patterns (TLP, DLP or PLP). An example of the analysis results, as seen in the MAPS IDE, is shown in Fig. 5.4. The original code corresponds to the small application from Fig. 5.2a. The code and the line profile can be visualized in a normal C source editor, as shown on the left-hand side of Fig. 5.4 (see the percentages to the right of the line numbers). The information collected for the loop in the main function is shown in a yellow box. The CG of the application is shown in the middle of the figure, with two edges, representing the two call sites to function `foo`. The left-hand side of the figure shows a portion of the DFG of the main function. It is possible to identify the loop entry node, a portion of the for condition and the switch node corresponding to the control flow introduced by the loop.

Graph analysis collects information at the level of loops and functions, which is important for the parallelism extraction and the code generation phases. In the case of loops, the following annotations are added to the loop entry nodes:

1. Loop type: This field indicates whether or not the loop is *well structured*. That is, if the loop has only one entry, one exit point and only one backward edge, i.e., no `break`, `continue` nor `return` statements.

2. Induction variables: This field indicates which variables are used to control the iterations of the loop. Loop carried dependencies due to these variables are therefore ignored during parallelism extraction.
3. Profiling: This field records how often the loop was executed and the average trip count. This information is important for assessing the benefit of parallelization strategies such as DLP or PLP. Note, for example, that the efficiency of a pipeline depends on the amount of iterations.
4. Loop carried dependencies: These dependency edges are explicitly recorded in the header, so that later phases can access the information rapidly.
5. I/O dependencies: These are dependencies that reach and leave the loop.

Functions are analyzed for *reentrancy*, i.e., whether or not they can be executed partially and then be re-executed with a correct completion. Reentrant functions are easily parallelizable, since several copies can run simultaneously. The following conditions are checked to determine if a function is reentrant:

1. The function must not modify global data.
2. The function must not have internal state (no static variables).
3. The function must not call other non reentrant functions.

All this information is reported in the form of log files to the user and flows into the parallelism extraction phase, discussed in the next section. In the case of library functions that are not visible to the flow, a *whitelist* of system functions that are known to be reentrant is provided to the analysis.

5.3 Partitioning and Parallelism Extraction

Until now, this chapter described how the sequential application model is constructed, which accounts for the first step of the sequential problem in Sect. 2.4.2. The remaining two steps of the problem definition are discussed in this section, namely how to obtain a parallel annotated model and how to select a suitable parallel implementation. Section 5.3.1 briefly reviews the graph clustering approach of the MAPS framework. Thereafter, Sect. 5.3.2 describes how parallelism patterns (TLP, DLP and PLP) are identified and annotated to the function graphs (second step of the problem definition). Section 5.3.3 then explains how the parallelism extraction at the level of functions is treated at the level of a complete application (third step of the problem definition).

5.3.1 Graph Clustering

The clustering approach in the MAPS framework has been reported in [47]. This section briefly discusses its main ideas.

The goal of the clustering process in the MAPS framework is to fragment a sequential application into blocks of a given granularity suitable for parallelism extraction.

Typical compiler granularities are either too fine or too coarse and highly depend on the programming style. On the one hand, statements represent the lowest granularity level, which is clearly too fine-grained. On the other hand, the granularity of basic blocks and functions is determined by the programming style. These observations motivated the definition of a new entity, called *Coupled Block* (CB).

Definition 5.1 A **coupled block** (CB^f) is a subgraph of the graph representation of a function f at the statement level (CDFG or DFG) that has a single entry and a single exit and is *densely connected.* More formally, given a graph $DFG^f = (S^f_{\text{stmt}}, E^f_c \cup E^f_d, \text{var}_{\text{size}})$ and profiling information function π^A, a cluster $C = \{s_1, \ldots, s_n\} \subseteq S^f_{\text{stmt}}$ with induced subgraph $G = (C, E_C)$ is a coupled block if:

1. Single entry/exit: There are two nodes in the subgraph that dominate and postdominate every node in the cluster, and no node outside the cluster is dominated and postdominated by these nodes, i.e.,

$$\begin{array}{c} \exists s_{\text{entry}} \in C, \text{pred}(s_{\text{entry}}) \cap C = \emptyset \wedge \forall s \in C,\ s_{\text{entry}} \text{ dom } s \\ \exists s_{\text{exit}} \in C, \text{succ}(s_{\text{exit}}) \cap C = \emptyset \wedge \forall s \in C,\ s_{\text{entry}} \text{ pdom } s \\ \forall s \in S^f_{\text{stmt}}, (s_{\text{entry}} \text{ dom } s \wedge s_{\text{entry}} \text{ pdom } s) \Rightarrow s \in C \end{array} \tag{5.2}$$

3. Densely connected: The amount of data that is transported during runtime must be high, compared to a given threshold T. This is modeled by

$$\frac{w_d \cdot \sum_{e \in E^f_d} \text{var}_{\text{size}}(e) \cdot \pi^A(e) + w_c \cdot \sum_{e \in E^f_c} \pi^A(e)}{t_C} > T \tag{5.3}$$

where t_C is the sequential time estimation for the execution of the statements in C. Notice that the control edges are also considered in the ratio, since they introduce synchronization overhead. The normalizing factor in the denominator (t_C) serves to scale communication in a similar way as it is done in ratio-cut approaches. The programmer can try out different values of the threshold T to tune the granularity of the clusters.

MAPS determines a good partitioning into CBs by using a hierarchical clustering algorithm (see Definition 2.32) based on *Density Boundary Scan* (DBSCAN) [80]. Further details can be found in [47]. After the partitioning process, all the functions are represented by partitioned graphs in the sense of Definition 2.33.

5.3.2 Discovering Parallelism Patterns

The initial clustering approach of the MAPS framework heavily relied on the user to produce the final parallel implementation. It was up to him to decide if two clusters which were loosely connected by data dependencies would actually feature a good

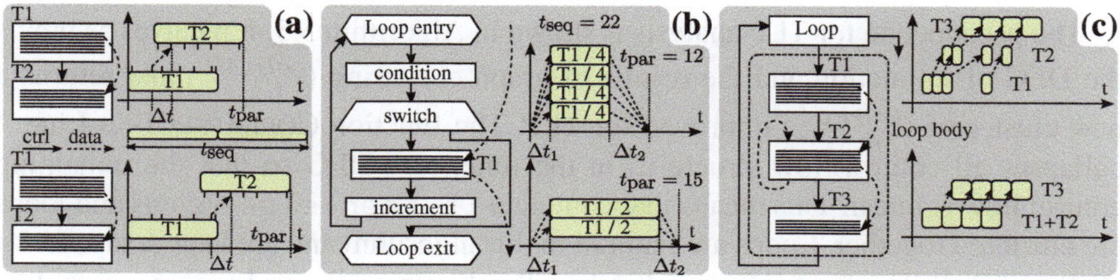

Fig. 5.5 Patterns in a partitioned DFG and sample Gantt Charts. *Lines* in the nodes represent program points. **a** Examples of TLP with different data dependencies. **b** Example of DLP with two different levels of parallelism. **c** Example of PLP with and without balancing

degree of TLP. Besides, DLP and PLP are not explicitly addressed by a plain clustering approach. This section presents new heuristics that expose all different kinds of parallelism, and reduce the amount of user interaction needed to obtain a parallel implementation.

Figure 5.5 shows the parallelism patterns introduced in Sect. 2.4.1.5, as seen in a dependence flow graph. Discovering those patterns from a DFG is a straightforward process. It is more challenging to decide when a given pattern exposes relevant parallelism. The way this is done is described in Sects. 5.3.2.1–5.3.2.3. The discussion in those sections is based on a measure of the efficiency of the parallel execution.

Definition 5.2 The **parallel efficiency** of a portion of code with sequential execution time t_{seq} and parallel execution time t_{par} when running on n_{PE} cores is defined as:

$$\eta = \frac{t_{\text{seq}}}{t_{\text{par}} \cdot n_{\text{PE}}} \tag{5.4}$$

The sequential execution time can be obtained by using some of the performance estimation functions on any given processor type. The parallel execution time must be estimated, depending on the type of parallelism, communication costs and other factors. The goal of the parallelism extraction process is to reduce t_{par} while retaining an acceptable parallel efficiency, as close as possible to unity. For acceptable results, $\eta \in (1/n_{\text{PE}}, 1)$. That is, a parallelization scheme for which $t_{\text{par}} > t_{\text{seq}}$ is not acceptable.

5.3.2.1 Analysis for TLP

As mentioned before, clusters in a partitioned graph explicitly express potential TLP. In general, TLP is characterized by nodes which feature few data dependencies, or no dependency at all. The benefit of actually creating tasks for each cluster depends on several factors. This is shown graphically in Fig. 5.5a for two different configurations of a two-clusters graph. As the figure shows, the efficiency of the parallel execution depends on the program points that create the dependencies and the time it takes to communicate the data (Δt in the figure). If the data is produced *late* and needed *early* in the clusters, the parallel execution could be even longer than the sequential one.

The pseudocode for TLP analysis is shown in Algorithm 5.3. It receives as inputs the DFG of a function and the results of graph clustering ($\mathscr{C}_i^{DFG^f}$) and returns a new clustering ($\mathscr{C}_o^{DFG^f}$) improved for TLP. The function COLLAPSECF in Line 2 collapses all control flow structures in the function's DFG so that the remaining control flow is linear. This means that complete `if-then-else` regions and loops are clustered together. Loops are ignored in this algorithm and are later analyzed for DLP and PLP. The loop in Lines 4–11 walks the linear list and merges every two consecutive clusters that display a parallel efficiency lower than a threshold (η^* in Line 8). The target-dependent sequential time t_{seq} is computed using Eq. 5.1 for all the basic blocks in clusters C and C'. The parallel execution time t_{par} is obtained from an ASAP scheduling of the two clusters on any given PE, using the fastest communication primitive to estimate data communication costs. The computation of t_{seq} and t_{par} is illustrated on the right-hand side of Fig. 5.5a. The Gantt Charts show the results of the ASAP schedules for two sample situations.

Algorithm 5.3 Improving TLP

1: **procedure** TLPFIND($DFG^f, \mathscr{C}_i^{DFG^f}$)
2:     COLLAPSECF($\mathscr{C}_i^{DFG^f}$)
3:     $C \leftarrow$ FIRST($\mathscr{C}_i^{DFG^f}$), $\mathscr{C}_o^{DFG^f} \leftarrow \emptyset$
4:     **while** succ(C) $\neq \emptyset$ **do**
5:         $C' \leftarrow$ succ(C)   *// Next cluster in the linear control flow*
6:         $t_{\text{par}} \leftarrow$ ASAP($\{C, C'\}, DFG^f$), $t_{\text{seq}} \leftarrow$ SEQTIME($\{C, C'\}, DFG^f$)
7:         $\eta \leftarrow t_{\text{seq}}/(2 \cdot t_{\text{par}})$
8:         **if** $\eta < \eta^*$ **then** $C \leftarrow C \cup C'$ // $\eta^* \geq 0.5$
9:         **else** $\mathscr{C}_o^{DFG^f} = \mathscr{C}_o^{DFG^f} \cup \{C\}, C \leftarrow C'$
10:         **end if**
11:     **end while**
12:     **if** $C \notin \mathscr{C}_o^{DFG^f}$ **then** $\mathscr{C}_o^{DFG^f} \leftarrow \mathscr{C}_o^{DFG^f} \cup \{C\}$
13:     **end if**
14:     **return** $\mathscr{C}_o^{DFG^f}$
15: **end procedure**

Although the example in Fig. 5.5a has only one dependency edge, it is easy to understand how this applies to cases with more data dependencies. Since the purpose of this phase is to expose all available parallelism in the application, ASAP scheduling is used, which ignores resource constraints. Whether or not two clusters will actually be exported as tasks is decided later. For this reason, Algorithm 5.3 ignores other costs associated with the parallel execution, e.g., the time needed for task creation.

5.3.2.2 Analysis for DLP

The DLP graph pattern is shown on the left-hand side of Fig. 5.5b. It is characterized by a loop body that has no loop carried dependencies, other than those generated by the loop induction variables. The loop has input and output dependency edges, but the memory locations that create these dependencies are disjoint across loop iterations. This means that the body of the loop can be copied several times and can work in parallel on different input and output data containers. The following restrictions are also checked before marking a loop as DLP:

- There must be one induction variable.
- The loop body must have no side effects, such as calls to non-reentrant functions.
- Incoming and outgoing data edges must refer to a single array.
- The loop must be well structured.

All this information is contained in the annotations produced by the graph analysis phase at the level of loops and functions (see Sect. 5.2.4). Since no data dependencies have to be considered as in the case of TLP, the efficiency computation is simpler for DLP. The parallel time is modeled by $t_{par} = t_{seq}/n_{PE} + t_{comm}(n_{PE})$, where t_{comm} measures the communication and synchronization overhead, depending on the amount of cores n_{PE} that are used. Note that if $\forall n_{PE}, t_{comm}(n_{PE}) \rightarrow 0$, then $\eta \rightarrow 1$. In reality, the synchronization overhead grows with the number of tasks that are created. This is illustrated in the sample Gantt Charts on the right-hand side of Fig. 5.5b. The efficiency of the 4-PE and 2-PE configurations is $22/(12 \cdot 4) = 0.46$ and $22/(15 \cdot 2) = 0.73$ respectively.

Algorithm 5.4 DLP Annotation

1: **procedure** DLPANNOT($DFG^f, C, \bar{c}, n$)
2: $PA^C \leftarrow (PA^C_{type} = \emptyset, X^{PA^C} = \emptyset, V^{PA^C} = \{v_1\})$ *// Initial empty parallel annotation*
3: $D^{PA^C}_{v_1} \leftarrow \{1, \ldots, \bar{c}\}$ *// $\bar{c}$: Average trip count*
4: **if** $\bar{c} < K \cdot n$ **then return** PA^C *// K: minimum ratio trip count – loop executions*
5: **end if**
6: $t_{seq} \leftarrow$ SEQTIME($\{C\}, DFG^f$)
7: $PA^C_{type} \leftarrow$ DLP
8: $m \leftarrow 0$
9: **repeat**
10: $m \leftarrow m + 1$
11: $t_{comm}(m) \leftarrow$ COMMCOST(DFG^f, C, m), $t_{par} \leftarrow t_{seq}/m + t_{comm}(m)$
12: $\eta \leftarrow t_{seq}/(m \cdot t_{par})$
13: **until** $(\eta < \eta^*) \vee (m \geq m_{max})$ *// m_{max}: e.g., number of processors*
14: $D^{PA^C}_{v_1} \leftarrow D^{PA^C}_{v_1} \cap \{1, \ldots, m\}$
15: **return** PA^C *// With associated domain $D^{PA^C}_{v_1}$*
16: **end procedure**

The heuristic for DLP is quite simple, as shown in Algorithm 5.4. It receives as input a cluster C from a function's DFG that represents the body of a loop that meets the conditions for DLP stated above. It also receives the average trip count ($\bar{c}$) and the amount of times the loop was executed (n). These two values are annotated in the loop header after graph analysis. The function returns a parallel annotation in the sense of Definition 2.34. Recall the definition of a parallel annotation for DLP $PA^C = (\text{DLP}, \emptyset, V^{PA^C} = \{v_1\})$, where the variable v_1 determines the number of data parallel tasks to create from the parallel loop. Algorithm 5.4 iterates over the possible number of tasks m in Lines 9–13 and stops once the efficiency falls below a threshold η^* or m exceeds a maximum value m_{max}, e.g., the number of PEs in the platform. The domain of the variable is then restricted to $\{1, \ldots, \min(\bar{c}, m)\}$ in Line

14. The code in Lines 4–5 is used to discard loops that provide DLP but would produce an overall low gain. This is determined by comparing the amount of times the loop is started with the average trip count. For example, a constant $K = 10$ enforces that a loop has to have at least 10 times more iterations than the amount of times the loop is instantiated. This helps to reduce the task creation overhead which is ignored in the computation of t_{par}.

5.3.2.3 Analysis for PLP

The PLP pattern is shown on the left-hand side of Fig. 5.5c, where the DFG loop nodes were omitted for the sake of clarity. PLP is characterized by a loop with several clusters and mostly forward dependencies. Loop carried dependencies are allowed, as long as they are not between the last and the first clusters.

Algorithm 5.5 PLP Annotation

1: **procedure** PLPANNOT($DFG^f, C, \bar{c}, n, \mathscr{C}_{i-1}^{DFG^f}$)
2: $PA^C \leftarrow (PA_{\text{type}}^C = \emptyset, X^{PA^C} = \emptyset, V^{PA^C} = \{v_1, v_2\})$ *// Initial empty parallel annotation*
3: $X^{PA^C} \leftarrow$ DECLUSTER($C, \mathscr{C}_{i-1}^{DFG^f}$)
4: $X^{PA^C} \leftarrow$ COLLAPSECF(X^{PA^C}); $m <\leftarrow |X^{PA^C}|$ *// X^{PA^C} from previous partition*
5: **if** $(\bar{c} < K \cdot n) \vee m < 2$ **then return** PA^C
6: **end if**
7: $D_{v_1}^{PA^C} \leftarrow \{2, \ldots, m\}$, $PA_{\text{type}}^C \leftarrow$ PLP, $t_{\text{seq}} \leftarrow \bar{c} \cdot$ SEQTIME($\{C\}, DFG^f$), $D_{v_2}^{PA^C} \leftarrow \emptyset$
8: **for** $i \in D_{v_1}^{PA^C}$ **do**
9: $(D, t_{\text{stage}}) \leftarrow$ BALANCEPIPE(DFG^f, X^{PA^C}, i)
10: $t_{\text{par}} \leftarrow (\bar{c} + i - 1) \cdot (t_{\text{stage}} + t_{\text{comm}})$, $\eta \leftarrow t_{\text{seq}} / (i \cdot t_{\text{par}})$
11: **if** $\eta > \eta^*$ **then** $D_{v_2}^{PA^C} \leftarrow D_{v_2}^{PA^C} \cup D$
12: **end if**
13: **end for**
14: **return** PA^C *// With associated variable domains*
15: **end procedure**

16: **procedure** BALANCEPIPE($DFG^f, \mathscr{L}, n_{\text{stage}}$)
17: $t_{\text{budget}} \leftarrow$ SEQTIME($\mathscr{L}, DFG^f$)$/n_{\text{stage}}$ *// Optimal time per stage*
18: $C \leftarrow$ FIRST($\mathscr{L}$), $t_{\text{stage}} \leftarrow$ SEQTIME(C, DFG^f), $s \leftarrow 1$, $t_{\text{stage}}^* \leftarrow t_{\text{stage}}$, $D \leftarrow \{(n_{\text{stage}}, C, s)\}$
19: **while** succ$(C) \neq \emptyset$ **do**
20: $C' \leftarrow$ succ(C), $t' \leftarrow$ SEQTIME(C', DFG^f)
21: **if** $t' + t_{\text{stage}} > K \cdot t_{\text{budget}}$ **then** $s \leftarrow \min(s+1, n_{\text{stage}})$, $t_{\text{stage}} \leftarrow 0$ *// K: Imbalance factor*
22: **end if**
23: $D \leftarrow \{(n_{\text{stage}}, C', s)\}$, $t_{\text{stage}} \leftarrow t_{\text{stage}} + t'$
24: **if** $t_{\text{stage}}^* < t_{\text{stage}}$ **then** $t_{\text{stage}}^* \leftarrow t_{\text{stage}}$
25: **end if**
26: **end while**
27: **return** (D, t_{stage}^*)
28: **end procedure**

The pseudocode in Algorithm 5.5 describes how the parallel annotations for PLP are created. The algorithm receives the same inputs as for DLP in Algorithm 5.4, together with the previous clustering in the hierarchy ($\mathscr{C}_{i-1}^{DFG^f}$). The function returns a PLP parallel annotation, $PA^C = (\text{PLP}, X^{PA^C}, V^{PA^C} = \{v_1, v_2\})$, where X^{PA^C} are the partitions that constitute the loop body, v_1 is a variable that determines the number of pipeline stages and v_2 is a function that maps clusters to pipeline stages.

The partitions that constitute the loop body (C) are retrieved from the clustering $\mathscr{C}_{i-1}^{DFG^f}$ by the function DECLUSTER in Line 3. Thereafter, the control flow is collapsed in Line 4, in the same way it was done for TLP in Algorithm 5.3. The code in Lines 5–6 discards the loop in a similar way as it was done for DLP. In addition to the execution condition, the loop is rejected if the loop body has no partition. The amount of clusters $m = |X^{PA^C}|$ defines the maximum number of pipeline stages, and therefore, restricts the domain of the first variable of the parallel annotation (see Line 7). The speedup of a pipeline cannot be measured per iteration as done for DLP, but requires a complete execution of the loop. For this reason, the sequential estimation returned for C is multiplied by the average trip count $\bar{c}$ in Line 7.

The loop in Lines 8–13 determines the domain of the second variable v_2 by iterating over all possible pipeline lengths. Note that this variable defines a mapping of clusters to pipeline stages depending on the number of stages to use. This can be modeled as a relation on $D_{v_1}^{PA^C} \times X^{PA^C} \times D_{v_1}^{PA^C}$, where a triple ($x \in D_{v_1}^{PA^C}$, $y \in X^{PA^C}$, $z \in D_{v_1}^{PA^C}$) means that in a x-stage pipeline, the cluster y is mapped to the z-th stage. The actual mapping happens within the function BALANCEPIPE, which returns a mapping D for a given number of stages n_{stage}. The function also returns the time of the largest pipeline stage t^*_{stage}. The algorithm in BALANCEPIPE corresponds to the greedy *first-fit* solution of the *bin packing problem* [89]. The *first-fit decreasing* variant is not used in order to retain the original control flow. The constant $K \in (1.0, 1.5)$ in Line 21 allows some imbalance. A value of $K = 1.0$ often results in a very unbalanced last stage.

After obtaining a mapping of the clusters to pipeline stages in Line 9, the time of the parallel execution is estimated. For the estimation, the time of the largest pipeline stage is considered. The time needed to initialize and flush the pipeline is approximated by $(i - 1) \cdot t_{\text{stage}}$, so that the total execution of the pipeline is $2 \cdot (i - 1) \cdot t_{\text{stage}} + (\bar{c} - (i - 1)) \cdot t_{\text{stage}} = (\bar{c} + i - 1) \cdot t_{\text{stage}}$. The communication cost is also added to the computation time, as shown in Line 10. Only if the efficiency of the pipeline configuration is above the threshold η^*, it is added to the domain of variable v_2. For the example in Fig. 5.5c, the domain of variable v_2 would then be $D_{v_2}^{PA^C} = \{(2, T1, 1), (2, T2, 1), (2, T3, 2), (3, T1, 1), (3, T2, 2), (3, T3, 3)\}$.

5.3.3 Global Analysis

The algorithms described in Sects. 5.3.2.1–5.3.2.3 work at the level of functions (and loops). After they are run, a parallel annotated graph model of the application is available that exposes all kinds of parallelism (see Definition 2.35). The purpose of global analysis is to finally select a suitable parallel implementation option in the sense of Definition 2.37. In order to judge how relevant the parallelism within a function is, the whole application has to be considered in the global analysis.

Algorithm 5.6 Complete parallelism extraction algorithm

1: **procedure** PARALLELISMEXTRACTION($\mathscr{M}^A = (\mathscr{SE}^A, CG^A, \pi^A), L = \{f_1, \ldots, f_n\} \subseteq S_f^A$)
Require: Results from graph analysis
2: **for** $f \in L$ **do**
3: $(\mathscr{C}_1^f, \ldots, \mathscr{C}_l^f) \leftarrow$ GRAPHCLUSTER(DFG^f) *// Briefly introduced in Sect. 5.3.1*
4: $\mathscr{C}_l^f \leftarrow$ TLPFIND($DFG^f, \mathscr{C}_l^f$) *// from Algorithm 5.3*
5: **for** $C \in \mathscr{C}^f$ a loop body **do**
6: Get $\bar{c}, n$ from analysis
7: DLPANNOT($DFG^f, C, \bar{c}, n$) *// from Algorithm 5.4*
8: PLPANNOT($DFG^f, C, \bar{c}, n, \mathscr{C}_{l-1}^f$) *// from Algorithm 5.5*
9: **end for** *// Parallel-annotated graph ready:* $DFG_{\text{par}}^f = (C, E, \mathscr{PA}^{C'})$
10: **end for** *// Parallel-annotated application model ready:* $\mathscr{M}_{\text{par}}^A = (\mathscr{SE}^A, CG_{\text{par}}^A, \pi^A)$
11: $L_{\text{hotspots}} \leftarrow$ REVERSETOPOLOGICALSORT(CG_{par}^A, L)
12: **for** $f \in L_{\text{hotspots}}$ **do** *// Walking the list in order*
13: **if** AMDAHL(f) $< T$ **then**
14: REMOVEANNOT(DFG^f), **continue**
15: **end if**
16: GETIMPLEMENTATION(DFG_{par}^f)
17: UPDATETIME($\mathscr{M}_{\text{par}}^A$)
18: **end for** *// Parallel implementation ready:* $\mathscr{PI}^A = (\mathscr{SE}^A, CG_{\text{pi}}^A, \pi^A)$
19: **return** $\mathscr{PI}^A$
20: **end procedure**

The pseudocode for the complete parallelism extraction, including global analysis, is shown in Algorithm 5.6. The loop in Lines 2–10 performs graph clustering and parallelism discovery, as discussed in the previous sections, over a list of user-specified functions L. The graph clustering process in Line 3 returns a hierarchical clustering of each DFG (see Definition 2.32). The TLP, DLP and PLP annotation processes are run over the clustering of each function. At the end of this loop (in Line 10), all parallelism has been exposed. With this information, the loop in Lines 12–18 selects the final parallel implementation. Only functions for which the parallel execution reports a considerable *gain* are analyzed further (see Lines 13–15). This gain is computed according to Amdahl's law [5], as:

$$\text{AMDAHL}(f) = \frac{t_{\text{seq}}(CG^A)}{t_{\text{seq}}(CG^A) - t_{\text{seq}}(f) + t_{\text{par}}(f)} \tag{5.5}$$

where $t_{\text{seq}}(CG^A)$ and $t_{\text{seq}}(f)$ are the sequential execution times of the application and of the function respectively. $t_{\text{par}}(f)$ is the estimation of the parallel execution of the function, considering the maximum speedups for TLP, DLP and PLP analyzed before. If the value is below a threshold T, all parallel annotations are removed (see Line 14).

The functions in the list L_{hotspots} are reverse-topologically sorted, as indicated in Line 11. This is important, since selecting a parallel implementation for a function

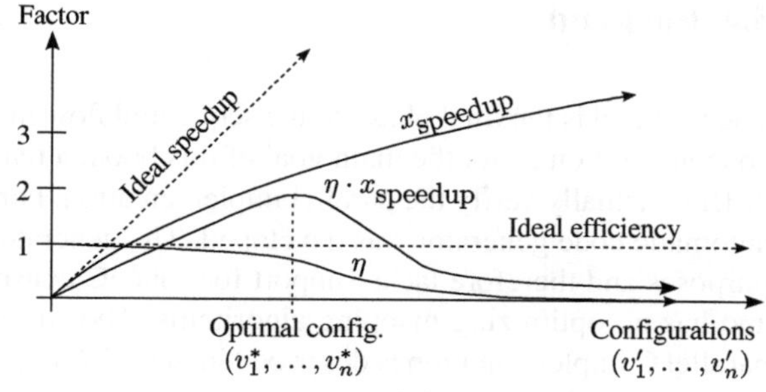

Fig. 5.6 Graphical view of the tradeoff efficiency-speedup

influences the timing of caller functions, i.e., predecessors in the CG. After an implementation has been selected, the function UPDATETIME is in charge of updating the execution time of the functions and recalculating the speedups.

The function GETIMPLEMENTATION in Line 16 is the core of the global analysis. It performs the following tasks:

1. It removes annotations within functions for which the Amdahl value is below the threshold T.
2. If a loop has two parallel annotations (DLP and PLP), the DLP annotation is selected per default, since DLP usually incurs less communication overhead.
3. It assigns values to the variables of every parallel annotation in the function.

For the variable assignment, the function GETIMPLEMENTATION uses the parallel efficiency and the speedup computed in the previous analysis phases. From these values, it selects the combination that maximizes the efficiency-speedup product, i.e., $\eta \cdot x_{\text{speedup}}$. Recall the general definition of a parallel annotation for a cluster C, $PA^C = (\{\text{DLP}, \text{PLP}\}, X^{PA^C}, V^{PA^C})$ from Definition 2.34. In this step, each variable $v_i \in V^{PA^C}$ is assigned a value v_i^*, such that

$$(v_1^*, \ldots, v_n^*) = \underset{V=(v_1',\ldots,v_n') \in \times_{v_i \in V^{PA^C}} D_{v_i}^{PA^C}}{\text{argmax}} (\eta^V \cdot x_{\text{speedup}}^V) \tag{5.6}$$

where η^V and x_{speedup}^V are the efficiency and speedup values that correspond to the configuration described by the variable assignment $V = (v_1', \ldots, v_n')$. This optimization process is intuitively illustrated in Fig. 5.6. The x-axis represents all the possible configurations, sorted in ascending order according to their speedup factor (in the y-axis). Given the small number of variables, the optimization problem in Eq. 5.6 is solved exhaustively.

5.4 Backend

The backend is the final phase of the sequential flow in Fig. 5.1. Although automatic code generation is not the main goal of this book, a real backend is indispensable in order to actually verify the parallel implementation. For this reason, an experimental automatic code generator was developed. The generator serves only for verification purposes and therefore lacks support for some C constructs, e.g., nested structures, and has no optimizing mapping algorithms. The strategy followed to generate the parallel C implementation is described in Sect. 5.4.1. The hints for migration to CPN are discussed in Sect. 5.4.2.

5.4.1 C Backend

For verification purposes, two main goals steered the design of the C backend. First, the parallel implementation must use a distributed memory model for communication in order to easily detect missing dependencies. A shared memory model could work fine in the verification host but break in the target architecture. Second, the generated code must retain the original structure as much as possible, e.g., variable names and control structures. This eases later modifications by the user.

To achieve the first goal, the generated code uses Pthreads for task management and MPI for data communication. Communication is implemented in a blocking fashion without buffering. To this end, synchronous MPI functions are used, with the prototypes shown in Listing 5.2.

```
int MPI_Ssend(void *buf, int count, MPI_Datatype datatype,
     int dest, int tag, MPI_Comm comm);
int MPI_Recv(void *buf, int count, MPI_Datatype datatype,
     int source, int tag, MPI_Comm comm, MPI_Status *status)
```

Listing 5.2 MPI functions used for code generation

The second goal is more difficult to achieve. Commonly, code generation approaches produce C code from the internal compiler IR. As a consequence, variable names are lost, several new temporary variables are added and the control structures are replaced by `goto` statements. The LLVM framework includes a C backend that suffers from these problems. An alternative to code generation from the IR, is to directly use the AST in the compiler frontend. Performing modifications at the AST level is however cumbersome. Besides, it requires to run the C pre-processor, which clutters the code with macro expansions and include-directives for system headers. In this book, code generation has been implemented by using the *Rewriter* class of the Clang frontend [160]. This class provides direct access to the locations in the source code (line and column numbers) for elements in the AST. Additionally, the class provides functions to insert, move and delete portions of code while updating the source locations accordingly. This functionality is key for producing human-readable code, as demonstrated in the following examples for TLP, DLP and PLP.

5.4.1.1 Parallel Main Function

In the parallel implementation, the original main function is removed and is replaced by a general main function. The structure of this new *parallel main* is shown in Listing 5.3. The parallel main initializes all the parallel threads by calling the application-dependent `initialize` function (Line `20`). It then triggers execution of the corresponding task (Line `23`) and waits until everything is done (Line `26`). An example of the initialization code is provided in Lines `31–37`.

```
1 /* Target include libraries */
2 #include "TARGET/mpi.h"  // ...
3
4 /* Handlers for worker threads and external initializers */
5 typedef int (*TaskPtr)();
6 extern TaskPtr TaskPtrs[];
7
8 /* More externs (application dependent) */
9 extern void initialize();
10 extern int numberOfCores;
11
12 int main () {
13   int argc; char ** argv;
14   int numprocs, rank, namelen, level = 0;
15
16   MPI_Init_thread(&argc, &argv, MPI_THREAD_MULTIPLE, &level);
17   MPI_Comm_size(MPI_COMM_WORLD, &numprocs);
18   MPI_Comm_rank(MPI_COMM_WORLD, &rank);
19   MPI_Get_processor_name(processor_name, &namelen);
20   initialize();
21
22   if (rank < numberOfCores && TaskPtrs[rank] != NULL) {
23     TaskPtrs[rank]();
24   }
25
26   MPI_Barrier(MPI_COMM_WORLD); MPI_Finalize();
27   return 0;
28 }
29
30 /* Initialization code (in different compilation unit) */
31 TaskPtr TaskPtrs[4] = { NULL };
32 int worker0(); int worker1(); int worker2(); int worker3();
33
34 void initialize() {
35     TaskPtrs[0] = &worker0; TaskPtrs[1] = &worker1;
36     TaskPtrs[2] = &worker2; TaskPtrs[3] = &worker3;
37 }
```

Listing 5.3 Main module of the parallel C code

5.4.1.2 Generation for TLP

For TLP, the original code of every cluster is wrapped into a function and the variables are privatized. Thereafter read and write function calls are inserted before every use and after every definition of an external dependency. This is illustrated by the example in Listing 5.4. The names of the new functions are extended with a number that serves to identify the line in the original source file where the code originates from. In the example, `worker4` in Line `14` wraps the cluster defined between Lines `4–6`. Similarly, `worker7` in Line `22` corresponds to the cluster in Lines `7–9`.

```
1  /************************** Original **************************/
2  void tlp_f() {
3    int a, b, c, t;
4    a = foo();    // start first cluster
5    b = bar();
6    c = bar()+b; // used somewhere else
7    t = foo();    // start second cluster
8    t = a;
9    a++;
10   //...
11 }
12
13 /************************** Generated *************************/
14 void worker4() {
15   int a, b, c; // Private
16   a = foo();
17   MPI_Ssend(&a,4,MPI_BYTE,2,0,MPI_COM_WORLD);
18   b = bar();
19   c = bar()+b;
20   MPI_Ssend(&c,4,MPI_BYTE,3,1,MPI_COM_WORLD);
21 }
22 void worker7() {
23   int a, t; // Private
24   t = foo();
25   MPI_Recv(&a,4,MPI_BYTE,1,0,MPI_COM_WORLD);
26   t = a;
27   a++;
28 }
```

Listing 5.4 Code generation for TLP

5.4.1.3 Generation for DLP

For DLP, the entire loop is copied to a new function. The data flow information in the loop header indicates which variables have to be communicated before and after the loop. As in TLP, variables are privatized in the function. An example of code generation for DLP is in Listing 5.5. Two new functions are created, one that spawns the threads and collects the results (`dlp_f_spawn` in Line `11`) and a function that is used to implement the copies of the for loop (`parLoop5` in Line `31`). In the example, the loop is split into four copies.

5.4.1.4 Generation for PLP

The generated code for PLP is similar to that of TLP and DLP, as can be seen in the example in Listing 5.6. In the example, the code generator exported a two-stage configuration of the pipeline due to the load distribution in the loop (as discussed in detail in Sect. 5.5.1). The code in Line `7` is mapped to the first stage (Lines `16–22`), while the code in Lines `8–10` is mapped to the second stage (Lines `23–34`).

```
/************************** Original **************************/
void dlp_f() {
  int i; float A[400];
  // ...
  for (i = 0; i < 400; i++) // loop at line 5
    A[i] = foo();
  //...
}

/************************** Generated **************************/
void dlp_f_spawn() {
  pthread_t parFor5;
  int parFor5_params_1[3] = { 0, 100, 0 };
  pthread_create(&parFor5, NULL, parLoop5, parFor5_params_1);
  int parFor5_params_2[3] = { 100, 200, 0 };
  pthread_create(&parFor5, NULL, parLoop5, parFor5_params_2);
  //...
  float A[400];

  MPI_Ssend(&A[0], 400, MPI_BYTE, 0, 5, MPI_COMM_WORLD);
  // ...
  MPI_Ssend(&A[300], 400, MPI_BYTE, 3, 5, MPI_COMM_WORLD);
  MPI_Recv(&A[0], 400, MPI_BYTE, 0, 5, MPI_COMM_WORLD,
    MPI_STATUS_IGNORE);
    // ...
  MPI_Recv(&A[300], 400, MPI_BYTE, 3, 33, MPI_COMM_WORLD,
    MPI_STATUS_IGNORE);
  pthread_join(parFor5, NULL);
  return 0;
}
void parLoop5(void *argument) {
  int MPI_thread_tag = 5;
  int lowerLimit = ((int *)argument)[0];
  int upperLimit = ((int *)argument)[1];
  int commID = ((int *)argument)[2];

  float A[400];
  MPI_Recv(&A[lowerLimit], ((upperLimit - lowerLimit) * 4),
     MPI_BYTE, commID, 5, MPI_COMM_WORLD, MPI_STATUS_IGNORE);
  for (int i = 0; i < 400; i++)
    if ((i >= lowerLimit) && (i < upperLimit)) {
      A[i] = foo();
    }
  MPI_Ssend(&A[lowerLimit], ((upperLimit - lowerLimit) * 4),
    MPI_BYTE, commID, 5, MPI_COMM_WORLD);
}
```

Listing 5.5 Code generation for DLP

5.4.2 CPN Hints

CPN hints are detailed comments intended to help to migrate C code into a parallel CPN specification. Together with the recommendations of the analysis phase to improve parallelism, the CPN hints are used during the manual CPN conversion shown in the lower right corner of Fig. 5.1c. This book does not include an automated conversion to CPN. A *safe* conversion to CPN would require more powerful data flow and program analysis or introducing restrictions to the input C specification. For restricted cases, e.g., nested loop programs (NLPs), some of the frameworks discussed in Sect. 3.2.2 could be used.

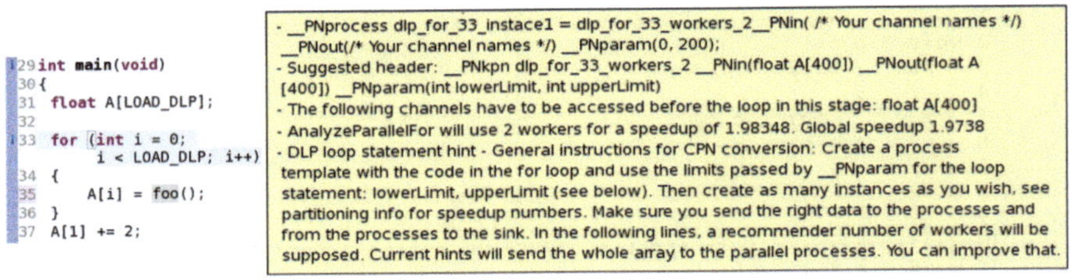

Fig. 5.7 Example of CPN hints in the MAPS IDE

```
/************************** Original **************************/
#define N    400
void plp_f() {
  int i;
  float x = 0.0, y = 0.0, z = 0.0, a = 9.2;
  for (i = 0; i < N; i++) { // loop at line 6
    x = foo();
    y = bar(x / 2);
    y = y + a;
    z = z + bar_(y / 4);
  }
  return z;
}

/************************** Generated **************************/
void pipe6Worker0(void *params) {
  float x; int i;
  for (i = 0; i < N; i++) {
    x = foo();
    MPI_Ssend(&x, 4, MPI_BYTE, 1, 6, MPI_COMM_WORLD);
  }
}
void pipe6Worker1(void *params) {
  float a, x, y, z; int i;
  MPI_Recv(&z, 4, MPI_BYTE, 0, 6, MPI_COMM_WORLD, /*...*/);
  MPI_Recv(&a, 4, MPI_BYTE, 0, 6, MPI_COMM_WORLD, /*...*/);
  for (i = 0; i < N; i++) {
    MPI_Recv(&x, 4, MPI_BYTE, 0, 6, MPI_COMM_WORLD, /*...*/);
    y = bar(x / 2);
    y = y + a;
    z = z + bar_(y / 4);
  }
  MPI_Ssend(&z, 4, MPI_BYTE, 0, 104, MPI_COMM_WORLD);
}
```

Listing 5.6 Code generation for PLP

Figure 5.7 shows an example of the hints exported for a DLP loop, similar to the one in Listing 5.5. The text in the yellow box indicates how to declare the process template and how to instantiate different workers, by using the `__PNparam` keyword to define the limits of the for loop. The last item in the yellow box contains a detailed description about how to write the CPN code. For TLP and PLP similar hints are exported to the MAPS IDE.

5.5 Case Study

This section contains two examples that demonstrate how the sequential tool flow is used to obtain a parallel implementation. The examples are not fully-fledged case studies, as the ones presented by Ceng in [47]. They are rather intended to showcase the new parallelism extraction features presented in Sect. 5.3.2. Section 5.5.1 uses a synthetic example to demonstrate the complete tool flow, down to a running implementation on a target architecture. The example in Sect. 5.5.2 describes the parallelization process for the real-life audio filter application mentioned in Sect. 2.1.

```
1 #define M1  500
2 #define M2  10
3 void main() {
4   int i;
5   float x = 0.0, y = 0.0, z = 0.0, a = 9.2;
6   for (i = 0; i < 400; i++) {
7     x = foo();              // Load varies linearly with M1
8     y = bar(x / 2);         // Load varies linearly with 2*M2
9     y = y + a;
10    z = z + bar_(y / 4); // Load varies linearly with M2
11  }
12  printf("z=%f\n", z);
13 }
```

Listing 5.7 Code with PLP

5.5.1 Synthetic Example

For the synthetic example, the application in Listing 5.7 is used (also in Listing 5.6). The load of the functions `foo`, `bar` and `bar_` is controlled by the values of the macros `M1` and `M2` as suggested by the comments in Lines 7–10. By varying the load of the functions, it is possible to test the heuristics presented in the previous sections to obtain a suitable parallel implementation. PLP was selected for the synthetic example, since it is the most involved form of parallelism. The final implementation is tested on the *Pitaya* platform simulator, described in Sect. 5.5.1.1. The results are then discussed in Sect. 5.5.1.2.

5.5.1.1 The Pitaya Platform

Pitaya is a virtual platform developed in the context of the EURETILE Project [81]. It is a regular 2-dimensional mesh of processing tiles interconnected by an NoC for inter-tile communication. Each tile contains an LTRISC PE, a local memory, a router, an optional floating point unit and other peripherals. Inside the tile, the elements are connected with a simple bus. The platform is modeled using the SystemC TLM 2.0 standard. The baseline Pitaya platform was made heterogeneous by replacing some of the LTRISC processors by LTVLIW processors. A schematic of the platform is shown in Fig. 5.8.

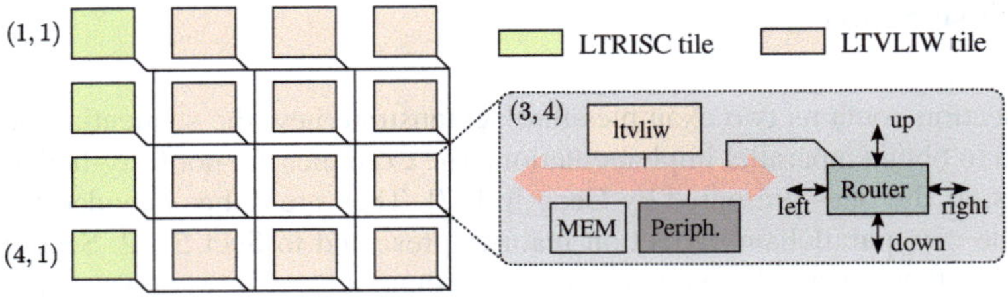

Fig. 5.8 The modified Pitaya platform

The simulator itself was also extended with a thin monitoring layer that uses the breakpoint API of the processor simulators. This layer produces a trace that records execution events, like entering or leaving a function. The events can be defined by a file that is passed to the simulator. This enables an accurate, non-intrusive, measurement of the time of the parallel implementation. The Pitaya platform comes with a thin software stack that implements a reduced set of the MPI API. This set includes the functions used by the backend (see Listing 5.2). However, the platform does not support Pthreads and as a consequence, every partition is mapped to a different processor.

5.5.1.2 Results

The results of the parallelism extraction phase for two different configurations are shown in Fig. 5.9. The partition results are shown as colored lines in the MAPS C editor on the left-hand side of Fig. 5.9. The right-hand side of the figure contains the dynamic call graphs of the two different versions of the application. The effect of the load configurations can be observed in the graphs by the intensity of the colors in the nodes and the number of calls to the nodes in the bottom (`maps_cos` and `maps_sin`). In both configurations, the functions `foo`, `bar` and `bar_` are called 400 times. Functions `bar` and `bar_` make 10 calls to the bottom nodes in the first configuration and 15 calls in the second, corresponding to the values of `M2=10` and `M2=15`. The value of `M2` controls the accuracy of a series implementation of the *sine* and the *cosine* functions in `maps_sin` and `maps_cos`.

In the first configuration in Fig. 5.9a, the parallelism extraction produced only two pipeline stages (see colored regions in the source code editor). This is due to the load distribution in the loop shown by the line profile. For this example, the estimated speedup of a three-stage configuration was of 1.39x, while that of a two-stage pipeline was of 1.59x. Consequently, the parallel efficiencies were of 0.46 and 0.79 respectively. Clearly, the two-stage configuration is more beneficial for this setup (recall Eq. 5.6). An excerpt of the generated code for this configuration can be seen in Listing 5.6.

(a)

(b)

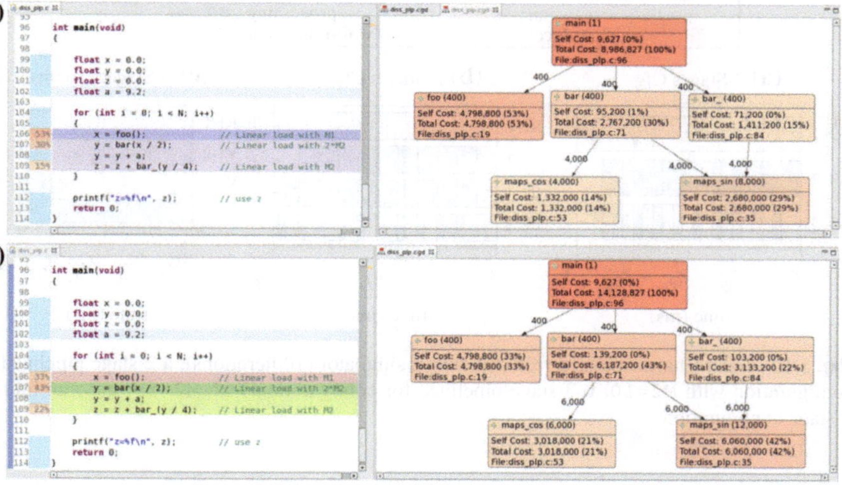

Fig. 5.9 Results of the parallelism extraction phase for the synthetic example. **a** Configuration with `M1=500` and `M2=10`. **b** Configuration with `M2=15`

In the second configuration, the load of `foo` was kept constant, while the load of `bar` and `bar_` was incremented. This produced a redistribution of the load in the main loop as shown in Fig. 5.9b. In this case, a three-stage pipeline reported a speedup of 1.80x and a corresponding efficiency of $\eta = 0.6$. A two-stage configuration reported a speedup of 1.38x, with $\eta = 0.69$. The efficiency-speedup product is then 1.08 and 0.95 for both configurations. For this reason, the parallel implementation contains 3 pipeline stages as shown on the left-hand side of Fig. 5.9b.

Figure 5.10 shows the actual results obtained with the Pitaya simulator for the first 10 iterations of the main loop. In both configurations, the main function runs on the RISC processor at position (1, 1) in the mesh (*R1-1* in the figure). This processor is responsible for initializing the application and triggering the execution. It then remains in a blocked state until the worker threads finish executing (see upper plot in Fig. 5.10a, b). The worker threads, in turn, are mapped to the VLIW processors on the first row (*V1-2–V1-3* in the figure). Compared to a sequential execution on a single VLIW, the measured speedup of the two different configurations was of 1.32x and 1.68x. These values correspond to a deviation of 5.3 and 7.1 % with respect to the speedup predicted by the sequential flow (1.39 and 1.80x). Note that the sequential flow produced the parallel implementations in less than a minute, while simulating the first 10 iterations on the cycle-accurate Pitaya simulator took around an hour. The time effort of a manual exploration with the simulator would therefore be at least two orders of magnitude higher.

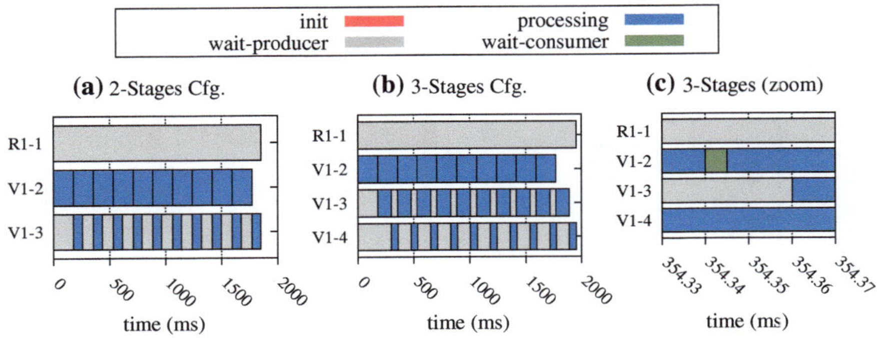

Fig. 5.10 Gantt Charts obtained from the Pitaya simulator (10 iterations). **a** 2-stage pipelined for configuration with `M2=10`. **b** 3-stage pipelined for configuration with `M2=15`. **c** Zoom into the 3-stage configuration

5.5.2 Audio Filter Application

The previous section demonstrated the applicability of the parallelism extraction algorithms proposed in this book on a simple application. In this section, the sequential flow is used to obtain a CPN specification from the C implementation of the audio filter application (LP-AF) from the motivational example in Sect. 2.1. This application has a higher complexity than the synthetic example, with above 500 lines of code and 39 different functions. A simplified version of the most relevant functions is presented in Listing 5.8–5.10. The following sections describe the parallelization process from the point of view of a programmer with little application knowledge.

```
int main()
{
  int t = 0, i = 0, rem = 0, len = 0, loop_cnt = 0;
  short buffer[BLOCK_LEN - NET_LEN][2] = { 0 };
  short src_data[BLOCK_LEN][2];
  short flt_data[BLOCK_LEN][2];
  FILE *in, *out;
  char inname[150] = "input_xs.wav";
  char outname[150] = "output_data_seq_c.wav";
  wave_param_t *wave_param = (wave_param_t *) \
                      malloc(sizeof(wave_param_t));

  // generate bit reverse index vector for fft
  bitRevVec_10B();

  // init the filter
  init_filt();

  in = fopen(inname, "r");
  out = fopen(outname, "wb");
  parse_file(in, wave_param);
  write_header(out, wave_param);
  len = wave_param->data_length / 4;
  rem = len % NET_LEN;
  loop_cnt = (int)((len - rem) / NET_LEN);

  for (t = 0; t < loop_cnt; t++) {
    for (i = 0; i < NET_LEN; i++) {
      src_data[i][0] = read_word(in);
      src_data[i][1] = read_word(in);
    }
    // zero pad
    for (i = NET_LEN; i < BLOCK_LEN; i++) {
      src_data[i][0] = 0;
      src_data[i][1] = 0;
    }

    complex freq_coef[BLOCK_LEN][2];
    complex flt_sink[BLOCK_LEN][2];
    fft_1024_dual(src_data, freq_coef);
    low_pass(freq_coef, flt_sink);
    ifft_1024_dual(flt_sink, flt_data);

    // write first 30 samples with overlap buffer added
    for (i = 0; i < (BLOCK_LEN - NET_LEN); i++) {
      write_word(out, flt_data[i][0] + buffer[i][0]);
      write_word(out, flt_data[i][1] + buffer[i][1]);
    }
    // write the next 964 samples unaltered
    for (i = (BLOCK_LEN - NET_LEN); i < NET_LEN; i++) {
      write_word(out, flt_data[i][0]);
      write_word(out, flt_data[i][1]);
    }

    // store the last 30 samples into overlap add buffer
    for (i = NET_LEN; i < BLOCK_LEN; i++) {
      buffer[i - NET_LEN][0] = flt_data[i][0];
      buffer[i - NET_LEN][1] = flt_data[i][1];
    }
  }
  /* ... process remainder */
  return 0;
}
```

Listing 5.8 LP-AF: Main function

```
void fft_1024_dual(const short time_coef[BLOCK_LEN][2],
                   complex freq_coef[BLOCK_LEN][2])
{
  int n1, d1, k1, m1;
  complex w1, wd1, t1, x1;
  bitReverse_to_complx2(time_coef, freq_coef);
  for (n1 = 1; n1 <= 10; n1++) {
    d1 = d_fac[n1 - 1];
    wd1 = twiddle_fac[n1 - 1];
    for (k1 = 0; k1 < BLOCK_LEN; k1 += d1) {
      w1.real = w_0.real;
      w1.imag = w_0.imag;
      for (m1 = 0; m1 < (d1 / 2); m1++) {
        t1 = compMul(w1, freq_coef[m1 + k1 + (d1 / 2)][0]);
        x1 = freq_coef[m1 + k1][0];
        freq_coef[m1 + k1][0] = compAdd(x1, t1);
        freq_coef[m1 + k1 + (d1 / 2)][0] = compSub(x1, t1);
        w1 = compMul(w1, wd1);
      }
    }
  }
  for (n1 = 1; n1 <= 10; n1++) {
    d1 = d_fac[n1 - 1];
    wd1 = twiddle_fac[n1 - 1];
    for (k1 = 0; k1 < BLOCK_LEN; k1 += d1) {
      w1.real = w_0.real;
      w1.imag = w_0.imag;
      for (m1 = 0; m1 < (d1 / 2); m1++) {
        t1 = compMul(w1, freq_coef[m1 + k1 + (d1 / 2)][1]);
        x1 = freq_coef[m1 + k1][1];
        freq_coef[m1 + k1][1] = compAdd(x1, t1);
        freq_coef[m1 + k1 + (d1 / 2)][1] = compSub(x1, t1);
        w1 = compMul(w1, wd1);
      }
    }
  }
}
```

Listing 5.9 LP-AF: FFT implementation

5.5.2.1 Application Tracing and Profiling

For the first phase of the sequential flow, a raw audio file of around 2 MB is used to analyze the application's runtime behavior. The analysis provides target-dependent application information that serves to gain a better understanding of the hot spots of the application. The results are displayed in the MAPS IDE as discussed in Sect. 5.2.4 and demonstrated in Fig. 5.9 for the synthetic example. The function profile of the LP-AF application is summarized in Fig. 5.11, where the functions are sorted according to their cost. The figure only shows the profile of the 20 most time consuming functions.

The dynamic call graph of the application is shown in Fig. 5.12. In the figure, the profiling and cost information is directly annotated in the nodes and in the edges of the graph. Note that the cost estimation is reported for two different processor types, namely IRISC and LTVLIW. The pair of numbers on the edges represents how often the function was called from a given call site and the line number from which the function was called.

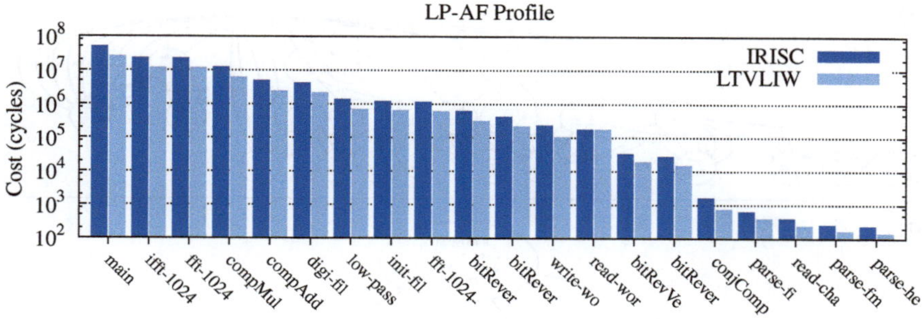

Fig. 5.11 Function profile for the low-pass audio filter application

The information provided by the function and the dynamic call graph allows to focus the parallelization efforts. After a short inspection of the code, four functions can be identified, namely `main`, `ifft_1024_dual`, `fft_1024_dual` and `low_pass`, corresponding to the first three and the seventh functions in the profile. The fourth and the fifth functions in the profile (`comMul` and `compAdd`) contain small operations on complex numbers that offer no coarse parallelism. The sixth and the eight functions are only executed once (see Fig. 5.12b), so that the impact of a performance improvement would diminish as the size of the input audio file grows.

5.5.2.2 Parallelism Extraction

The entire parallelization extraction process is depicted in Fig. 5.13. Two possible algorithmic representations of the application are shown in Fig. 5.13a. These representations correspond to typical *pencil-and-paper* specifications for these kinds of DSP applications. Therefore, they serve as *target* for the parallelization process.

The `main` function is the first one to be analyzed. The sequential flow focus on the main loop, which has an structure similar to the one shown in Fig. 5.13b (see also Listing 5.8). The tool reports no DLP for this loop due to a loop carried dependency on a variable called `buffer`. This variable is actually an array that allows to overlap the first and the last samples of consecutive blocks used by the algorithm. Despite the loop carried dependency, the tool reports several configurations for PLP with up to five stages. The efficiencies computed by the tool were of 0.65, 0.36, 0.22, 0.15 for 2, 3, 4 and 5-stage configurations. The selected configuration has therefore two stages with a speedup of 1.5x over 10 iterations of the main loop. The result of the extraction in the main function can be represented as a two-process KPN, shown on the left of Fig. 5.13e. The tool assigns input processing, FFT processing and the actual filtering to the first process and IFFT processing as well as data write-back to the second process.

Due to their similar structure, the parallelization process for the functions `fft_1024_dual` and `ifft_1024_dual` is identical. The nested loops that

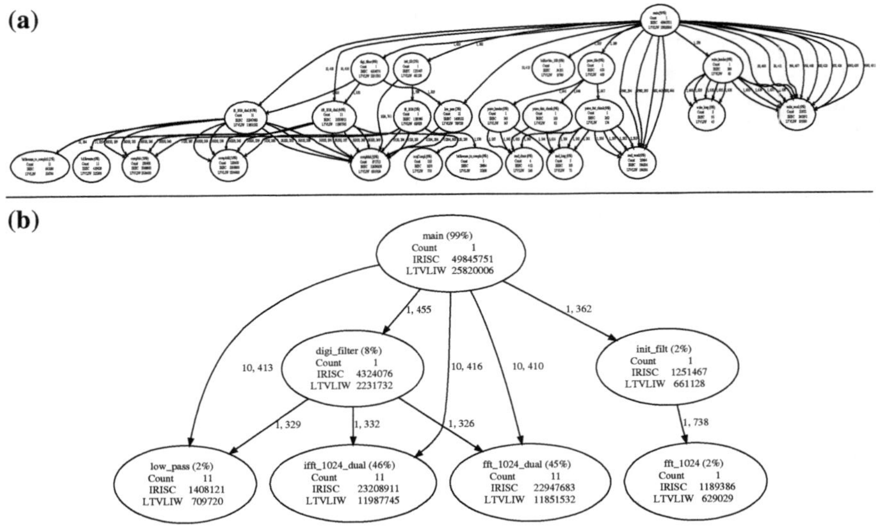

Fig. 5.12 Dynamic call graph of the LP-AF application. **a** Complete call graph. **b** Relevant portion of the call graph

implement the (I)FFT processing feature neither DLP nor PLP. However, the tool reported relevant TLP on the two main outer loops as depicted in Fig. 5.13c (see also Listing 5.9). These loops actually process the left and the right audio channels separately, which explains the decision of the tool. The result of the extraction for the FFT and IFFT functions is represented by the second and the third KPN in Fig. 5.13e.

Finally, the parallelism extraction for the `low_pass` function reported a loop with DLP. This loop has the structure shown in Fig. 5.13d (see also Listing 5.10). Although this parallelization produces a function speedup of 1.74x, it is discarded by the global analysis, since the evaluation of Eq. 5.5 reports a global speedup of less than 2 %. The parallelization hints are however useful to split the filtering process into left and right as it was done for the FFT/IFFT processing.

The final KPN representation obtained with the sequential flow is shown on the right-hand side of Fig. 5.13e. Note that the structure is very close to the algorithmic representation of Fig. 5.13a. The main difference between both specification lies on the filter processing, which is merged with the FFT processing due to pipeline balancing. The semi-automatic conversion to a CPN specification took about half a day, where most of the time was spent implementing the code modifications suggested by the CPN hints. Producing the CPN hints itself took less than 3 min (including application tracing, trace analysis and parallelism discovery).

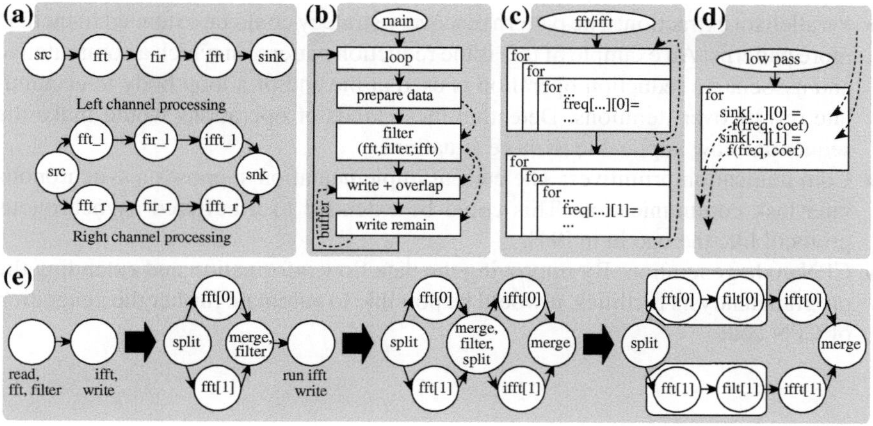

Fig. 5.13 Parallelism extraction for the LP-AF application. **a** Algorithmic representations (mixed and separate processing of audio channels). **b** Parallelism in `main`. **c** Parallelism in `fft_1024` and `ifft_1024`. **d** Parallelism in `low_pass`. **e** Step-wise derivation of a CPN implementation

5.6 Limitations and Outlook

The tool flow for sequential C code presented in this chapter has limitations and drawbacks. Some of them are inherent to the approach, while other are implementation restrictions. The following items represent the major potential extensions for future work:

- Input language: The sequential flow supports all features of C99 but function pointers and recursive functions.
- Handling structures: It is a common approach to use data structures to collect results during the application execution. While the framework can determine the exact field in the structure, it cannot partition the structure in order to expose more parallelism.
- Tracing: The *flat* trace file produced during application tracing typically contains repeated sections induced by loops in the application. As a consequence, the size of the file grows rapidly with the execution time of the application on the host. This issue prevents the flow from being applied to more complex benchmarks. To compress this file, some of the post-processing logic can be migrated to the uTracer runtime library.
- Safety: A fundamental problem of dynamic DFA is that the accuracy of the analysis depends on the inputs fed to the application during tracing. This is the reason why the sequential tool flow requires the programmer to analyze the results and judge if the proposed partitions are safe. To improve safety, a better integration of state-of-the-art static analyses is required. Introducing code coverage to the sequential flow would also help to better assess the safety of a parallel implementation.

- Parallelism extraction: The pattern discovery strategy could be extended to include more patterns. An example of this is the reduction pattern, in which a commutative and associative reduction operation is used at the end of a loop body to accumulate results over iterations. Detecting these kinds of operations would make the sequential flow applicable to more situations.
- Communication primitives: The current implementation supposes a synchronous inter-task communication. This could be extended to a buffered asynchronous protocol like the one in in [94].
- CPN code generation: By improving the data flow information and extending the program analysis facilities, it would be possible to automate further the generation of CPN code.

5.7 Synopsis

This chapter presented a tool flow that solves the sequential problem introduced in Sect. 2.4. The chapter described four new algorithms (Algorithms 5.3–5.6) that complement the graph clustering approach by discovering parallelism patterns that offer a significant speedup. A new experimental C backend was also introduced that allows to verify the parallel implementation on the host or on a target MPSoC.

The main goal of the sequential flow is to help a programmer to get a parallel specification from a C sequential application. This was demonstrated in this chapter by means of simple examples. By using this semi-automatic approach, it is possible to obtain a CPN version of the application that can be then used as an input to the parallel flow, as shown in Sect. 5.5.2. How such a parallel version is finally mapped to the target MPSoC is discussed in the next chapter.

```
void low_pass(const complex src_sig[BLOCK_LEN][2],
              complex sink_sig[BLOCK_LEN][2])
{
  int i;
  for (i = 0; i < BLOCK_LEN; i++) {
    sink_sig[i][0] = compMul(src_sig[i][0], filt_coef_freq[i]);
    sink_sig[i][1] = compMul(src_sig[i][1], filt_coef_freq[i]);
  }
}
```

Listing 5.10 LP-AF: Filter implementation

Chapter 6
Parallel Code Flow

As discussed in Chap. 1, not only sequential but also parallel programming must be supported in order to close the productivity gap. With a parallel specification, programmers can use their application knowledge to explicitly express parallelism that can be otherwise difficult to extract automatically. For the reasons described in Sect. 2.5.1.4, the KPN MoC was selected as the underlying formalism to specify parallel applications. This chapter describes a solution to the parallel code problem as stated in Definition 2.49.

This chapter is organized as follows. The overall tool flow for obtaining a valid runtime configuration and a final executable parallel implementation is introduced in Sect. 6.1. Sections 6.2–6.5 describe how KPN applications are analyzed and present the algorithms proposed in this book for computing an application mapping. Code generation for an OSIP-based target platform is discussed in Sect. 6.6. The algorithms are benchmarked in Sect. 6.7, followed by an analysis of the deficits of the approach in Sect. 6.8. The chapter ends with a summary in Sect. 6.9.

6.1 Tool Flow Overview

An overview of the parallel flow is shown in Fig. 6.1. Besides the CPN code and the architecture model, the flow receives a file with user constraints and configuration options. The focus of this chapter is on the mapping and scheduling component in Fig. 6.1c, which outputs valid runtime configurations (see Definition 2.16).

The parallel flow is divided into four phases. (1) The token logging phase, in Fig. 6.1a, records the history of tokens in all the KPN channels by running a Pthreads implementation of the application on the host. (2) The KPN tracing phase, in Fig. 6.1b, uses the channel logs to profile each process in an isolated manner. This phase shares similarities with the tracing phase of the sequential flow. (3) The mapping and scheduling phase, in Fig. 6.1c, computes a valid runtime configuration using the

J. Castrillón Mazo and R. Leupers, *Programming Heterogeneous MPSoCs*, DOI: 10.1007/978-3-319-00675-8_6,

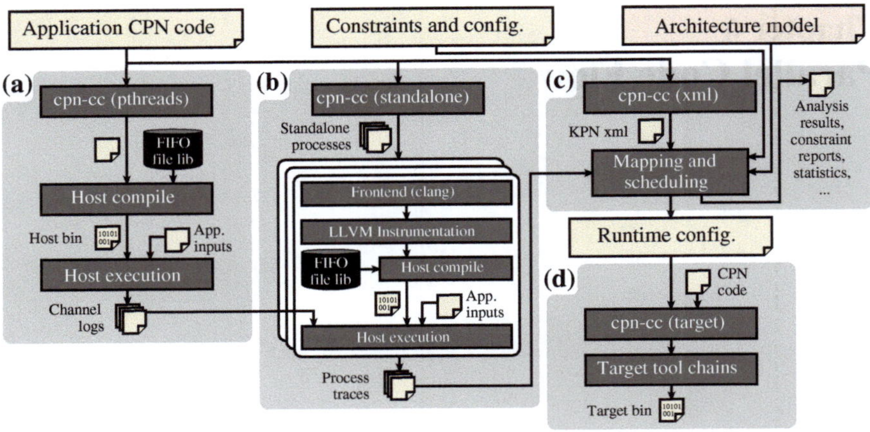

Fig. 6.1 Parallel tool flow. **a** Token logging phase. **b** KPN tracing phase. **c** Mapping and scheduling phase. **d** Backend phase

information collected during tracing. (4) Finally, the binary images for the target platform are generated in the backend phase, as shown in Fig. 6.1d.

The parallel flow uses the Clang and LLVM projects described in Sect. 5.1.1. Additionally, the flow uses the CPN C compiler (*cpn-cc*) of the MAPS framework, which is briefly introduced in Sect. 6.1.1. A functional description of the components of the parallel flow is given in Sect. 6.1.2. This section ends with a detailed presentation of the formats used for the flow's input and output files in Sect. 6.1.3.

6.1.1 The CPN C Compiler

cpn-cc is a retargetable source-to-source compiler implemented as an extension of the Clang frontend. It takes as input the CPN application code together with a runtime configuration and produces C code. In the output code, the CPN constructs are lowered into C constructs and API calls that vary depending on the target architecture. In this work, a few extensions were added to *cpn-cc* (see Sect. 6.6).

The CPN compiler has several different backends. Besides code generation for the target architecture, it can generate different implementations for the host which are intended for functional verification and code analysis. As shown in Fig. 6.1, the parallel flow uses four different *cpn-cc* options.

- Pthreads: This backend produces a Pthreads implementation of the KPN application that can be executed on the host. It allows the programmer to test the functionality of the application while using the more comfortable environment of the host.

- Standalone: This backend generates a standalone C application for every process that can be compiled and run on the host. This allows to debug and analyze each process independently of the rest of the KPN application.
- Xml: This backend produces an XML file with the topology of the KPN application and meta-information relevant to the mapping and scheduling phase.
- Target: This is the actual backend for the target architecture. Therefore, it is the only backend that receives the runtime configuration as input.

6.1.2 Tool Flow Components

The parallel flow of Fig. 6.1 is composed of several components. This section describes their functionality in order to provide a general understanding of the flow.

6.1.2.1 Token Logging Phase

The purpose of this phase is to collect the tokens in all the channels of the KPN specification by running the application. The output channel logs in Fig. 6.1a serve to analyze each of the processes separately during the tracing phase.

The Pthreads implementation is compiled and linked against a custom FIFO library. The FIFO write function in the library stores every token into a binary file. After host execution, a file for every channel is created that contains the history of all the tokens that were written to the channel. While executing, the application's threads are scheduled by the host OS. Note that the way the OS schedules the threads is irrelevant, since KPN applications are determinate, as discussed in Sect. 2.5.1.3.

6.1.2.2 KPN Tracing Phase

The purpose of the tracing phase, in Fig. 6.1b, is to profile the behavior of each process in the application. This is done in a similar way as in the sequential flow described in Sect. 5.2. Each standalone process produced by *cpn-cc* is instrumented and executed to obtain a model $P^A = (\mathcal{SE}^{P^A}, CG^{P^A}, \pi^{P^A})$ (see Definition 2.44). In this phase, the channel logs obtained during the token logging phase are used as input for each consumer process. The read function in the FIFO library, linked to the instrumented object files, reads from the corresponding channel log instead of attempting to read from an actual FIFO. This phase outputs process traces that are used by the heuristics in the mapping and scheduling phase.

6.1.2.3 Mapping and Scheduling Phase

This phase constructs the parallel application model (see Definition 48) and uses heuristics to compute valid runtime configurations. This includes the buffer sizes for the channels, the mapping of processes and channels to platform resources and the scheduling policies to be used during runtime (see Sect. 2.5.4). Besides the runtime configuration, this phase also exports several files with application profiles, execution statistics and other reports that help the user to get a better understanding of the application.

6.1.2.4 Backend Phase

The purpose of this final phase is to produce an executable implementation of the KPN application for the target platform, according to the runtime configuration selected by the programmer. This configuration can be the one generated by the mapping and scheduling phase or a custom configuration produced by the user.

6.1.3 Input and Output Modeling

Besides the CPN code, the parallel flow receives three more inputs (architecture model, constraints and configuration files in Fig. 6.1). The formats of these input files as well as the format of the runtime configuration file are described in the following.

6.1.3.1 Architecture Model

The platform graph from Definition 2.5 can be difficult to specify explicitly, since it may contain a high number of edges. For that reason, a simpler XML model is introduced, leaving the actual model construction to the compiler (see Sect. 6.3).

```
1 <Platform>
2   <Processors      List="irisc0 ltvliw0 ltvliw1"/>
3   <Memories        List="loc_mem_0 loc_mem_1 loc_mem_2 sha_mem1 hw_fifo"/>
4   <Interconnects   List="Bus1"/>
5   <CommPrimitives  List="over_lm over_sm1 over_hw"/></Platform>
6 <Processor Name="irisc0"  CoreRef="ARM9"/>
7 <Processor Name="ltvliw0" CoreRef="C65"/>
8 <Processor Name="ltvliw1" CoreRef="C65"/>
9 <SharedM Name="sha_mem1" Size="8388608">
10  <Connect List="Bus1"/><VisibleFrom List="irisc0 ltvliw0 ltvliw1"/>
11 </SharedM>
12 <HwFifo Name="hw_fifo" Size="512" Source="ltvliw0" Target="ltvliw1"
13        Channels="1"/>
14 <LocalM Name="loc_mem_0" Size="524288" PERef="irisc0"/><!--... others -->
15 <CommPrimitive Name="over_lm">
16   <Cost Offset="100" StairX="4" StairY="1" />
17   <DefinedFor List="loc_mem_0 loc_mem_1 loc_mem_2"/>
18 </CommPrimitive> <!--... other CPs -->
19 <Core Name="ARM9" CoreType="ARM926EJS_AHB_Model" Category="RISC">
20   <MultiTaskingInfo MaxNumberOfTasks="-1">
21     <ContextSwitchInfo LoadTime="100" StoreTime="100" />
22     <SchedulingPolicies List="FIFO PriorityBased"/></MultiTaskingInfo>
23   <CostTable>
24     <Operation Name="mul">
25       <VarType Name="Char" Cost="1"/> <VarType Name="Float" Cost="2" />
26       <!-- ...-->
27 </CostTable></Core> <!--... other cores -->
```

Listing 6.1: Example of an architecture description file

An example of an XML platform description is shown in Listing 6.1. The text in Lines 1–5 describes the available hardware resources and the communication primitives. Each PE is linked to a processor type (`CoreRef` in Lines 6–8). The characteristics of each type are described later in the file (see Lines 19–27), including information about the context switch execution time (200 cycles), the maximum number of tasks that can be mapped (unlimited), the supported scheduling policies and the operation costs needed for table-based performance estimation. The text in Lines 9–14 describes the memories in the platform. Shared memories include a list of processors that can access the memory, while local memories specify a single processor. Hardware FIFOs specify the producer and the consumer processor. Besides the total memory size, HW FIFOs are restricted to a single logical channel (see `Channels` in Line 13).

The sample file includes a description of a simple communication primitive over local memory in Lines 15–18. The description contains the constants that define the communication cost model presented in Sect. 2.3.1.3 (see Eq. 2.3). It also contains a list of communication resources over which this API can be executed (`DefinedFor` in Line 17). This allows a compact representation of all the communication primitives without explicitly enumerating them as required by Definition 2.4.

6.1.3.2 Application Constraints

The application constraints (in Definition 2.6) that define the constraint satisfaction problem are specified in an XML file as well. An example of such file is shown in Listing 6.2. The kinds of constraints that can be specified are:

```
1 <Latency MaxCycles="150000" Path="P1 P2 P3" Class="hard"/> <!-- or soft -->
2 <Throughput Process="P1" MaxPeriod="150000"  Class="hard"/> <!-- or soft -->
3 <PreMapping>
4   <MapConstraint Process="P1" Processor="irisc0"  />
5   <MapConstraint Process="P2" Processor="ltvliw0" /></PreMapping>
6 <PlatformSubset Processors="irisc0 ltvliw0"/>
7 <TimeTrigger>
8   <TimedProcess Process="P4" Period="8000"/></TimeTrigger>
9 <LogicChanBounds>
10   <FifoConstraint Fifo="Channel_P1-P2" Constraint="3" /></LogicChanBounds>
11 <MemoryConstraint size="8000"/>
```

Listing 6.2: Example of an application constraints

- Latency: This constraint specifies a maximum time allocated for the execution of a single process or a path in the KPN application. For example, Line `1` expresses that the path P1–P2–P3 should execute within 150000 cycles.
- Throughput: This constraint specifies a list of processes and their required throughput (the reciprocal of the period). In Listing 6.2, process P1 has a throughput constraint determined by a period of 150000 cycles (in Line `2`).
- Fixed mapping: This constraint restricts the mapping of some processes to processing elements. In the example, process P1 is mapped to `irisc0` and P2 to `ltvliw0`. Such a constraint allows the user to specify tasks that must be mapped to a specific processor, e.g., tasks for which an accelerator is available.
- Platform subset: This constraint allows the user to restrict the processing elements that will be considered in the mapping process. In the example in Listing 6.2, the mapping is restricted to `irisc0` and `ltvliw0` (see Line `6`).
- Time trigger: This constraint specifies which processes, if any, are triggered by a timer. This allows to model inputs from the environment. In the example, process P4 is time-triggered with a period of 8000 cycles (in Line `8`).
- Logical channel bounds: This constraint lets the user set the maximum size of some channel buffers. In the example, the logical channel between `P1` and `P2` will be constrained to three elements (in Line `10`).
- Memory: This constraint bounds the amount of memory available for the FIFO channels (see Line `11` of Listing 6.2).

As discussed in Sect. 2.5.3, timing properties, like throughput and latency, are not well defined for KPNs. This motivated the inclusion of the time checkpoint in the KPN application model (see Definition 2.45). However, since processes can have arbitrary behavior, there are some restrictions on the type of paths the user can define constraints on. As stated in Definition 2.6, the average case is analyzed for soft real-time applications, while a more strict constraint test is performed for hard real-time applications. How exactly the time is measured to check constraint compliance and what kinds of paths are supported is discussed in Sect. 6.3.1.2.

6.1.3.3 Tool Flow Configuration

The tool flow has several configuration options that allow the user to steer the optimization engine. The main configuration options allow to select the search algorithm

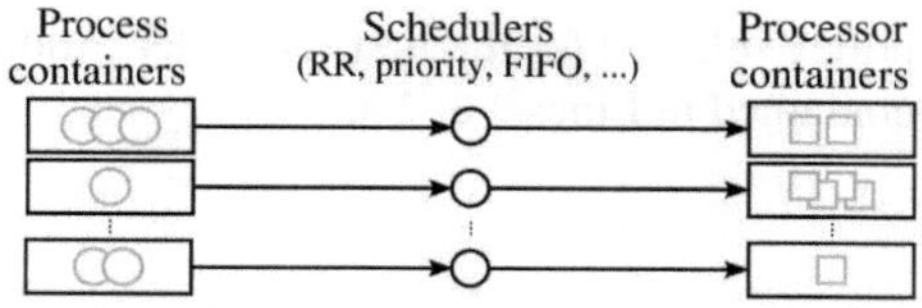

Fig. 6.2 Conceptual view of a schedule descriptor

(among those discussed in in Sect. 6.4) and the dynamic scheduling policy, among FIFO scheduling, priority-based, Round-Robin (RR) or RR With Skipping (RRWS).[1] The scheduler parameters, such as process priorities and time slot duration, are computed during the mapping and scheduling phase.

6.1.3.4 Schedule Descriptors

The main output of the mapping and scheduling phase is the runtime configuration, as shown in Fig. 6.1. This configuration is exported in the form of a schedule descriptor in an XML file. The general concept behind a descriptor is shown in Fig. 6.2. Application processes are mapped to a scheduler, and this in turn is assigned to a group of processors. Although more restricted, this structure is similar to the configuration of the OSIP runtime (see Fig. 4.1). The assignment of processes to schedulers is decided in the mapping and scheduling phase of the parallel flow.

```
1 <Scheduler Name="RR0" SchedulingPolicy="RoundRobin">
2   <Param TimeSlot="1000"> <ProcessContainer List = "P1 P2 P3" /> </Param>
3 </Scheduler> <!-- ... other schedulers -->
4 <Mapper>
5   <ProcessorContainers> <!-- Which processors belong to which groups -->
6     <PeGroup Name="RISCS" Processors="irisc0" />
7     <PeGroup Name="VLIWS" Processors="ltvliw0 ltvliw1" />
8   </ProcessorContainers>
9   <ProcessMapping> <!-- Connect schedulers to processor groups -->
10    <Connect Scheduler="RR0" Group="RISCS" /> <!-- ... other connects -->
11  </ProcessMapping>
12  <ChannelMapping>
13    <Map Fifo="Channel_P1-P2" Primitive="over_shared" Resource="sha_mem1"/>
14    <!--... other mappings -->
15 </ChannelMapping></Mapper>
16 <Buffersizing>
17   <FifoBound Fifo="Channel_P1-P2" Bound="4"/> <!-- ... other channels -->
18 </Buffersizing>
```

Listing 6.3: Example of an schedule descriptor

A simplified example of the descriptor file is shown in Listing 6.3. This descriptor contains most of the elements that compose a runtime configuration $RC^A = (\mu_p, \mu_c, \mu_a)$ from Definition 15. Process mapping (μ_p) is specified in Lines 1–11 and channel mapping (μ_c) in Lines 12–15. Variable mapping (μ_a) is specified in different locations of the file. As an example, consider the round-robin scheduling

[1] Even if a task is not ready to execute, the RR scheduler reserves its slot. In RRWS instead, if a task is not ready at the beginning of its slot, the next ready task would use it.

policy selected for the first scheduler. Additionally, the result of buffer sizing (β) is contained in Lines `16–18`.

6.2 Token Logging and KPN Tracing

This section provides further details about the first two phases of the parallel flow in Fig. 6.1. Since the tracing approach is close to the one of the sequential flow, the implementation details of the instrumentation process are omitted in this section. Instead, the section focuses on the definition and the format of the KPN traces in Sect. 6.2.1, and on how timing is added to traces in Sect. 6.2.2.

6.2.1 KPN Traces

As introduced with the motivational example in Sect. 2.5.3 (see Fig. 2.10), in order to predict the effect of a mapping, it is important to understand the control flow within every process in the KPN application. While in static dataflow programming models the behavior of a process is described in the specification itself, in a CPN specification it is hidden in the C implementation of the processes. At runtime, a process may follow different paths in its internal CFG. Along these paths, the process may access (read or write) any of its channels. In order to characterize the behavior of a KPN application, every process is analyzed in isolation with the sequential tracing flow described in Sect. 5.2.1. As mentioned before, this analysis is performed on the standalone implementation of the processes and using the channel logs produced by the token logging phase.

Recall that sequential tracing produces a file that records the control flow followed by the application and the memory accesses (see Fig. 5.2b). This information is used to build the model of each process according to Definition 2.44 in the same way it was done for a sequential application (see Sect. 5.2.2). Note, however, that for the purpose of scheduling, only three types of runtime events are relevant: a read access to an input channel, a write access to an output channel and a call to the time checkpoint. Channel accesses are important because they can potentially change the state of processes in the KPN application, i.e., a read from an empty channel and a write to a full channel will block a process, while a read from a full channel and a write to an empty channel may awake blocked processes. The time checkpoint event is needed to assess whether or not timing constraints are met. In order to make it easier to recognize these events in the execution trace, the instrumentation process was slightly modified.

The execution of a process is then characterized by a sequence of events, separated by arbitrary computations. The sequence is called a *process trace* and the arbitrary computations are denoted *segments*. More formally,

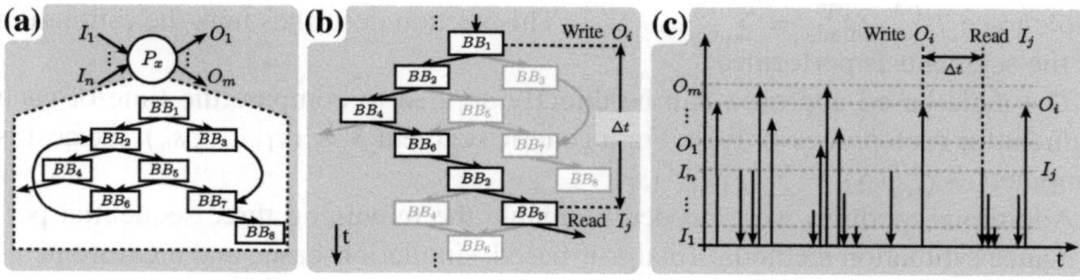

Fig. 6.3 Process trace example. **a** General process and sample control flow graph. **b** Sample execution trace. **c** Graphical representation of a process trace

Definition 1 A **segment** (S^{P^A}) of a process $P^A = (\mathscr{S}\mathscr{E}^{P^A}, CG^{P^A}, \pi^{P^A})$ is a sequence of statements $S^{P^A} = (s_1, \ldots, s_n)$ that determine a path in its CFG, where s_1 follows immediately from a synchronization event and s_n is the only statement in the sequence that generates a synchronization event. Naturally, $\forall s_i \in S^{P^A}, s_i \in \mathscr{S}\mathscr{E}^{P^A}$. The set of all segments in an application A is denoted $\mathscr{S}^A$.

Definition 2 A **process trace** (T^{P^A}) of a process P^A is a sequence of segments $T^{P^A} = (S_1^{P^A}, \ldots, S_m^{P^A})$ observed during the tracing process. The set of all process traces of an application A is denoted $\mathscr{T}^A$.

The concepts of segments and traces are illustrated in Fig. 6.3. Figure 6.3a shows a generic process P_x with n inputs and m outputs together with its CFG. A portion of the execution trace is shown in Fig. 6.3b, with the execution path marked in black solid lines. This path corresponds to a segment that starts after a write to channel O_i, ends with the statement that reads from channel I_j and contains all the statements in between from basic blocks BB_1, BB_2, BB_4, BB_6, BB_2 and BB_5. Finally, Fig. 6.3c shows a graphical representation of the process trace, in which read and write events are marked by arrows. The time elapsed in between synchronization points, i.e., the duration of a segment, depends on the processor the process is mapped to. This time (Δt in the figure) is obtained via performance estimation, as discussed in the next section.

6.2.2 Sequential Performance Estimation Revisited

The parallel flow requires time estimates of the sequential portions between synchronization events, i.e., for segments. This represents a coarser granularity than the one required in the sequential flow (see Sect. 5.2.3). For this reason, all the techniques introduced in Sect. 2.2.2 are enabled in the parallel flow. The purpose of this estimation process is to determine the duration of every segment $S^{P^A} \in \mathscr{S}^A$, $\zeta_{\text{seg}}^{\text{PT}}(S^{P^A})$. Provided the estimation for a segment, it is possible to define the estimation for a whole trace T^{PA}, $\zeta^{PT}_{trace} = \sum_{S \in T^{PA}} S^{PA}$. This section describes how the estimation for the segments is performed.

The table-based approach can be directly applied to compute the time of each segment for each processor type. For a generic segment $S = (s_1, \ldots, s_n)$, the cost is computed as $\zeta^{PT}_{seg}(S) = \sum_{i=1}^{n} \zeta^{PT}_{tb}(s_i)$.

Additional modules were needed to enable the remaining three sequential performance estimation methods: Totalprof-based, simulation-based and measurement-based. To enable Totalprof, a new plugin was implemented that instruments the channel access functions and the time checkpoint function. At every event, the plugin exports a time stamp into a text file. For simulation-based, a set of scripts and monitoring classes were implemented that directly connect to the processor ISSs that comes with Synopsys PA. These classes monitor the execution of each process and export time stamps as well. Finally, a simple postprocessing tool was implemented that analyzes the execution logs from actual HW boards to produce a process trace. Each of these three profiling flows works on the output of the standalone *cpn-cc* backend, and produces a file like the one in Listing 6.4. The file records all the events that occurred during the profiling run. In the example, the process with identifier 5 is entered (see Line 1), it then writes to channel 0 after 1,446 cycles from the source code line 193 (see Line 2), writes to channel 3 after 10 cycles from source line 194 (see Line 3), reads from channel 2 after 5,685 cycles from source line 365 (see Line 4) and hits a time checkpoint after 5 cycles (see Line 5).

```
1 e 5              // enter process with id 5
2 w 0 1446 193 // Write channel 0 after 1446 cycles, line 193
3 w 3   10 194 // ...
4 r 2 5685 365 // Read channel 2 after 5685 cycles, line 365
5 f       5     // Reached time check point after 5 cycles.
```

Listing 6.4: Sample output of coarse-grained time estimation for process

6.3 Building and Evaluating the Model

Until now, this chapter described how the behavior of processes is captured in the form of process traces. Before analyzing the heuristics built on these traces, this section explains how the different models of the parallel code problem of Definition 2.49 are constructed from the input files in Sect. 6.3.1. In order to asses the validity of a runtime configuration, the parallel flow includes a high-level simulator that allows to evaluate the model and verify constraint compliance. The simulator is described in Sect. 6.3.2.

6.3.1 Model Construction

The first step in the mapping phase of the parallel flow in Fig. 6.1c consists in creating the actual models for the target architecture and the application. This includes not only the underlying models of the application behavior ($\mathcal{M}^A$), but also the variables and the constraints that define the constraint satisfaction problem.

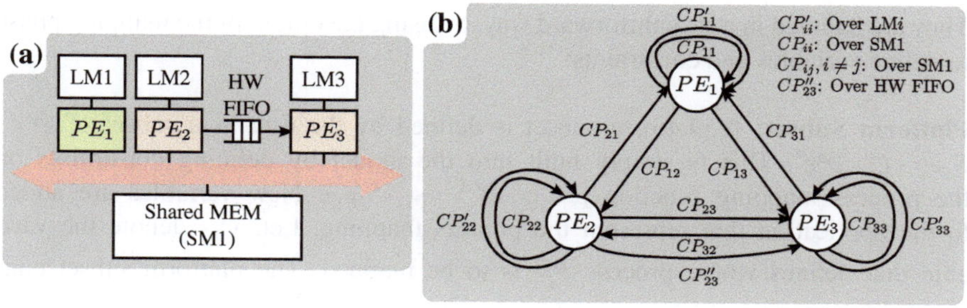

Fig. 6.4 Platform graph construction. **a** Platform schematic. **b** Platform graph

6.3.1.1 Building the Platform Graph

The process of obtaining the platform graph of Definition 2.5 from the architecture file introduced in Sect. 6.1.3.1 is straightforward. Every processor in the file defines a node in the platform graph, whereas every communication primitive in the file defines a set of edges in the platform graph. More specifically, it defines an edge between every two processors that can access a communication resource used by a given communication API.

This is demonstrated with the example in Fig. 6.4, which corresponds to the architecture file of Listing 6.1. As shown in Fig. 6.4a, there is a local memory for each PE and a shared memory for inter-processor communication. Additionally, processors PE_2 and PE_3 are interconnected with a HW FIFO. Figure 6.4b shows the corresponding platform graph. In this example, it is supposed that there are only two APIs for communicating over memory, one over the local and one over the shared memory. Additionally, there is an API to communicate using the HW FIFO. As a result, every PE can communicate with itself over both the local and the shared memories, indicated by two self-loops for every PE in the figure. The attributes and variables of the processor types and the communication resources are extracted directly from the XML file.

6.3.1.2 Generating Application Constraints

Recall the general definition of an application from Definition 2, $A = (\mathcal{M}^A, V^A, \mathcal{K}^A)$. The parallel application model $\mathcal{M}^A = (\mathcal{PAE}^A, KPN_A)$ from Definition 2.48 is built from the XML representation of the KPN application produced by *cpn-cc* (see Fig. 6.1c). This section addresses the remaining application elements, namely V^A and $\mathcal{K}$.

The first condition of a valid runtime configuration is determined by the application constraints (see Definition 2.16). The constraints introduced in Sect. 6.1.3.2 can be classified into two categories, *timing* and *non-timing* constraints. Non-timing constraints include fixed mapping, platform subset, logical channel bounds and memory.

They are handled in a straightforward way, by using the output of the mapping phase to define variables and constraints:

Platform Subset: A platform subset is defined by the user as a subset of $\mathscr{PE}$, $S_{\mathscr{PE}} \subset \mathscr{PE}$. This is simply built into the model by defining constraints on the process mapping function $\mu_p : \mathscr{P}^A \to S_{\mathscr{PE}}$. New variables are added to the application that represent the process mapping. Let, $\nu^i_{\mu_p}$ denote the variable that defines where process P_i^A is to be mapped. The platform subset constraint is defined as $K^A_{\text{subset}} = (S = \{\nu^1_{\mu_p}, \ldots, \nu^{|\mathscr{P}^A|}_{\mu_p}\} \subset V^A, R_{\text{subset}})$, with $R_{\text{subset}} = \times_{i \in \{1,\ldots,|\mathscr{P}^A|\}} S_{\mathscr{PE}}$. Note that pruning PEs from the platform automatically constrains the available communication primitives, so that the channel mapping function μ_c is also constrained. As a consequence a similar constraint is defined on the channel mapping function μ_c.

Fixed Mapping: These constraints represent a special case of the previous constraint, in which the subset $S_{\mathscr{PE}}$ is reduced to a single element for selected processes.

Logical Channel Bounds: To add logical channel bounds to the model, a constraint is defined on the variables in the domain of the buffer sizing function, $\beta : V^A_{\text{size}} \to \mathbb{N}$ (see Definition 2.48 and Sect. 2.5.4). Let, $S = \{\nu^1_\beta, \ldots, \nu^m_\beta\} \in V^A_{\text{size}}$ be the set of variables for which the user has defined a size constraint. The constraint is then defined as $K^A_{\text{lcb}} = (S = \{\nu^1_\beta, \ldots, \nu^m_\beta\} \subset V^A, R_{\text{lcb}})$ and $R_{\text{lcb}} = \{(n_1, \ldots, n_m) \in \mathbb{N}^m, n_i \leq k_i\}$, with k_i the constraint defined by the user as the maximum amount of tokens.

Memory: This imposes a global constraint on the buffer sizing function. The definition is similar to the previous one, $K^A_{\text{mem}} = (S = V^A_{\text{size}} \subset V^A, R_{\text{mem}})$ and $R_{\text{mem}} = \{(n_1, \ldots, n_{|V^A_{\text{size}}|}) \in \mathbb{N}^{|V^A_{\text{size}}|}, \sum_{i=1}^{|V^A_{\text{size}}|} n_i \cdot \text{var}_{\text{size}}(C_i^A) \leq k_{\text{mem}}\}$, with k_{mem} the global constraint defined by the user with the maximum amount of memory. var_{size} is the function that returns the amount of memory required by a token (see Definition 2.46).

Timing constraints are more complex to define than the non-timing constraints discussed above. As mentioned before (see Definition 2.45), time checkpoint events are used to measure latency and throughput. This is done by defining variables that represent the time stamps of each event after the mapping process. In order to accomplish this, the artificial concept of a *process iteration* is introduced, as a sequence of segments between time checkpoints. More formally,

Definition 3 The **process iterations** ($\mathscr{I}^P$) of a given process P modeled by an infinite trace $T^P = (S_i^P), i \in \mathbb{N}$ is a possibly infinite sequence of finite subsequences from T^P, $\mathscr{I}^P = (I_i^P) = ((S_1^P, S_2^P, \ldots, S_{f_1}^P), (S_{f_1+1}^P, \ldots, S_{f_2}^P), \ldots) = (S_{Q_i}^P)$. The indexes f_i correspond to the segments in the trace that finish with a time checkpoint. They define a partition of $\mathbb{N}$, $\mathcal{Q}(\mathbb{N}) = \{\{1, \ldots, f_1\}, \{f_1 + 1, \ldots, f_2\}, \ldots\} = \{\mathcal{Q}_i\}$.

For convenience, let N_P^{it} denote the number of iterations observed in the process trace of process P.

As an example, consider a trace $T^P = (S_1, S_2, S_3, S_4, S_5, \dots)$ with time checkpoints at S_2 and S_5. The sequence of auxiliary indexes would be $(f_j) = (2, 5, \dots)$ and the partition $\mathcal{Q}(\mathbb{N}) = \{\{1, 2\}, \{3, 4, 5\}, \dots\}$. Finally $\mathcal{I}^P = ((S_1, S_2), (S_3, S_4, S_5), \dots)$, meaning that the first iteration is composed of the first two segments, the second iteration of the next three segments and so on.

The execution time of each iteration can be estimated by adding the sequential execution time of the individual segments within the iteration and the time spent due to blocking and waiting times. The blocking time depends on the state of the FIFOs buffers, whereas the waiting time depends on the scheduler. For complex runtime systems, it is difficult to obtain a closed form to model the time of each iteration. This requires a simulation of the parallel execution, as discussed in the Sect. 6.3.2. For the moment, assume that there is a method to assign a time stamp to each iteration after the mapping process. Let $v_i^{I^P}$ denote such time stamp for the i-th iteration of process P. For convenience, the duration of the i-th iteration is defined as:

$$\Delta v_i^{I^P} = \begin{cases} v_i^{I^P} & \text{if } i = 1 \\ v_i^{I^P} - v_{i-1}^{I^P} & \text{if } i > 1 \end{cases} \tag{6.1}$$

With these variables, it is possible to define the constraints, as follows:

Throughput: For every throughput constrained process P, a constraint is defined as $K_{\text{th}}^P = (S = \{v_1^{I^P}, \dots v_o^{I^P}\} \subset V^A, R_{\text{th}}^P)$, where $o = N_{\text{it}}^P$. The region of the constraint is defined as $R_{\text{th}}^P = \{(n_1, \dots, n_o) \in \mathbb{N}^o, n_1 \leq k^P \wedge \forall i \in \{2, \dots, o\}, (n_i - n_{i-1}) \leq k^P\}$, with k^P the constraint defined by the user with the maximum number of cycles per iteration. For soft real-time applications a weaker constraint is defined by averaging the iteration times, i.e., $R_{\text{th}}^P = \{(n_1, \dots, n_o) \in \mathbb{N}^o, \frac{1}{o}(n_1 + \sum_{i=2}^{o}(n_i - n_{i-1})) = \frac{n_o}{o} \leq k^P\}$.

Latency: For every latency constraint defined along a path $P_{i->j} = (P_i, \dots, P_j)$, a constraint is defined as $K_{\text{path}}^{i->j} = (S = \{v_1^{P_i}, \dots, v_o^{P_i}, v_1^{P_j}, \dots, v_1^{P_j}\} \subset V^A, R_{\text{path}}^{i->j})$, where $o = N_{\text{it}}^{P_j}$ and $q = N_{\text{it}}^{P_j}$. The region of the constraint is defined as $R_{\text{path}}^{i->j} = \{(n_1, \dots, n_o, m_1, \dots, m_q) \in \mathbb{N}^{o+q}, m_1 \leq k^{i->j} \wedge \forall i \in \{2, \dots, o\}, (m_i - n_{i-1}) \leq k^{i->j}\}$, with $k^{i->j}$ the path constraint defined by the user. Similarly, the average is used for soft real-time applications.

In general, it is not possible to define a path latency in a KPN, since there is no way to relate events of different processes (recall from Sect. 2.5.3 that KPN is an untimed MoC). For this reason, the constraint definition above is restricted to paths in which the head and the tail have the *same* number of iterations (actually, $N_{\text{it}}^{P_j} = N_{\text{it}}^{P_j} + 1$). Notice also, that the latency constraint is determined by the time checkpoints of the

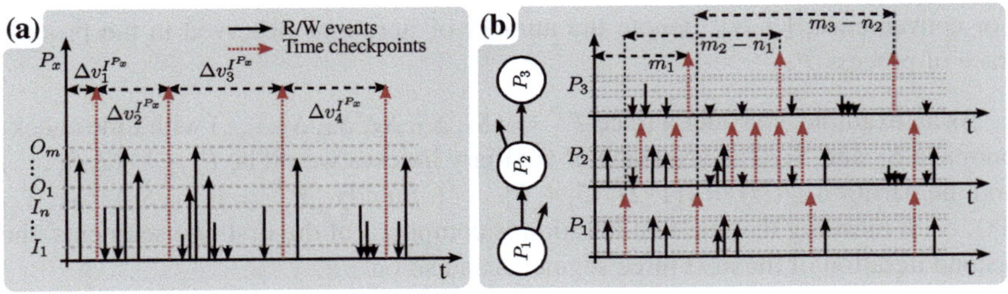

Fig. 6.5 Example of how timing constraints are measured. **a** Throughput constraint on a process P_x. **b** Latency constraint on the path (P_1, P_2, P_3)

tail and head processes. Additionally, during model construction, it is checked that between these two checkpoints, at least one time checkpoint is observed for every other process in the path. These kinds of path constraints are typical in iterative algorithms, e.g., iterative receivers, where a process produces tokens into a subnetwork that iterates for an unknown amount of times before producing the final result to the sink process.

The above definitions are clarified by means of the example in Fig. 6.5, where the time stamps of all the events after the mapping process are displayed. The measurement for a throughput constraint is shown in Fig. 6.5a. This consists in simply subtracting the time stamps of consecutive time checkpoint events. Notice, that each iteration in the example is composed of a different number of segments. The first, second and third iterations contain 2, 5 and 8 segments, respectively. The latency constraint in Fig. 6.5b demonstrates how the checkpoints of the head and the tail processes are used. Notice that in this simple example, the process in the middle (P_2) is allowed to iterate several times.

This section covered all application constraints, but the *time trigger* constraint defined in Sect. 6.1.3.2. This constraint is interpreted as a latency constraint on a single process. Additionally, time-triggered processes are treated differently during model evaluation, since their activation depends not only on the status of the channels but also on a timer.

6.3.1.3 Generating Architecture Constraints

The second condition for a valid runtime configuration in (Definition 16) regards constraints imposed by the attributes of the target platform. Three different attributes define constraints on the mapping process: the size of the communication resources (e.g., memories), the amount of channels supported by a communication resource and the amount of tasks supported by a processor type. These attributes are all described in the architecture file (see Listing 6.1).

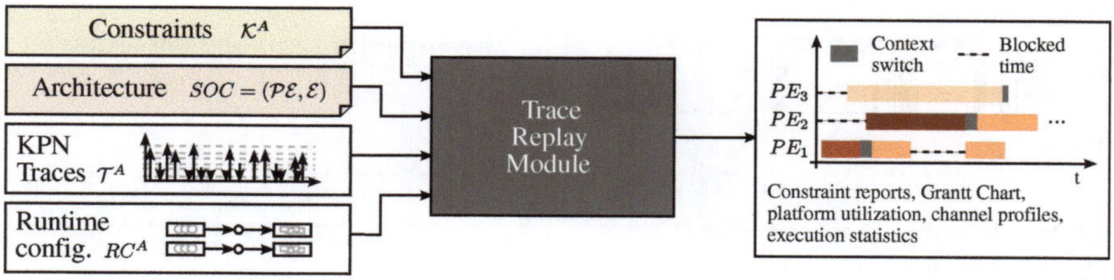

Fig. 6.6 Trace replay module

Memory Size: The amount of memory, due to logical channels mapped to a given communication resource, cannot exceed the amount of memory available. Let S^{CR} denote the set of application channels mapped to $CR = (x_{\mathrm{MEM}}^{\mathrm{CR}}, x_{\mathrm{CH}}^{\mathrm{CR}})$, that is $S^{\mathrm{CR}} = \{C^A \in \mathscr{C}^A, \mu_c(C^A) = CP = (PE_i, PE_j, S, \mathscr{CM}^{\mathrm{CP}}), \mathrm{CR} \in S\}$. For every communication resource, the inequality $\sum_{C^A \in S^{\mathrm{CR}}} \beta(C^A) \cdot \mathrm{var}_{\mathrm{size}}(C^A) \leq x_{\mathrm{MEM}}^{\mathrm{CR}}$ must hold. If there is no global memory constraint set, it is set to the maximum available memory in the platform, i.e., $k_{\mathrm{mem}} = \sum_{\mathrm{CR} \in \mathscr{CR}} x_{\mathrm{MEM}}^{\mathrm{CR}}$.

Supported Channels: The amount of channels mapped to a communication resource must not exceed the maximum allowed. This constraint applies to hardware channels, such as HW FIFOs. For every communication resource $CR = (x_{\mathrm{MEM}}^{\mathrm{CR}}, x_{\mathrm{CH}}^{\mathrm{CR}})$, the inequality $|S^{\mathrm{CR}}| \leq x_{\mathrm{CH}}^{\mathrm{CR}}$ must hold.

Supported Tasks: The amount of tasks mapped to a given processing element must not exceed the maximum allowed. This applies to hardware accelerators. Let S^{PE} denote the set of processes mapped to a processor PE, i.e., $S^{PE} = \{P^A \in \mathscr{P}^A, \mu_P(P^A) = PE\}$. For every processor PE_i^{PT}, the inequality $|S^{PE_i^{PT}}| \leq x_{\mathrm{tasks}}^{\mathrm{PT}}$ must hold.

6.3.2 Model Evaluation

Model evaluation refers to the process of assessing the quality of a runtime configuration. This is required in the mapping phase in order to decide whether a runtime configuration is valid or not. For this purpose a lightweight discrete-event simulator is included in the flow, called *Trace Replay Module* (TRM).

The TRM emulates the runtime system of the target platform and generates a Gantt Chart of the execution of the application by *replaying* the traces of the processes. The way the segments of the traces are scheduled is controlled by the given runtime configuration, as illustrated in Fig. 6.6. The TRM uses the cost models of the platform to estimate the time of the segments and the delays introduced by data communication. After replaying the traces, every segment of every trace is annotated with a time stamp. These time stamps are used to verify constraint compliance, as defined

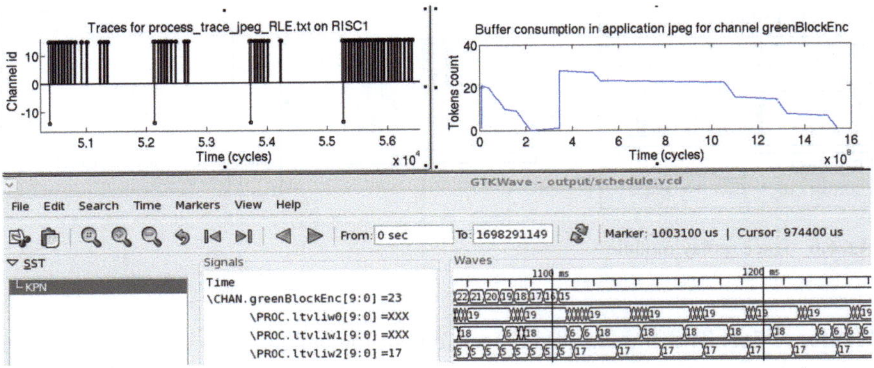

Fig. 6.7 Sample visualization of analysis results for the JPEG application

in Sect. 6.3.1.2. Apart from a report about the timing constraints, the TRM produces the following files with analysis results:

- Gantt Chart: The emulated schedule of the application is exported as a *value change dump* (vcd) file. This file includes the schedule and the buffer utilization.
- Channel profiles: The occupation of each one of the application channels is exported as a function of time. This allows the user to identify under-utilization of buffers as well as potential bottlenecks.
- Platform utilization: The utilization of each one of the processors of the platform is also exported as a function of time.
- Statistics: Several statistics are exported which can be visualized in the MAPS IDE. This includes process blocking time and overhead due to context switches.

Figure 6.7 gives an example of the analysis results for the JPEG application from the motivational example in Sect. 2.1. On the upper left corner, the figure shows the trace of the RLE process (see also Listing 2.1). Lines pointing downwards represent read events and lines pointing upwards represent write events. As expected for a run-length encoder, the amount of write events between read events varies, depending on the content of the input tokens. The plot on the upper right corner of Fig. 6.7 shows the profile of the channel containing encoded green color components (greenBlockEnc). Finally, the Gantt Chart of the application on a platform with three VLIW processors is shown on the bottom of Fig. 6.7. The first signal is a different representation of the greenBlockEnc channel profile. The remaining signals show the process identifier that runs at a given time on each of the processors. In this example, the identifiers 17–19 correspond to the IDCT processes for the three color components, which take most of the computation time. The value XXX in a signal expresses that the processor is idle, i.e., all the processes mapped to it are in the blocked or finished state.

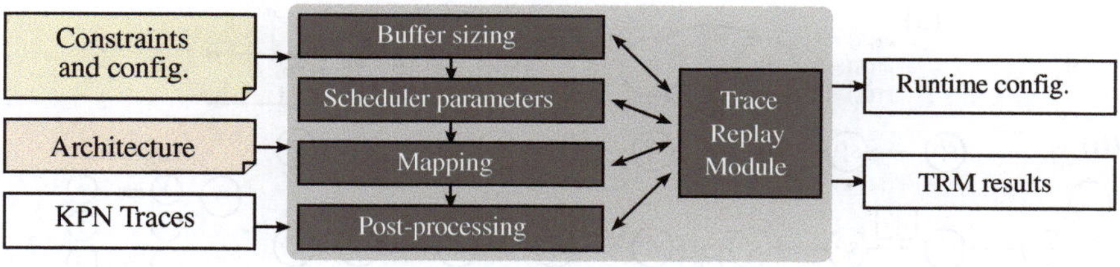

Fig. 6.8 Flow of the mapping and scheduling phase

6.4 Best-Effort Mapping and Scheduling

This section introduces the general structure of the mapping and scheduling phase of the parallel flow from Fig. 6.1. It also describes a set of heuristics for computing a runtime configuration for best-effort applications, i.e., applications with no timing constraints. For these applications, the optimization target consists in minimizing the application makespan. Irrespective of the method, the computed runtime configuration always respects the non-timing constraints, defined in Sect. 6.3.1.2.

The flow of the mapping phase is depicted in Fig. 6.8. It is composed of four different steps, which are presented in this section. The heuristic for buffer sizing is described in Sect. 6.4.2. Those for setting the parameters of the schedulers are treated in Sect. 6.4.3. Sections 6.4.4–6.4.5 present the algorithms for process and channel mapping. Finally, the post-processing step is described in Sect. 6.4.6. Before presenting the heuristics, the next section introduces a graph representation of the process traces which is used by some heuristics.

6.4.1 Trace Graph

For the purpose of analysis, the process traces described in Sect. 6.2.1 can be represented as a graph. Intuitively, given a KPN application A with traces $\mathscr{T}^A$, its trace graph would contain a node for every segment of every process. Segments would be connected by dependencies, e.g., the segment that starts after reading the n-th token from a given channel is only ready after the segment writing that n-th token was executed. Since the trace graph must model communication costs and account for blocking read/write semantics, its actual construction is more complex.

Definition 4 The **trace graph** ($\mathscr{TG}^A$) of a KPN application $KPN^A = (\mathscr{P}^A, \mathscr{C}^A, \text{var}_{\text{size}})$ with traces $\mathscr{T}^A$ is a directed graph $\mathscr{TG}^A = (V = \mathscr{S}^A \cup RE^A \cup \{v_s, v_e\}, E \subseteq V \times V)$, where $\mathscr{S}^A$ is the set of segments in the application traces, RE^A is a set that contains all read events in the traces and $\{v_s, v_e\}$ are auxiliary nodes. Edges in the graph represent dependencies.

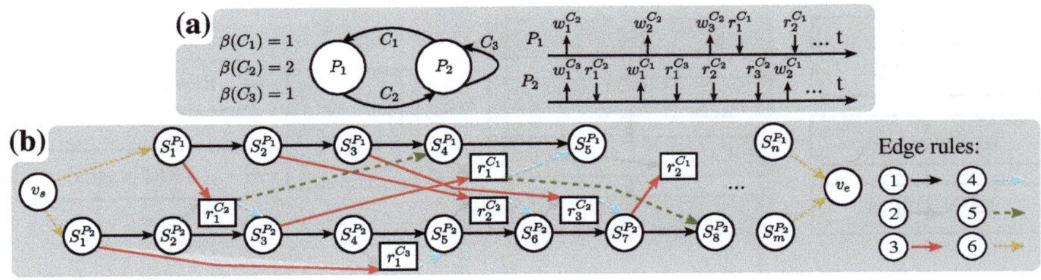

Fig. 6.9 Example of a trace graph. **a** KPN traces. **b** Corresponding trace graph

Let $r_i^{C^A}$ denote the i-th read access to channel C^A, so that $\forall C^A \in \mathscr{C}^A, i \in \mathbb{N}, (r_i^{C^A} \in RE^A)$. For convenience, let $S_{ri}^{C^A} \in \mathscr{S}^A$ be the segment that finishes with read event $r_i^{C^A}$ (note that $S_{ri}^{C^A} \in T^{\text{dst}}(C^A)$). Similarly, let WE^A be the set of all write events and $S_{wi}^{C^A} \in \mathscr{S}^A$ the segment that finishes with write event $w_i^{C^A}$ ($S_{wi}^{C^A} \in T^{\text{src}}(C^A)$). Finally, let nextseg$(S^P)$ denote the segment coming directly after S^P in the trace of process P, i.e., $S_{i+1}^P = \text{nextseg}(S_i^P)$. Note that every channel access may consume or produce several tokens. However, to ease the presentation of the trace graph, assume that every channel access corresponds to a single token. The extension to multi-token access follows the same concepts presented here. With this simplification, the edges of the trace graph are constructed following six rules:

1. Sequential order: $\forall S_i^P \in \mathscr{S}^A, (S_i^P, S_{i+1}^P) \in E$.
2. Read after compute: $\forall r_k^{C^A} \in RE^A, (S_{rk}^{C^A}, r_k^{C^A}) \in E$, i.e., the read event happens only after the segment has finished.
3. Block-reads: $\forall w_k^{C^A} \in WE^A, (S_{rk}^{C^A}, r_k^{C^A}) \in E$, i.e., the k-th token can be read only after it has been written.
4. Unblock-read: $\forall r_k^{C^A} \in RE^A, (r_k^{C^A}, \text{nextseg}(S_{rk}^{C^A})) \in E$, i.e., after the read event issued by a segment is done, the next segment in the trace is unblocked.
5. Block-writes: $\forall w_k^{C^A} \in WE^A, (r_{k-\beta(C^A)}^{C^A}, \text{nextseg}(S_{rk}^{C^A})) \in E$, i.e., due to buffer sizing (β), the k-th token can only be written if at least $(k - \beta(C^A))$ tokens have been read.
6. Single root, single leaf: There is an edge from v_s to every root node and from every leaf node to v_e.

Figure 6.9 illustrates the construction of a trace graph for a simple KPN. The application topology and the initial portion of its traces are shown in Fig. 6.9a,b shows the corresponding trace graph, in which the six different kinds of edges can be seen (see keys in the lower right corner). As an example, consider the *block-write* edge $(r_1^{C_2}, S_4^{P_1})$. In this case, since two tokens fit in channel C_2 ($\beta(C_2) = 2$), the third write would block if no token has been read.

6.4.2 Graph-Based Buffer Sizing

Theoretically channels in a KPN are unbounded [135]. In practice, channels have to be bounded, which may lead to artificial deadlocks [210]. This section presents a heuristic to solve the buffer sizing problem, i.e., determining the buffer sizing function $beta : \mathscr{C}^A \rightarrow \mathbb{N}$ for a KPN application $KPN^A = (\mathscr{P}^A, \mathscr{C}^A, \text{var}_{\text{size}})$. The graph-based heuristic is similar to classical simulation-based buffer sizing approaches [18, 210] (see also Sect. 3.3.2.2). However, instead of using a simulator, the heuristic uses the trace graph introduced in Sect. 6.4.1. The goal of this buffer sizing heuristic within the mapping flow in Fig. 6.8 is to determine the minimum required buffer sizes that allow the application to run without deadlocks. Later in the flow, buffers can be enlarged in order to improve the throughput of constrained processes (see Sect. 6.5).

In the trace graph, changes in the buffer sizes modify the *block-write* edges. Given fixed buffer sizes (β), a trace is deadlocked if its corresponding trace graph has a cycle. It is therefore possible to incrementally construct a trace graph until no cycle is observed. This is equivalent to incrementing the buffer sizes once a the simulation enters a deadlock in simulation-based approaches. The trace graph offers more information about the reason of the deadlock, so that size increments can be tailored better.

The pseudocode of the graph-based buffer sizing heuristic is shown in Algorithm 6.1. The function INITBURSTY in Line 2 performs the initialization of the buffer sizes. This is based on a local, per-channel analysis of the process traces. It assigns every channel a size that equals the biggest *burst* access (read or write) observed over a window of time. Since there is no mapping information, the time of the segments is assumed to be the average among all processor types. The window of time is determined by the average cost of a context switch. The function BUILDTRACEGRAPH in Line 2 performs the graph construction described in Sect. 6.4.1. The loop in Lines 4–15 iteratively increases the buffer sizes of channels that lead to artificial deadlocks. These correspond to edges created due to blocking writes in the cycle of the trace graph (edges in S_{cycle} in Line 7). The function UPDATEGRAPH in Line 14 updates only the *block-write* edges affected by changes in β^{tg}.

Algorithm 6.1 Graph-based Buffer Sizing

procedure BSTG($A = ((\mathscr{PAE}^A, KPN^A), V^A, \mathscr{K}^A), \mathscr{T}^A, SOC$)
  $\beta^{tg} \leftarrow$ INITBURSTY($\mathscr{T}^A, SOC$)
  $\mathscr{TG}^A \leftarrow$ BUILDTRACEGRAPH($\mathscr{T}^A, KPN^A, \beta^{tg}$)
  while ISCYCLIC($\mathscr{TG}^A$) **do**
    if GETMEMORY(β^{tg}, var_{size}) $> k_{mem}$ **then error:** not enough memory; **return** $\emptyset$
    end if
    $S_{cycle} \leftarrow$ GETCYCLEEDGES($\mathscr{TG}^A$)
    for $C^A \in S_{cycle}$ **do**
      if C^A is block write **then**
        $s \leftarrow \beta^{tg}(C^A)$
        $\beta^{tg} \leftarrow \beta^{tg} \cup (C^A, s+1)$
      end if
    end for
    $\mathscr{TG}^A \leftarrow$ UPDATEGRAPH($\mathscr{TG}^A, \beta^{tg}$)
  end while
  return β^{tg}
end procedure

6.4.3 Heuristics for Setting Scheduler Parameters

Depending on the configuration file passed to the mapping flow, different scheduling policies may be used for dynamic scheduling. The purpose of this phase is to fix the different parameters of the schedulers. This book proposes methods for setting the time slot used by time-slicing policies and the relative importance of processes. The latter is used to sort processes in several stages of the flow and can also be directly used as process priority in the case of priority-based schedulers. The different techniques are described in the following.

6.4.3.1 Computing a Time Slot

The time slot duration plays an important role for the efficiency of time-slicing policies. Small time slots allow applications to progress continuously, but incur in high context switching overhead. Big time slots, in turn, reduce the context switch overhead, but may introduce unnecessary blocking times.

There are no reported strategies to compute the time slot for a KPN application. In this book, a simple heuristic is followed, which takes into account the average cost of a context switch and the time between channel accesses in the process traces. For structured processes, i.e., processes with time checkpoints, the processes iterations from Definition 6.3 are used. More precisely, for every process P a time slot t^P_{slot} is computed by averaging the maximum execution time of its iterations, or its segments if it is not structured, i.e.,

$$t_{\text{slot}}^{P} = \begin{cases} \frac{1}{N_{\text{it}}^{P}} \sum_{i=1}^{N_{\text{it}}^{P}} \sum_{S \in I_i^P} \max_{PT \in \mathscr{PT}} (\zeta_{\text{seg}}^{PT}(S)) & \text{if } N_{\text{it}}^{P} \geq 1 \\ \frac{1}{|T^P|} \sum_{S \in T^P} \max_{PT \in \mathscr{PT}} (\zeta_{\text{seg}}^{PT}(S)) & \text{otherwise} \end{cases} \tag{6.2}$$

The global time slot is then fixed by comparing the average time slot per process (t_{slot}^{-P}) with a multiple of the average context switch costs (x_{cs}^{-PT}):

$$t_{\text{slot}} = \max(t_{\text{slot}}^{-P}, k_{\text{slot}} \cdot x_{\text{cs}}^{-\text{PT}}) \tag{6.3}$$

The constant k_{slot} is a parameter in the flow, with a default value of 10. With this arbitrary value, the context switch overhead is restricted to less than 10 %. The rationale behind the time slot computation in Eqs. 6.2–6.3 is to avoid context switches before potential channel writes, which usually accumulate at the end of iterations.

6.4.3.2 Process Importance

The importance of processes is modeled as a function $\phi : \mathscr{P}^A \rightarrow \mathbb{R}$ that associates a number with every process. The higher the number, the more important it is. Several implementations of this function are available in the flow as discussed in the following.

Topology based: The topology-based heuristic defines the importance as $\phi^{\text{top}}(P^A) = -l^{dfs}$, where l^{dfs} is the level in a depth-first search, with $l^{dfs} = 1$ for all the source processes. This policy is well suited for simple graphs, i.e., graphs with few feedback edges or even acyclic. It induces a data-driven execution with less context switches. However, it can produce higher latency for internal nodes. The alternative implementation $\phi^{\text{top}}(P^A) = l^{dfs}$ induces a demand-driven execution. This simple heuristic is however not suitable for graphs with very irregular computation patterns and several feedback edges. Therefore, the parallel flow includes other measures that in special cases provide better results. This is the heuristic used in the case studies presented in this chapter.

Output Rate: The output rate of a process measures the average amount of time that the process requires to generate tokens to its outputs. For convenience, let $\bar{\zeta}(T^{P^A})$ denote the average cost among all processor types, i.e., $\bar{\zeta}(T^{P^A}) = \sum_{PT \in \mathscr{PT}} \zeta_{\text{trace}}^{\text{PT}} / |\mathscr{PT}|$. Given a process with trace T^{P^A}, its importance is computed as:

$$\phi^{\text{or}}(P^A) = \bar{\zeta}(T^{P^A}) / \sum_{C^A \in \mathscr{C}^A : \text{src}(C^A) = P^A} |WE_{C^A}^{A}| \tag{6.4}$$

Recall that $WE_{C^A}^{A}$ is the set with all read events to channel C^A. The rationale behind this measure is that a process that needs more time to produce its outputs should be allowed more computation time.

Execution Weight: Similarly to the output rate, a process that needs more computation should be allowed more computational resources. The execution weight of a process P_x^A is defined as:

$$\phi^{\text{ew}}(P_x^A) = \bar{\zeta}(T^{P_x^A}) / \max_{P^A \in \mathscr{P}^A} (\bar{\zeta}(T^{P^A})) \tag{6.5}$$

Weighted Output Rate: This heuristic is a combination of the two previous measures:

$$\phi^{\text{wor}}(P^A) = \phi^{\text{ew}}(P^A) \cdot \phi^{\text{or}}(P^A) \tag{6.6}$$

6.4.4 Heuristics for Decoupled Process and Channel Mapping

As mentioned in Sect. 3.3.2, the mapping heuristics used in literature mostly address process mapping, and perform channel mapping once the processes have been mapped (see also [53]). This approach suffices for simple platforms with a single shared memory and simple communication primitives. For the platforms addressed in this book, this approach falls short, as will be discussed in Sect. 6.7. This section presents a greedy decoupled mapping heuristic. It first computes a process mapping, discussed in Sect. 6.4.4.2, followed by a channel mapping, discussed in Sect. 6.4.4.3.

6.4.4.1 Definitions

To ease the presentation of some the algorithms in the upcoming sections, let M^X denote the set of KPN elements mapped to a resource X. Inversely, an assignment set $\mathbb{W}^e$ is the set of resources to which a KPN element can be mapped. More formally,

Definition 5 A **mapping set** of a platform element $e \in (\mathscr{PE} \cup \mathscr{CR} \cup \mathscr{CP})$ is denoted M^e and contains the KPN elements that are mapped to it, i.e., $M^{PE} = \{P^A \in \mathscr{P}^A, \mu_p(P^A) = PE\}$, $M^{CR} = \{C^A \in \mathscr{C}^A, \mu_c(C^A) = CP = (PE_i, PE_j, CR, \mathscr{CM}^{CP})\}$ and $M^{CP} = \{C^A \in \mathscr{C}^A, \mu_C(C^A) = CP\}$.

Definition 6 An **assignment set** of a KPN element $e \in \mathscr{PAE}^A$ is denoted $\mathbb{W}^e$ and contains several platform resources to which the element can be mapped.

The algorithm in Sect. 6.4.5 uses groups of resources, so it is convenient to extend the mapping set to a group of resources.

Definition 7 A **group mapping set** of a set of platform resources $G \subseteq (\mathscr{PE} \cup \mathscr{CR} \cup \mathscr{CP})$ is denoted M^G and contains all the KPN elements that are mapped to the elements in G, i.e., $M^G = \bigcup_{e \in G} M^e$.

6.4.4.2 Process Mapping Heuristics

Let L be a list with all the processes of a KPN application, sorted according to their importance (see Sect. 6.4.3.2). The following heuristics for *process-only* mapping ($\mu_p : \mathscr{P}^A \rightarrow \mathscr{PE}$) have been adapted from the literature to the trace-based mapping approach proposed in this book:

Computation Balancing: This heuristic traverses the list L and finds the processor with the least current load for every process. The load of a PE_j of type PT, $\lambda(PE_j)$, is defined as the amount of computation already mapped to it. With M^{PE_j} the partial mapping to processor PE_j (see Definition 6.5), the current load can be computed as $\lambda(PE_j) = \sum_{P^A \in M^{PE_j}} \zeta_{\text{trace}}^{PT}(T^{P^A})$. The mapping is then defined by:

$$\mu_{\text{p}}^{\text{cb}}(P^A) = \underset{PE_j \in \mathscr{PE}}{\text{argmin}} \; (\lambda(PE_j)) \tag{6.7}$$

In the ideal case of an application in which processes do not communicate, computation balancing would produce a speedup close to the number of processors used. However, in the presence of communication, balancing could introduce a high communication overhead, for example, by mapping two communicating processes to processors that can only communicate with a slow primitive.

Affinity: This heuristic assigns processes to the more *affine* processor type, i.e., the processor type that executes the process the fastest. While traversing the list L, this heuristic assigns the fastest processor type to every process, i.e.,

$$\mu_{\text{p}}^{\text{aff}}(P^A) = PE_j^u \tag{6.8}$$

where $u = \underset{PT \in \mathscr{PT}}{\text{argmin}} \; (\zeta_{\text{trace}}^{PT})$. In the presence of multiple instances of the same processor type, the index j represents the processor with the lowest load.

Random Walk: This heuristic consists in randomly selecting a processor for every process and evaluating the result using the TRM. This process is repeated several times and the best observed mapping is returned as solution. Due to the complexity of the mapping process, such randomized mapping have been used even for the simpler DAG mapping problem, e.g., in [279].

Simulated Mapping: This heuristic uses the ETF scheduler introduced in Sect. 2.2.1 at the segments-granularity, meaning that segments are dynamically mapped to the processor that reports the earliest finishing time. In the case of two or more ready segments, the list L determines the order in which segments are assigned. In an ETF scheduler, segments can freely migrate from processor to processor. The simulated mapping procedure gradually reduce the amount of migrations by diminishing the migration probability over time, so that in the end, processes settle in on a processor.

This heuristic is inspired by the well-known simulated annealing technique.
Let $t_{\text{free}}^{PE_j^u}$ be the time after which PE_j^u is free at a given simulated instant, and let $R = \{S_i^{P_1}, \ldots, S_l^{P_k} \ldots\}$ be the set of ready segments at that instant. The heuristic traverses the ready set and assigns

$$\mu_p^{\text{sim}}(P_k) = PE_j^u \tag{6.9}$$

with $PE_j^u = \underset{PE_m^v \in \mathscr{PE}}{\text{argmin}} (t_{\text{free}}^{PE_m^v} + \zeta_{\text{seg}}^v(S_l^{P_K}))$. If the selected PE differs from the processor the process was mapped to, the process is migrated only if $\epsilon > t_{\text{free}}^{PE_m^v} / \sum_{P^A \in \mathscr{P}^A} \bar{\zeta}(T^{P^A})$, where ϵ is a random variable taken from a uniform distribution $U(0, 1)$.

6.4.4.3 Independent Channel Mapping (ICM)

ICM refers to the process of finding a channel mapping $\mu_c : \mathscr{C}^A \to \mathscr{CP}$, given a pre-computed process mapping $\mu_p : \mathscr{P}^A \to \mathscr{PE}$, so that (1) The KPN mapping (μ_p, μ_c) is logically valid (see Definition 2.16) and (2) the application makespan is minimized.

A greedy heuristic was devised in order to solve this problem. It starts by creating an assignment set with all possible communication primitives for every KPN channel, i.e., $\mathbb{W}^{C^A} = \{CP = (PE_i, PE_j, CR, \mathscr{CM}^{CP}) \in \mathscr{CP}, PE_i = \mu_p(\text{src}(C^A)) \wedge PE_j = \mu_p(\text{dst}(C^A)) \wedge \beta(C^A) \cdot \text{var}_{\text{size}}(C^A) \le x_{\text{MEM}}^{\text{CR}}\}$. Recall that $x_{\text{MEM}}^{\text{CR}}$ is the maximum amount of memory available for communication in the underlying communication resource. Then the channels are sorted, in decreasing order, according to the amount of traffic observed during tracing. While traversing the sorted list, every channel C^A with total traffic X_{C^A} is assigned the fastest available communication primitive, i.e., $\mu_C(C^A) = CP^* = (PE_i, PE_j, CR^*, \mathscr{CM}^{CP})$, with

$$\begin{gathered} CP^* = \underset{CP \in \mathbb{W}^{C^A}}{\text{argmin}} (\zeta^{CP_{ij}}(X_{C^A})), \ s.t. \\ \beta(C^A) \cdot \text{var}_{\text{size}}(C^A) + \textstyle\sum_{C \in M^{CR^*}} \beta(C) \cdot \text{var}_{\text{size}}(C) \le X_{\text{MEM}}^{\text{CR}^*} \wedge |M^{CR^*}| < X_{\text{CH}}^{\text{CR}^*} \end{gathered} \tag{6.10}$$

Additionally, a simple backtracking was implemented for CPs over HW FIFOs. This improved the success rate of ICM on platforms with restricted communication links. If at any point it is not possible to assign a CP, the algorithm fails.

6.4.5 Joint Process and Channel Mapping: The GBM Algorithm

The mapping heuristics described in the previous section share similarities with commonly used algorithms. However, as discussed in Sect. 1.1.1, new communication architectures have made channel mapping equally important to process mapping. This section presents an algorithm that maps KPN processes and channels in no predefined order, but depending on the application and the target platform.

The goal of a joint mapping process is to compute both μ_p and μ_c simultaneously, obtaining a valid KPN mapping with minimum makespan. To solve this problem, a *Group-Based Mapping* (GBM) algorithm is proposed with two main underlying goals: (1) Narrow the mapping space while avoiding early selection of specific HW resources, (2) analyze jointly processes and channels in no predefined order.

To achieve the first goal the algorithm was split into two phases. In the first one, KPN elements are iteratively mapped to *groups* of HW resources, i.e., assignment sets from Definition 6.6. This reflects the fact that different resources may display the same timing characteristic for a given KPN element. In the second phase, a sort of homogeneous mapping is performed in which the actual HW resources are selected. To achieve the second goal, the algorithm was designed so that processes and channels are selected according to an improvement measure. This measure is not biased to processes or channels, thus, they are selected in no prescribed order.

The pseudocode of the GBM heuristic is shown in Algorithm 6.2. As mentioned before, the first phase in Lines 2–12 works on assignment sets of the KPN elements ($\mathcal{W}^e$). These sets are reduced iteratively by calls to the function MAKEPROPOSAL until no more reductions are possible. The function MAKEPROPOSAL selects a KPN element e^* and reduces its set from $\mathcal{W}^{e^*}$ to $\mathcal{W}^*$. The first call to ASSESS in Line 5 checks whether the reduction is feasible. If it is not, the proposed new set $\mathcal{W}^{e^*}$ is removed from the previous set $\mathcal{W}^{e^*}$ and the feasibility is checked anew (second call to ASSESS in Line 7). At the end of the first phase, every KPN element e has an optimized assignment set $\mathcal{W}^e$. The second phase then performs homogeneous mapping on these groups (see Line 13). Further details of the GBM algorithm are provided in the following sections.

Algorithm 6.2 Group-Based Algorithm

procedure GBM($\mathcal{M}^A = (\mathcal{PAE}^A, KPN^A = (\mathcal{P}^A, \mathcal{C}^A, \text{var}_{\text{size}})), SOC = (\mathcal{PE}, \mathcal{CP}), \beta$)
  $\forall P^A \in \mathcal{P}^A, \mathcal{W}^{P^A} \leftarrow \mathcal{PE}, \forall C^A \in \mathcal{C}^A, \mathcal{W}^{C^A} \leftarrow \mathcal{CP}$
  while CANMAKEPROPOSALS($\{(e \in \mathcal{PAE}^A, \mathcal{W}^e)\}$) **do**
    $(e^*, \mathcal{W}^*) \leftarrow$ MAKEPROPOSAL($\{(e \in \mathcal{PAE}^A, \mathcal{W}^e)\}$)
    $(f, \{(e, \mathcal{W}'^e)\}) \leftarrow$ ASSESS($\{(e \in \mathcal{PAE}^A, \mathcal{W}^e)\}, (e^*, \mathcal{W}^*)$)
    if $f =$ `False` **then**
      $(f, \{(e, \mathcal{W}'^e)\}) \leftarrow$ ASSESS($\{(e \in \mathcal{PAE}^A, \mathcal{W}^e)\}, (e^*, \mathcal{W}^{e^*} - \mathcal{W}^*)$)
      if $f =$ `False` **then fatal error; return** $(\emptyset, \emptyset)$
      end if
    end if
    $\forall e \in \mathcal{PAE}^A, \mathcal{W}^e \leftarrow \mathcal{W}'^e$
  end while
  return DOHOMOGENEOUSMAPPING($\{(e \in \mathcal{PAE}^A, \mathcal{W}^e)\}$) *// returns the KPN mapping* (μ_p, μ_c)
end procedure

procedure ASSESS($\{(e, \mathcal{W}^e)\}, (e^*, \mathcal{W}^*)$)
  $\{(e, \mathcal{W}'^e)\} \leftarrow$ PROPAGATE($\{(e \in \mathcal{PAE}^A, \mathcal{W}^e)\}, (e^*, \mathcal{W}^*)$)
  if $\exists e \in \mathcal{PAE}^A, \mathcal{W}'^e = \emptyset$ **then return** (`False`, $\emptyset$)
  end if
  $f \leftarrow$ LOADCONTROL($\{(e \in \mathcal{PAE}^A, \mathcal{W}'^e)\}$)
  if $f =$ `False` **then return** (`False`, $\emptyset$)
  end if
  if $e^* \in \mathcal{C}^A$ **then** *// extra check for communication channels*
    $\{(e, \mathcal{W}'^e)\} \leftarrow$ CONSISTCHECK($\{(e \in \mathcal{PAE}^A, \mathcal{W}'^e)\}, e^*$)
    if $\exists e \in \mathcal{PAE}^A, \mathcal{W}'^e = \emptyset$ **then return** (`False`, $\emptyset$)
    end if
  end if
  return (`True`, $\{(e, \mathcal{W}'^e)\}$)
end procedure

6.4.5.1 Making a Proposal

In this context, making a proposal refers to the process of selecting a KPN element and refining its assignment set. The algorithm for making proposals uses the trace graph $\mathcal{TG}^A$ introduced in Sect. 6.4.1. Since the buffer sizing process has already taken place, the trace graph is acyclic. Proposals are generated by analyzing the *Dominant Sequence* [237] of the trace DAG, i.e., the critical path of the partially mapped graph.

At every call to MAKEPROPOSAL, the critical path of the trace DAG is determined. This is done by using the timing provided by the slowest resource in the assignment set $\mathcal{W}^e$, $e \in \mathcal{PAE}^A$. The algorithm compares the impact of alternative assignments for the nodes in the critical path (better mapping than the slowest resource). The KPN element corresponding to the trace graph nodes for which this impact is the highest is selected. The function then returns this element together with its assignment set reduced to the resource group that produced the highest improvement. Once all the elements in the critical path have been assigned to a group that contains only resources of the same type, no more proposals can be done. In this case, a call to the function CANMAKEPROPOSALS returns false (see Line 3 in Algorithm 6.2).

6.4.5.2 Mapping Propagation

After a proposal has been made, the function PROPAGATE in Line 16 of Algorithm 6.2 updates all the assignment sets accordingly. Note that if the assignment set of a KPN element is reduced, the assignment sets of other elements must be updated. As an example, consider a process whose assignment set is reduced from the set of all processors ($\mathbb{W}^P = \mathscr{PE}$) to a single processor ($W'^P = \{PE_i\}$). As a consequence, all ingoing and outgoing communication edges must be mapped to communication primitives that have the processor PE_i as destination and source, respectively. Again, as an example, consider a communication channel C whose assignment set is reduced to a single primitive ($\mathbb{W}^C = \{CP_{ij}\}$) over a HW FIFO. In such extreme case, the assignment sets of the source and destination processes become $\mathbb{W}^{\mathrm{src}(C)} = \{PE_i\}$ and $\mathbb{W}^{\mathrm{dst}(C)} = \{PE_j\}$. As it can be seen, the update process has to be propagated throughout the application graph. This process continues recursively, until no more change in the assignment sets is observed.

6.4.5.3 Load Control

The function LOADCONTROL in Line 19 of Algorithm 6.2 prevents proposals from always selecting the same group of resources. For this purpose, a measure of the occupation of a group has to be defined. This measure has to take into account that the *bigger* the KPN elements are, the more difficult it is to distribute them within a resource group. Consider for example an accumulated total memory requirement of 100 kB to be distributed on 3 memories of 40 kB each. While it is easy to distribute 20 *small* channels of 5 kB among the group, it is impossible to do it for 4 *bigger* channels of 25 kB. A similar observation holds for processes, where the *size* relates to the computation time. In the case of HW FIFOs, the function checks that there are no more KPN channels assigned than HW FIFOs in the group.

Consider a group of resources $G \subset (\mathscr{PE} \cup \mathscr{CR})$ with associated mapping set $M^G \subset \mathscr{PAE}^A$, i.e., a set of processes or channels mapped to the group G (see Definition 6.7). Generally speaking, if $|M^G| > |G|$, the load control ensures that the utilization of the group U stays below a variable threshold:

$$U < \frac{k_x \cdot |M^G|}{|M^G| + |G| \cdot (k_x/k_y - 1)} \tag{6.11}$$

The threshold is controlled by the amount of KPN elements and the amount of resources. The parameters $k_x \in (0, 1),\ k_y \in (0, k_x)$ provide further control. In the extreme cases $|M^G| = |G|$ and $|M^G| \gg |G|$ the utilization is compared against k_y

and k_x respectively. The values of the parameters were determined empirically. For memories, they are $k_x = 1,\ k_y = 0.5$ and for processors $k_x = 0.95,\ k_y = 0.75$.

The utilization U for a group of memories is determined by the ratio of the required and the available memory, i.e.,

$$U_{\text{mem}} = \frac{\sum_{C \in M^G} \beta(C) \cdot \text{var}_{\text{size}}(C)}{\sum_{\text{CR} \in \text{G}} x_{\text{MEM}}^{\text{CR}}} \tag{6.12}$$

For processors it is more involved, since the quantity *available time* is not as well defined as the available memory. The LOADCONTROL module uses an estimation of the makespan T to model the available time. This estimation is obtained by performing list scheduling on the trace DAG. With this, the utilization of a group of processors is defined as the ratio of the required computation (cumulative segment costs) and the total available computing time, determined by T cycles on each of the processors of the group. More precisely, for a group G of processors of type PT,

$$U_{\text{proc}} = \frac{\sum_{P \in M^G} \zeta_{\text{trace}}^{PT}(P)}{|G| \cdot T} \tag{6.13}$$

6.4.5.4 Consistency Check

When a KPN channel is assigned to a group of CRs, inconsistencies may appear. This is controlled by the function CONSISTCHECK in Line 23 of Algorithm 6.2. Consider a group of HW FIFOs G_{fifo} that sparsely connect processors within a group G_{proc}. In this case, it is not enough to perform local checks, i.e., whether all producers and consumers of the KPN channels mapped to G_{fifo} are mapped to G_{proc}. It has to be further checked that there actually exists at least one fixed feasible mapping. These checks are performed by the CONSISTCHECK module, with a time complexity of $O(|\mathscr{C}^A|^2 \cdot |\mathscr{C}\mathscr{P}| \cdot |G_{\text{proc}}|)$.

6.4.5.5 Homogeneous Mapping

After the heterogeneous phase, the assignment sets of the nodes in the critical path of the trace DAG consist only of resources of the same type. For these nodes, the problem is reduced to multiple smaller homogeneous mapping problems, where the main concern is to decide how to share hardware resources. Since sharing processors usually has a higher impact on the runtime than sharing communication resources (e.g., memories), this phase focuses on process mapping. Channel mapping can be done, in this case, as an afterthought, as long as the assignment sets are correct. The same holds for the KPN elements outside the critical path.

Any homogeneous mapping algorithm can be used in this phase, like those derived from the *bin packing* algorithm. The function in Line 13 of Algorithm 6.2 uses the information collected in the previous phase to compute the final mapping. First, note that due to constraits imposed by channel assignment sets, mapping a process $P^A \in \mathscr{P}^A$ will sometimes imply mapping a group of processes $G^{P^A} \subseteq \mathscr{P}^A$. Processes are sorted primarily by the size of these induced sets ($|G^{P^A}|$) and then by the size of the assignment sets ($|\mathcal{W}^{P^A}|$). The bigger the induced group and the smaller the assignment set, the earlier the process has to be mapped. The list of processes in the critical path of the trace DAG is sorted according to the aforementioned criteria. Every process P^* is mapped to the processor PE^* on which a local scheduler on the trace DAG reports the best finishing time. In order to reduce the complexity of this mapping phase, the local scheduler considers only segments of processes that are already mapped to the target processor PE^* and ignores other dependencies. More precisely, let $ALAP(S_i^{P^*})$ be the *As Late As Possible* time of segment $S_i^{P^*}$, obtained during the critical path computation on the trace DAG. Let $t^{PE}_{S_i^{P^*}}$ be the time computed by the local list scheduler for the same segment on every processor in the assignment set, i.e., $PE \in \mathcal{W}^{P^*}$. The final mapping is determined by $\mu_p^{\text{gbm}}(P^*) = PE^*$, with

$$PE^* = \underset{PE \in \mathcal{W}^{P^*}}{\operatorname{argmin}} \sum_{S_i^{P^*} \in T^{P^*}} \max(0, t^{PE}_{S_i^{P^*}} - ALAP(S_i^{P^*})) \tag{6.14}$$

The term $t^{PE}_{S_i^{P^*}} - ALAP(S_i^{P^*})$ can be seen as a penalty for processors that do not have free time slots that allow to schedule the segment before its theoretical ALAP time. After every fixed mapping decision, the tests described in Sects. 6.4.5.2–6.4.5.4 are performed.

6.4.6 Post-processing Phase

The final phase of the mapping and scheduling flow from Fig. 6.8 is in charge of performing final adjustments to the schedule descriptors. This includes checking for compliance with all non-timing constraints and for the logical validity of the configuration. Additionally, a pass is included that further tightens the buffer sizes based on the current schedule configuration.

6.5 Mapping and Scheduling with Timing Constraints

The previous section described the proposed algorithms for computing a valid configuration for applications with no timing constraints. In the presence of timing constraints, the objective of the optimization problems is shifted. Instead of attempting to produce a runtime configuration with the minimum makespan, a configuration is now sought which meets the timing constraints while using as few platform resources as possible. The approach followed is therefore to start from the smallest possible platform configuration and gradually use more resources (e.g., processors and memories) until a valid best-effort mapping is obtained. The initial platform configuration, $SOC^+ = (\mathscr{PE}^+ \subseteq \mathscr{PE}, \mathscr{CP}^+ \subseteq \mathscr{CP})$, is determined by user-defined mapping constraints. A similar approach is followed for more restricted application representations in [192, 245].

The pseudo code for the iterative mapping heuristic is presented in Algorithm 6.3. The algorithm receives the KPN application $A \in \mathscr{A}^{\mathrm{kpn}}$, the traces and the model of the architecture. The initial platform subset is generated by the function INITTARGETSUBSET in Line `2`. For this purpose, it uses the constraints specified by the user (platform subset and fixed mapping as defined in Sect. 6.3.1.2). If no platform constraint was specified, the function returns a randomly selected processor PE_i together with all the communication primitives that connect it with itself CP_{ii}. The outer loop in Lines `5–24` is in charge of gradually adding more processors to the mapping process (see INCRPLATFORM in Lines `19–23`). For every new platform configuration, a best-effort mapping is constructed following the steps presented in the previous section (see Lines `6–9`). Note that the logical bound constraints and the memory constraint are passed to the buffer sizing function. After the post-processing phase, every runtime configuration RC^A is logically valid and satisfies the platform constraints (see Definition 2.16). To check the remaining constraints, the TRM presented in Sect. 6.3.2 is used, as suggested by the call to TRMEVAL in Line `10`. This function returns a flag f_c that indicates if the constraints were met and several runtime information (*runinfo* in Line `10`), which includes the Gantt Chart, platform utilization, channel profiles and other execution statistics. If the timing constraints are not met (f_c = **`False`**), the inner loop in Lines `11–18` attempts to improve the timing behavior by increasing the sizes of the buffers. This is done until the memory constraint corresponding to the subset **SOC**$^+$ is surpassed. In the implementation, the inner loop is also exited in case no timing improvement is observed after several iterations. For every new set of buffer sizes β, the application is remapped and the constraints are checked (see Lines `15–17`). The algorithm returns the last computed runtime configuration together with the flag f_c that indicates whether the configuration is valid or not. Further details on the two new functions INCRBUFFERS and INCRPLATFORM are provided in the following.

Algorithm 6.3 Iterative algorithm for real-time applications

procedure ITERATIVEMAPPING($A = (\mathscr{M}^A = (\mathscr{PAE}^A, KPN^A), V^A, \mathscr{K}^A)$, $\mathscr{T}^A$, SOC)
  $SOC^+ \leftarrow$ INITTARGETSUBSET($SOC, \{K^A_{\text{fixedmap}}, K^A_{\text{subset}}\} \subset \mathscr{K}^A$)
  $f_c \leftarrow$ `False` // *flag for constraints met*
  $f_b \leftarrow$ `False` // *flag for buffer increment successful*
  repeat
    $\beta \leftarrow$ BSTG($A, \mathscr{T}^A, SOC^+, \{K^A_{\text{lcb}}, K^A_{\text{mem}}\} \subset \mathscr{K}^A$) // *from Algorithm 6.1, recall $\beta \subset \mu_a$*
    $\mu_a \leftarrow$ GETSCHEDPARAMS($\mathscr{T}^A, SOC^+$) // *described in Sect. 6.4.3*
    $(\mu_p, \mu_c) \leftarrow$ GBM($\mathscr{M}^A, SOC^+, \beta$) // *can also use other heuristic from Sect. 6.4.4*
    $RC^A = (\mu_p, \mu_c, \mu_a) \leftarrow$ POSTPROCESS(μ_p, μ_c, μ_a) // *described in Sect. 6.4.6*
    (f_c, runinfo) $\leftarrow$ TRMEVAL($\mathscr{K}^A, RC^A, \mathscr{T}^A, SOC^+$) // *TRM in Sect. 6.3.2, Fig. 6.6*
    while $f_c =$ `False` **do**
      $(f_b, \beta) \leftarrow$ INCRBUFFERS($\beta, SOC^+, \{K^A_{\text{lcb}}, K^A_{\text{mem}}\} \subset \mathscr{K}^A$, runinfo)
      if $f_b =$ `False` **then break** // *no more memory in current SOC^+*
      end if
      $(\mu_p, \mu_c) \leftarrow$ GBM($\mathscr{M}^A, SOC^+, \beta$)
      $RC^A \leftarrow$ POSTPROCESS(μ_p, μ_c, μ_a)
      (f_c, runinfo) $\leftarrow$ TRMEVAL($\mathscr{K}^A, RC^A, \mathscr{T}^A, SOC^+$)
    end while
    if $f_c =$ `False` **then** $SOC' \leftarrow$ INCRSOC(SOC^+, SOC)
      if $SOC' = SOC^+$ **then break**
      end if
      $SOC^+ \leftarrow SOC'$
    end if
  until ($f_c =$ `True`)
  return (f_c, RC^A)
end procedure

Incrementing Buffer Sizes: As mentioned before, several algorithms have been proposed in the literature to increment buffer sizes, either for resolving deadlocks [91, 210] or for improving performance [52]. The function INCRBUFFERS in Algorithm 6.3 increments the size of the buffers in a non uniform way. It does it by analyzing the channel profiles of the replayed execution as reported by the TRM. It then selects the channel that was full for the longer time and increments its size by the biggest burst access (see also Sect. 6.4.2).

Incrementing the Platform Subset: The fundamental idea behind the iterative approach in Algorithm 6.3 is to obtain a best-effort mapping that meets the constraints while using as few resources as possible. To achieve this, one would have to try all possible configurations with one processor, then all possible configurations with two processors and so forth. Note that for a platform with n processors, this procedure requires the mapping process to be run for $2^n - 1$ subgraphs of the platform graph SOC. To avoid this exponential complexity, the function INCRPLATFORM in Algorithm 6.3 enlarges the current subgraph SOC^+ by adding a processor at a time, together with the corresponding communication primitives.

This incremental procedure can be seen as a simple indexing over a set of subgraphs that are traversed linearly. Let $\mathscr{L}_{\text{proc}} = \{\mathscr{PE}_1, \mathscr{PE}_2, \ldots, \mathscr{PE}_n\} \subset \wp(\mathscr{PE})$ be the list of processor subsets that define platform subgraphs $\{SOC_1, SOC_2, \ldots,$

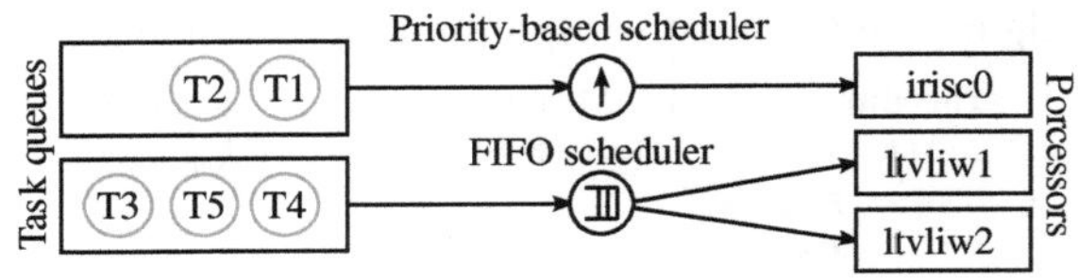

Fig. 6.10 Sample OSIP configuration for code generation

$SOC_n\}$, where $SOC_i = (\mathscr{PE}_i, \mathscr{CP}_i)$. The i-th call to function INCRPLATFORM then returns the subgraph SOC_{i+1}, with SOC_1 the subgraph returned by the function INITTARGETSUBSET. Calls to function INCRPLATFORM following the $(n-1)$-th call will all return the last subset (SOC_n), marking the end of the algorithm (see Line `20` in Algorithm 6.3).

6.6 OSIP Backend

The backend is the final phase of the parallel flow shown in Fig. 6.1. For the case studies presented in this book, a backend for OSIP-based MPSoCs was added to the CPN compiler. This section explains the structure of the generated code together with a debugging environment for virtual platforms.

6.6.1 OSIP Generated Code

As mentioned in Chap. 4, three different cores can be currently integrated in an OSIP-based MPSoC: An ARM processor, a small IRISC controller and an LTVLIW processor. The generated code is different for each of these processors, as explained in the following.

ARM Code: The ARM processor is used as host in an OSIP-based MPSoC. It is in charge of the system setup, including OSIP configuration, channel initialization and task creation. A simplified example of OSIP configuration code is shown in Listing 6.5 for the structure in Fig. 6.10. The code in Lines `2–7` tells OSIP about the available processors for this application and their interrupt signal identifiers. In the example `CLASS_1` represents the VLIW class with two processors (*ltvliw1*, *ltvliw2*). The code in Lines `9–12` configures two task queues with their corresponding scheduling policies and assigns them to the processing classes. Finally, the code in Lines `14–17` shows how tasks are created. In the example, `Task1` is assigned to the queue attached to *irisc0* with a priority of 1.

```
/* Platform (CLASS_0 = 0 (irisc0) CLASS_1 = 1 (ltvliws)) */
ProcClasses[CLASS_0].processing = 1;
ProcClasses[CLASS_0].InterID = pID0; //*pID0 = [0]
ProcClasses[CLASS_1].processing = 2;
ProcClasses[CLASS_1].InterID = pID1; //*pID1 = [1,2]
pParams->numProcClass = 2;
pParams->numProcResInstances = 3;
/*... Schedule configuration */
TaskQueues[CLASS_0].procClass     = CLASS_0;
TaskQueues[CLASS_0].entryNodeAlgo = OSIPSchedPriority;
TaskQueues[CLASS_1].procClass     = CLASS_1;
TaskQueues[CLASS_1].entryNodeAlgo = OSIPSchedFIFO;
/*... Task creation */
NewTaskParams.fpTask = (int (*)(void*))Task1;
NewTaskParams.Priority = 1;
NewTaskParams.EntryNodeRef = TaskQueueRefs[CLASS_0];
CreateTask(&NewTaskParams);
```

Listing 6.5: OSIP configuration code

Code for IRISC: The code for an IRISC processor is fairly simple. It starts by retrieving pointers to global channels initialized by ARM and then initializing local channels. Whether a channel is local or global depends on the runtime configuration produced by the mapping phase. The configuration also contains the size of the channels, so statically allocated arrays are used by the code generator. After initialization, the IRISC goes idle, awaiting OSIP interrupts. For the implementation of communication, the OSIP APIs are used so that the *wait* and *signal* primitives are communicated to the OSIP processor.

Code for LTVLIW: The LTVLIW performs a similar initialization process as the IRISC. However, after the initialization, the processor enters an infinite loop that constantly polls the HW proxy for tasks. This difference is due to the lack of interrupt support in the LTVLIW, as discussed in Sect. 4.2.3.1. Additionally, since Protothreads are used for multi-tasking, the code generator changes all process variables to static variables. In this way, the process context is maintained. The use of Protothreads also restricts the scheduling policy to FIFO, since preemption is not supported.

6.6.2 Debugging Environment

For the purpose of debugging, both real and virtual platforms offer standard interfaces that allow to connect traditional tools such as the *GNU Project Debugger* (GDB) [99]. After code generation, the high level application information is lost, so that debuggers are not aware of the original logical processes and communication channels of the application. In order to enable this kind of high level debugging, an extra debugging layer was implemented on top of the debugging facilities offered by Synopsys PA. The scripts are integrated within Synopsys *Virtual Platform Analyzer* (VPA) and extend its graphical controls.

The debugging layer consists of a set of scripts that receive high level application information exported by the CPN backend. The layer is customized for the OSIP runtime manager, so that several system events can be tracked, such as interrupts, context switches and OSIP API calls (task creation, wait, preemption and yield).

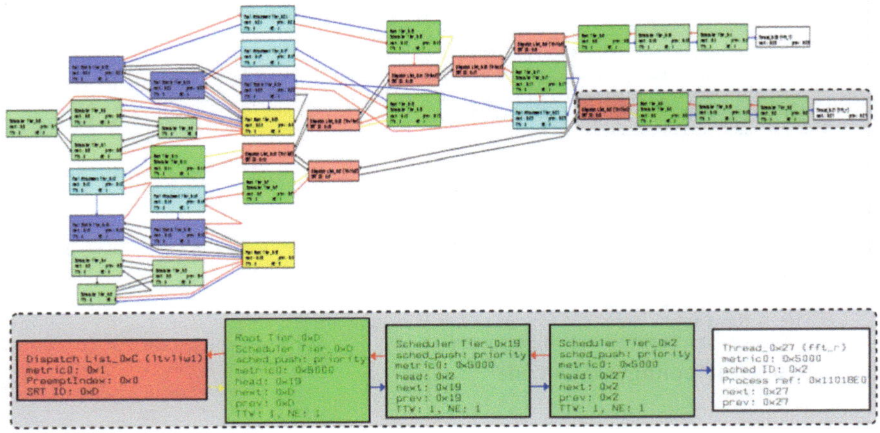

Fig. 6.11 OSIP state from debugger (zoomed-in region on the *bottom part*)

Provided with the application information, the debugging layer allows to graphically inspect the internal state of the OSIP scheduler at any moment (see Sect. 4.2.3.2). As an example, consider the snapshot of the internal state of OSIP shown in Fig. 6.11. The figure shows different colors for different elements in the scheduling hierarchy, including white thread identifiers. Figure 6.11 shows the details of a schedule path for the *ltvliw1* on the bottom part. In the example, the right FFT processing task is queued for scheduling, three levels down the hierarchy.

At the application level, the debugging environment allows to set breakpoints at channel accesses and time checkpoints among others. For setting the breakpoints, the programmer does not need to know the mapping of the application, or how logical channels were translated into actual C structures and APIs.

6.7 Case Study

This section presents results that help to assess the quality of the algorithms proposed in this chapter. It does it with the experimental setup described in Sect. 6.7.1. The algorithms for best-effort applications are benchmarked in Sect. 6.7.2. Section 6.7.3 presents the results for real-time applications and Sect. 6.7.4 validates the results on an actual hardware board featuring the TI Keystone platform.

6.7.1 Experimental Setup

The proposed algorithms are tested on both real and synthetic applications on two different platforms. Further details are provided in the following.

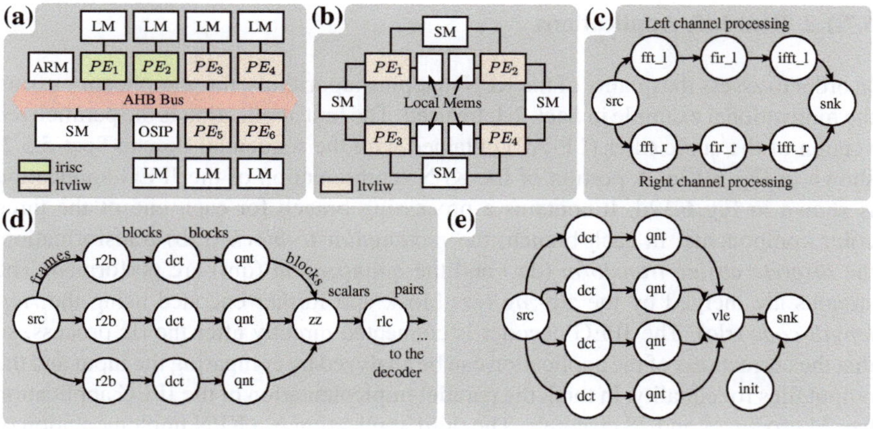

Fig. 6.12 Experimental setup **a** *Densely Connected Platform* (DCP). **b** *Sparsely Connected Platform* (SCP). **c** JPEG. **d** Audio filter. **e** MJPEG

6.7.1.1 Target Virtual Platforms

To test the algorithms, two virtual platforms are assembled that reflect the characteristics of current MPSoCs, described in Sect. 1.1.1 (see also Fig. 1.6). The first one is an OSIP-based MPSoC, containing an ARM host processor, 2 IRISCs and 4 LTVLIWs (see Sect. 4.2.3.1). Each processor has a local memory and can access a shared memory over an AHB bus, as shown in Fig. 6.12a. In this case study the functional model of OSIP is used (see Sect. 4.2.3.2). Due to its standard interconnect, this MPSoC is denoted *Densely Connected Platform* (DCP).

The DCP virtual platform serves also to test the generated code of the OSIP-backend for functional correctness. For timing analysis, the results from both the virtual platform and the TRM can be used. When using simulation-based KPN tracing, the TRM timing deviates only 3 % from the results of the actual virtual platform. This is due to the accuracy of OSIP's model and the rather simple architecture of the MPSoC. When using table-based and Totalprof for KPN tracing, the deviation increases. Note however, that this book is not concerned with improving the quality of the sequential performance estimates, but with methodologies that enable a programmer to choose which technique to use.

The second test MPSoC is a *Sparsely Connected Platform* (SCP), shown in Fig. 6.12b. It represents more specialized MPSoCs, that include dedicated, high bandwidth hardware communication channels. This channels are typically sparse, as in the case of so-called *pipelined MPSoCs* [39, 130] or the BlueWonder's BWC200 [67]. The platform includes four LTVLIW processors with local memories and four hardware channels, implemented as dual port memories. Due to its sparse interconnect, the SCP platform poses more challenges to the mapping process than the DCP platform. It is therefore used to validate the joint mapping proposed in Sect. 6.4.5.

6.7.1.2 Real-Life Applications

In order to assess the quality of the KPN mapping algorithms, three applications from the motivational example in Sect. 2.1 are used. The first application is the parallelized version of the audio filter (LP-AF) obtained with the sequential flow in Sect. 5.5.2, shown in Fig. 6.12c. A portion of the KPN representation of the JPEG application is shown in Fig. 6.12d. It contains a processing branch for each one of the three color components. In each branch, the *rectangular-to-block* (r2b) transformation, the *discrete cosine transform* (dct) and the *quantization* (qnt) are performed. The streams are merged by the *zig-zag* (zz) block and further encoded using the *run-length-code* (rlc). The JPEG decoder is connected directly after the rlc process, so that the correctness of the application can be analyzed by comparing the input and the output files for equality. In total, the parallel implementation of the JPEG application has 24 processes and 28 channels. The third application is a KPN implementation of the MJPEG standard, shown in Fig. 6.12e, presented in [235].

6.7.1.3 Random KPN Applications

Apart from the real applications, in order to effectively assess the quality of a mapping algorithm, many more tests have to be performed. For this purpose, several KPN graphs with their corresponding traces were randomly generated.

The random KPN generator uses the graph generation facility of the SDF3 tool [244]. The generated SDF graph is then modified by selecting actors and modifying their behavior to be *less static*. Given a random SDF, actors are turned into CSDF actors with a probability of 30 % and into KPN processes with a probability of 40 %. With a probability of 30 %, the SDF behavior of an actor is retained. The conversion from SDF to CSDF is done by dividing the behavior of the original SDF actor into *fragments*. As an example, consider an SDF actor with a single input of rate 4 and a single output of rate 3. The same behavior can be obtained with a CSDF actor with two phases, a first phase that consumes and produces two tokens ((2, 2)) and a second phase that consumes two and produces one token ((2, 1)). Many other alternatives are possible, e.g., two phases ((3, 3), (1, 0)) or three phases ((1, 2), (2, 1), (1, 0)). With this conversion, the consistency of the original SDF is conserved. The conversion to KPN process is more complex. This is achieved by first defining phases, as in the case of CSDF, and then introducing randomness in the way the phases are selected at *runtime*, i.e., during trace generation.

With the application topology and the behavior description of each node in the graph, artificial KPN traces are generated. Note that actual trace generation as described in Sect. 6.2.1 is not possible, since there is no source code to do profiling on. Once the artificial traces are created, the rest of the parallel flow can be used to produce a KPN mapping and evaluate it on the TRM. For the experimental evaluation, two sets of random graphs were used: a low-communication set (LC-RKPN) and a high-communication one (HC-RKPN). Each set contains 2 groups of 200 graphs, one with high and one with low variance (in terms of computation

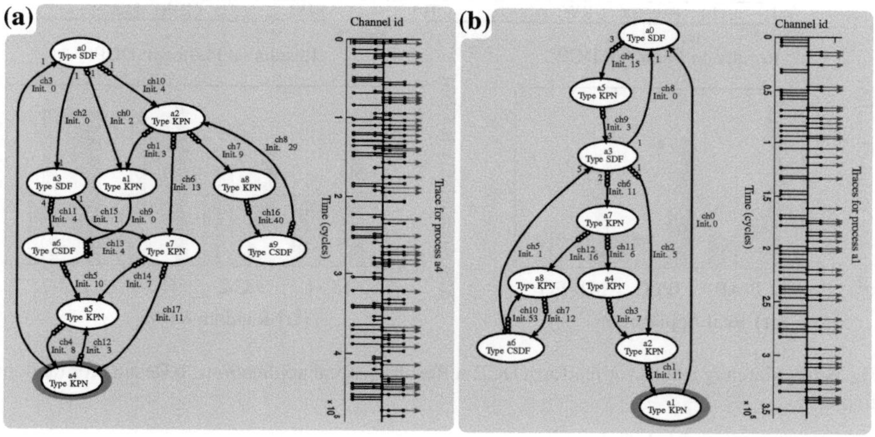

Fig. 6.13 Sample randomly generated KPN graphs and example execution trace for the highlighted KPN processes (on the bottom of the graphs). **a** A graph from the HC-RKPN set. **b** A graph from the LC-RKPN set

and communication times). The graphs in LC-RKPN and HC-RKPN have an average communication-to-computation ratio of 0.001 and 0.1 respectively. Two sample randomly generated graphs are shown in Fig. 6.13.

6.7.2 Best-Effort Results

The results for platform DCP are summarized in Fig. 6.14. The figure shows the achieved makespan for all the test KPNs with the different algorithms relative to the makespan obtained by GBM. The keys in the figure refer to the ICM heuristics introduced in Sect. 6.4.4, namely affinity (AFF), random walk (RW), computation balancing (CB) and simulated mapping (SIM).

The makespan of LP-AF, JPEG and MJPEG obtained by GBM was of 57.2, 156.9 and 336.7 Mcycles respectively. As shown in Fig. 6.14a, GBM produces results that are as good as the results produced by simulated mapping, which is the best process-only mapping heuristic. In average, GBM and SIM produced makespans that are 53, 36 and 38 % smaller than those produced by the other ICM heuristics for the three test applications, respectively. The makespans achieved by GBM correspond to a speedup of 3.76x, 3.82x and 3.02x compared to a solution in which all processes run on a single VLIW. For the JPEG application, with traces of around 25 MB, the runtime of the whole mapping flow was of 2.1s (On a AMD Phenom host processor running at 3.2 GHz with 8 GB of RAM).

The results for the 4 different groups of random KPNs are shown in Fig. 6.14b. For the low communication case (LC-RKPN1-2), GBM reported a speedup of 10 % with respect to average result of the remaining heuristics. For cases HC-RKPN1-2 the speedup was of 24 %. Also here, the benefit of GBM increases with the complexity

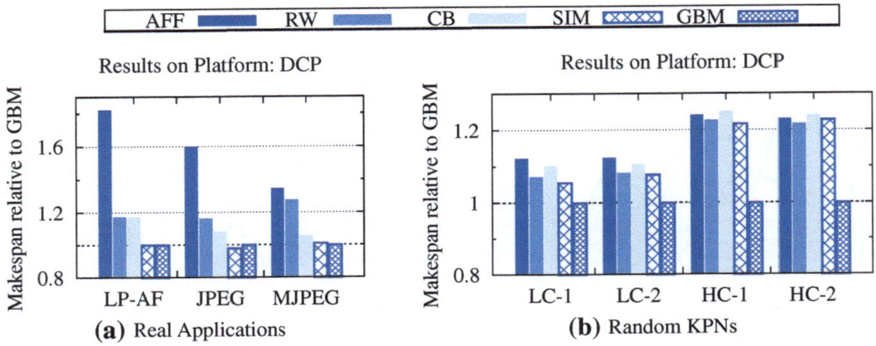

Fig. 6.14 Mapping results for platform DCP. **a** Results for real applications. **b** Results for random graphs

of the application. Notice also, that GBM was almost always the best algorithm, while the second best varied along the experiments. Finally, note that the results of heuristics like CB, SIM and AFF were similar to those of RW. This has not been the case in previous works which neglect or oversimplify the interconnect.

GBM performed well on platform DCP, which is not a very challenging platform for channel mapping. The benefits of a joint mapping algorithm are more evident on platform SCP. Mapping channels after processes failed for the three real applications. GBM, instead, found correct mappings with a makespan of 60.5, 206.4 and 345.3 Mcycles for LP-AF, JPEG and MJPEG respectively. A similar situation was observed in the case of random graphs. GBM managed to map all random graphs, whereas the ICM heuristics failed in over 96 % of the cases (see Table 6.1). This high failure rate was expected since the restricted interconnect of platform SCP introduces constraints on process mapping that are not considered in the process-only mapping heuristics of Sect. 6.4.4.2.

6.7.3 Results for Real-Time Applications

The iterative approach of Algorithm 6.3 is tested on the two real-time applications, LP-AF and MJPEG on platform DCP. Every execution of the sink process of the LP-AF application produces 1024 16-bit samples. Therefore, to achieve an audio rate of 192 kbit/s, it has to execute every 83.3 ms (as specified in Sect. 2.1). For MJPEG,

Table 6.1 Test results for 800 random KPNs on platform SCP

	Successes	Failures	Success Ratio (%)
GBM	800	0	100
Computation Balancing	30	770	3.8
Simulated mapping	30	770	3.8
Affinity	38	768	4.8
Random walk	26	776	3.2

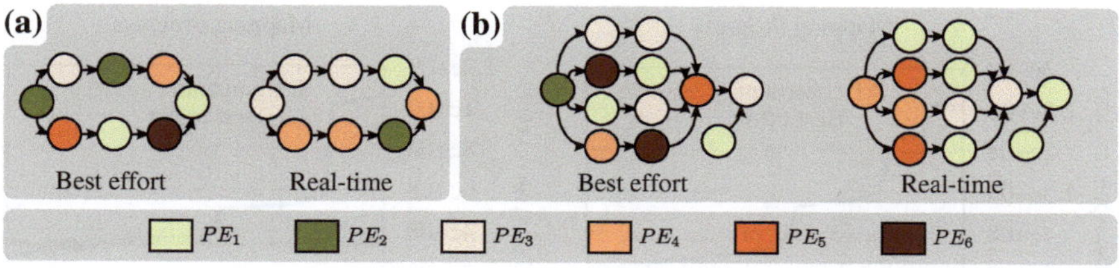

Fig. 6.15 Real-time mapping results. **a** Mapping results for LP-AF with and without real time constraints. **b** Same for MJPEG

the constraint of 10 frames-per-second requires that the sink process executes every 100 ms.

Given the relative small size of platform DCP, all platforms subsets are analyzed in this case study. For both applications, the algorithm was able to find a valid runtime configuration after 13 and 34 iterations, respectively. Finding a solution took the tool 60 s for LP-AF and 330 s for MJPEG. The computed mappings are shown in Fig. 6.15 together with the best-effort mapping. As expected, the computed mapping uses less resources. The result for LP-AF uses two IRISCs and two LTVLIWs (see Fig. 6.15a). Note that the best-effort mapping uses dedicated LTVLIW processors for the FFT and IFFT processes. Provided with the throughput constraint, the mapper decided to share the FFT processing with the filter and the source processes on PE_3 and the filter and the sink processes on PE_4. The additional load on the LTVLIW is compensated by mapping the IFFT processes to the IRISC processors. Note that merging the filter with the FFT matches the initial partition proposed by the sequential flow. The result for MJPEG uses one IRISC and three LTVLIWs, as shown in Fig. 6.15b.

Figure 6.16 shows the progress of the iterative mapping process for both applications. This figure shows how the makespan decreases by making more resources available to the mapper. Small improvements in the makespan are usually due to an increase of the buffer sizes in the inner loop of Algorithm 6.3. Notice however, that increasing the buffers not always improves sthe throughput since new delays might appear due to a new data mapping to a bigger, slower memory. The bigger steps in the figure are produced by adding processors to the mapping problem. The figure also shows the value below which the throughput constraint is fulfilled. It corresponds to 91.6 Mcycles for LP-AF and 330 Mcycles for MJPEG, since the sink processes are executed 11 and 33 times in each application respectively. The tool also allows to perform this kind of study for best-effort applications. In this way, the user can select the configuration with the best performance-utilization tradeoff.

6.7.4 Validation on TI's Keystone Platform

So far, the algorithms presented in this chapter have been tested on virtual platforms. With the timing information collected from simulation runs, the performance of the runtime configurations on the virtual platforms proved to match that of the internal TRM. In this section, the TI Keystone architecture (recall Fig. 1.6b) is used to vali-

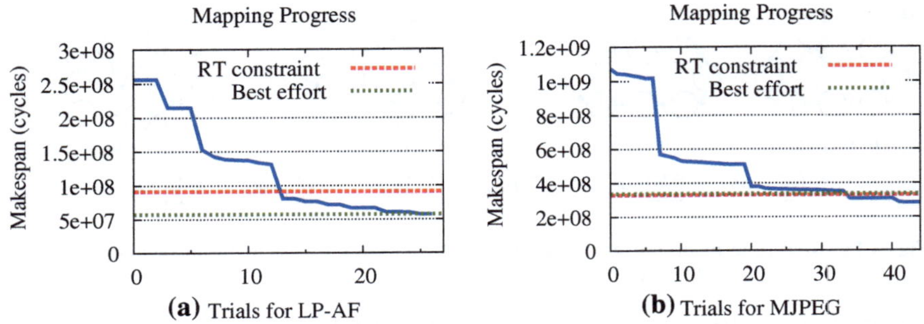

Fig. 6.16 Real-time iterative mapping. **a** For LP-AF. **b** For MJPEG

date the mapping results on a commercial platform. Before discussing the results in Sect. 6.7.4.2, the tool flow setup is briefly discussed in Sect. 6.7.4.1.

6.7.4.1 Keystone Tool Flow Setup

For the analysis in this section, the TMS320C6678 *Evaluation Module* (EVM) is used. This board features a C6678 multiprocessor chip of the Keystone product family, containing 8 fixed and floating-point VLIW DSPs. For code generation, this section leverages the newly developed MAPS backend for C6000 DSP platforms described in [205].

The tracing flow uses in this case a measurement-based approach (recall Sect. 2.2.2). By doing so, it avoids the inaccuracies due to high-level profiling and tracing (i.e., by table-based or Totalprof), allowing to focus the analysis on the quality of the algorithms and the accuracy of the TRM. For obtaining the information directly from the hardware, the execution of each process is monitored in isolation, in a similar way as it was done in [236]. After executing all the processes of an application, a simple tool translates the information exported by the monitor into the trace format of Listing 6.4. By having accurate numbers for the process traces, the sources for inaccuracies are restricted to the TRM and the communication costs. For the latter, different FIFO buffer implementations were designed and benchmarked to derive the cost model of the different communication primitives.

6.7.4.2 Keystone Results

To validate the results of the parallel flow on the Keystone architecture, the three applications introduced in Sect. 6.7.1.2 are used. For each of them, several mappings are produced with the tool flow by varying the platform subset. The predicted makespan reported by the TRM is then compared against the actual makespan measured directly on the board. This allows to measure the deviation from the speedup reported by the tool and the real speedup.

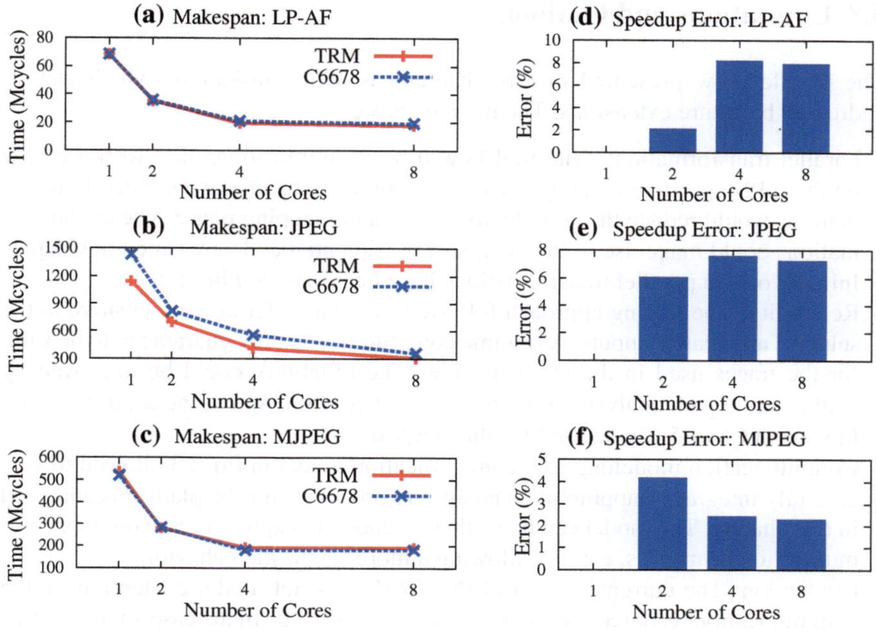

Fig. 6.17 Keystone results. **a–c** Makespan for test applications obtained with TRM and on the C6678 board. **d–f** Magnitude of the relative error of the speedup prediction

The results for the three applications are shown in Fig. 6.17. As shown in the horizontal axis of the figure, four different platform subsets were used, with one, two, four and eight processors. The left hand side of the figure (Fig. 6.17a–c) shows the makespan reported by the tool and the real one obtained from the C6678 board. The average makespan estimation error was of 4.2, 18.7 and 3.7 % for LP-AF, JPEG and MJPEG respectively. Note that the makespan of the LP-AF and MJPEG applications is considerably shorter on the Keystone than on platform DCP (see Sect. 6.7.2, Fig. 6.16). This is due to the highest computing power of the Keystone platform. In the case of the JPEG application, the execution on platform DCP featured lower values. This is due to several system calls in the code of the JPEG application that are emulated on the simulator (platform DCP) but consume actual time in the real board.

As shown in the makespan plots, the estimations provided by the tool flow are close to the real ones. However, an accurate prediction of the speedup is often more relevant than the absolute numbers, since programmers typically use the speedup to steer design decisions. For this study, the speedup prediction error is shown in Fig. 6.17d–f. Acceptable average errors of 4.59 % for LP-AF, 5.25 % for JPEG and 2.24 % for MJPEG were observed.

Most of the estimations were slightly under the real value obtained on the C6678. This deviation was tracked down to inaccuracies of the scheduler model inside the TRM, which does not fully emulate what the scheduler in the board does. This deviation has a regular impact in the prediction numbers, so that it is not reflected in the trend prediction as indicated by the plots in Fig. 6.17a–c.

6.8 Limitations and Outlook

The parallel flow presented in this chapter has some limitations which can be addressed by future extensions. The main ones are:

- Parallel transformations: The tool flow has no optimizations that work directly on the KPN specification, e.g., by merging or splitting processes. Such transformations would reduce the complexity of the later mapping phase. These transformations could make use of the analysis for sequential code presented in Chap. 5. Initial work on parallel transformations have been reported in [246].
- Reliability: The tracing approach followed in the parallel flow is sensitive to the selected application inputs. A runtime configuration is only guaranteed to be valid for the traces used in the mapping flow. The reliability could be improved by adding static path analysis of the processes' behavior. Again, the sequential tool flow from Chap. 5 can be used for this purpose.
- Communication modeling: The communication model in Eq. 2.3 allowed to successfully integrate mapping of channels and processes for the platforms analyzed in this chapter. The model can be further extended to capture more complex communication primitives, e.g., by allowing a more stochastic behavior.
- Contention: The current version of the TRM does not model contention on the communication resources. This proved not to be an issue in the case studies of this book, but may become relevant for other platforms and applications.

6.9 Synopsis

This chapter presented a solution to the parallel problem stated in Sect. 2.5. The solution approach is based on the analysis of process traces, for which a KPN tracing framework was proposed. Several heuristics were described for obtaining a valid runtime configuration. The simplicity of these heuristics, as opposed to evolutionary approaches, allows the parallel flow to be integrated in interactive programming environments, like the MAPS IDE. Furthermore, the infrastructure proposed in this book can be used as basis for further research on upcoming dynamic applications and complex platforms. The algorithms were benchmarked on several applications and different target platforms, including the recent TI Keystone architecture.

The parallel flow presented in this chapter is based on source code analysis, source-to-source transformation and, finally, traditional compilation. This approach falls short in the case of applications described as block diagrams, for which no source code is available. The approach also fails to discover potential hardware acceleration in the platform to implement a process specified using the C language. Both features are needed for demanding applications, like the MIMO-OFDM receiver in Sect. 2.1. This extensions are the matter of the next chapter.

Chapter 7
Extensions for Software Defined Radio

Chapter 1 explained how heterogeneous MPSoCs for high-end applications will continue to have hardware acceleration and specialized processors in order to meet tight energy constraints. For such platforms, the software compilation approach of the previous chapter would not leverage all the computing power. As a consequence, the performance of an automatically generated implementation would lag orders of magnitude behind an optimum manual design. This chapter presents an extension to the parallel flow that accounts for such optimized platforms. The extensions target applications from the SDR domain, thereby solving the SDR problem from Def. 2.57.

This chapter is organized as follows. Section 7.1 provides an overview of the programming flow. The main extensions to the parallel flow are described in Sects. 7.2–7.4, followed by a case study in Sect. 7.5. Section 7.6 discusses the deficits of the flow and Sect. 7.7 ends the chapter with a summary.

7.1 Tool Flow Overview

An overview of the SDR flow is shown in Fig. 7.1. This flow is an implementation of the Nucleus development concept introduced in Sect. 2.6.1. Recall that the main goal is to improve productivity by means of abstraction and retain performance by exposing hardware acceleration and specialized routines in form of a *flavor library*. Abstraction is provided by the CPN language and a mechanism that allows to mark processes as being taken from a *nucleus library*. The SDR flow is divided into three phases. (1) The construction phase, in Fig. 7.1a, builds the SDR application model (see Def. 2.52). (2) The nucleus matching phase, in Fig. 7.1b, generates *matching nucleus mappings* according to Def. 2.55. (3) The last phase in Fig. 7.1c computes valid runtime configurations and generates code for them. An overview of these phases is given in Sect. 7.1.1, followed by a description of the nucleus and flavor librares in Sect. 7.1.2.

J. Castrillón Mazo and R. Leupers, *Programming Heterogeneous MPSoCs*, DOI: 10.1007/978-3-319-00675-8_7,

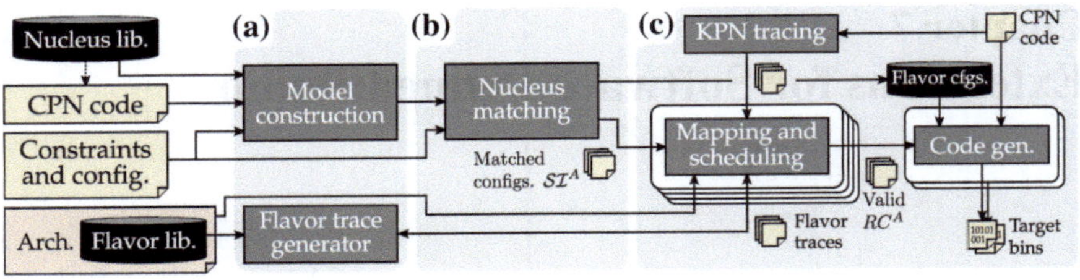

Fig. 7.1 SDR tool flow. **a** Model construction phase. **b** Nucleus matching phase. **c** Mapping, scheduling and code generation

7.1.1 Tool Flow Components

The following sections describe the purpose of the components in the three phases of the SDR flow of Fig. 7.1.

7.1.1.1 Model Construction Phase

The SDR model construction phase is similar to that of the parallel flow. It uses the XML option of *cpn-cc* to construct the application graph. The names of the process templates are used to identify processes that represent instances of nuclei from the library. With this information, it is possible to build the SDR application model from Def. 2.52 ($\mathcal{M}^A = (\mathcal{PAE}^A, KPN^A = (\mathcal{P}^A \cup \mathcal{N}^A, \mathcal{C}^A, \text{var}_{\text{size}}))$). The model is then annotated with constraints, as discussed in Sect. 6.3.1.2. This phase also includes a so-called *flavor trace generator*, which creates traces from a flavor specification for the mapping and scheduling phase.

7.1.1.2 Nucleus Matching Phase

This phase analyzes all possible nucleus mappings, $NC^A = (\mu_n, \mu_f)$ from Def. 2.53, i.e., all the possible ways to implement the functionality described in CPN with the flavors available in the target MPSoC. It then removes non *matching* mappings in the sense of Def. 2.55 and mappings that violate *non-timing* constraints. As shown in Fig. 7.1b, this phase produces a set of SDR implementation models (see Def. 2.56) which are further analyzed in the subsequent phase. Additionally, this phase produces configuration files that are used by the code generator to produce setup code for the target platform.

7.1.1.3 Mapping, Scheduling and Code Generation

This phase uses the parallel flow from Chap. 6 to compute valid runtime configurations for every matched configuration. If no valid configuration is found, the option is discarded. For all valid configurations, the flow produces binary executables as shown in Fig. 7.1c. The final implementation can be selected based on simulation.

```
1 <Nucleus name="FFT">
2  <parameter name="bitwidth">
3   <domain><int min="8" max="128"></int></domain></parameter>
4  <parameter name="points">
5   <domain><int min="1"></int></domain></parameter>
6  <property name="latency">
7   <domain><float></float></domain></property>
8  <!-- ... other parameters and properties -->
9  <input name="fft_in"/>
10 <output name="fft_out"/>
11 </Nucleus>
```

Listing 7.1: Example of a nucleus description

7.1.2 Input and Output Modeling

The flow in Fig. 7.1 has the same inputs as the parallel flow (see Sect. 6.1.3). In addition, the flow includes a platform-independent nucleus library and a platform-dependent flavor library. The formats of these inputs are described in the following.

7.1.2.1 Nucleus Library

This library contains a collection of nuclei, which represent algorithmic blocks that can be composed to implement wireless communication standards. How these nuclei are discovered and characterized is an ongoing research effort (see [142] as an example). Although at a finer granularity, the approach is similar to the application *dwarfs* discussed in [11] or Intel's *Recognition, Mining and Synthesis* (RMS) initiative in [70].

Recall the representation of a nucleus from Def. 2.50, $N^A = (P^{N^A}, V^{N^A}, \mathscr{K}^{N^A}, IN^{N^A}, OUT^{N^A})$. The functional description (P^{N^A}) of the nucleus is specified using the CPN language. This functional specification is intended for functional verification, e.g., using the Pthreads backend of *cpn-cc*. It is also used for the target platform, if the tool cannot find a matching flavor. The variables (V^{N^A}) as well as the input and output ports (IN^{N^A}, OUT^{N^A}) are described in the nucleus library. The constraints ($\mathscr{K}^{N^A}$) are defined by the programmer, restricting the domains of the nucleus variables. The set of all nuclei define the nucleus library ($\mathscr{N}$).

An example of a nucleus description for the FFT algorithm is given in Listing 7.1. This algorithm has a single input and a single output, `fft_in` and `fft_out` in Line `10`. In the description, variables are defined as either *parameters* (in Lines 2–5) or *properties* (in Lines 6–7). Parameters are variables that receive a single value, specified by the user according to the specific wireless application. Properties, in turn, are functions of the parameters. In both cases, the file describes the domains of the variables. As an example, consider the parameter variable `bitwidth` in Line 3 defined over a domain $\{8, 9, \ldots, 128\} \in \mathbb{N}$. When instantiating this nucleus in a receiver specification, the user can fix the `bitwidth` variable to a fixed number or further restrict the domain, e.g., to a value in {8, 16}.

```
1 <Flavor name="fft_HW" Nucleus="FFT">
2  <parameter name="bitwidth"><value>32.0</value></parameter>
3  <parameter name="points"><values List="32 64 128 256" /></parameter>
4  <property name="latency"><function>16*points+100</function></property>
5  <input name="fft_in"><port>input</port>
6   <DataType representation="fixed_point" format="Q31" DataWidth="32" />
7   <Interface type="buffer_flag_1of2">
8    <val name="size" val="8" /><val name="stride" />
9    <val name="cnt" val="64" />
10   <val name="fsize" val="4" /></Interface>
11  <Interface type="buffer_flag_2of2">
12   <addr name="addr" pool="in" min="0x04100000" max="0x041F0000" gran="8"
13          size="8" stride="fft_in__stride" cnt="64"/>
14   <addr name="faddr" pool="in" min="0x04100000" max="0x041F0000" gran="4"
15          size="4" cnt="1"/></Interface></input>
16 <output name="fft_out"><port>output</port>
17   <!-- ... --></output>
18 <Resource>
19  <PEs List="HACC_FFT0 HACC_FFT1"><Type>FFT_IP_v2.0</Type></PEs></Resource>
20  <!-- Other parameters and properties -->
21 <memory name="instance" pool="instance_fft" min="0" max="1"
22           gran="1" size="1" stride="1" cnt="1"/>
23 </Flavor>
```

Listing 7.2: Example of a flavor description

7.1.2.2 Flavor Library

This library is a collection of optimized kernels supported in the target platform. If the kernel is an optimized assembly routine, its description specifies how to call the routine and how to parametrize it. If the kernel is implemented in hardware, its description specifies how to trigger the hardware block, how to send and receive data from it and how to do synchronization. An example of a flavor characterization for the platform P2012 [242] is presented in [142].

Recall Def. 2.51, $F = (N, V^F, \mathcal{K}^F, IN^F, OUT^F, \mathcal{CM}^F, \mathcal{PE}^F)$. The fields of a flavor are better understood from the example in Listing 7.2. The flavor first specifies the nucleus it implements (N). In the example, flavor `fft_HW` is an implementation of the nucleus `FFT` from Listing 7.1 (see Line 1). This implementation contains the same variables (V^F) as the nucleus (`bitwidth`, `points` and `latency` in Lines 2–4). The variables and constraints of a matching flavor are a refinement of the nucleus variables. For example, the number of points of the flavor is restricted to {32, 64, 128, 256}, so that it cannot implement a 2048-point FFT nucleus.

The example in Listing 7.2 also shows the usage of flavor properties. The latency property of this flavor is defined as a function of the `points` parameter (see Line 4). In this case, the latency function is also the cost model ($\mathcal{CM}^F$) of the flavor. The flavor inputs and outputs (IN^F, OUT^F) match the algorithmic ports of the nucleus. In the case of a flavor, the ports have hardware-related annotations, as shown in Lines 5–15. It defines the exact data type in Line 6 and the hardware interfaces in Lines 7–15. This information is used during the matching process and in the code generation phase, as will be discussed later in this chapter. Finally, the set of resources that contains this flavor ($\mathscr{PE}^F$) is specified in Lines 18–20. The set of all flavors in the platform defines the flavor library ($\mathscr{F}^{SOC}$).

The format used for flavors and nucleus is intended as a prototype. Such a format could be later merged with standards for *Intellectual Property* (IP) descriptions such as the IEEE IP-XACT Standard [119].

7.2 Tracing for Software Defined Radio

In a similar way as in Chap. 6, the behavior of an SDR application is characterized by traces. This section explains how traces are generated for flavors and how the flow deals with performance estimation.

7.2.1 Flavor Trace Generation

In order to use the parallel flow presented in the previous chapter, the behavior of the application has to be characterized by means of execution traces. An SDR application has two types of nodes, $\mathscr{P}^A$ and $\mathscr{N}^A$. For non-nuclei algorithmic blocks, i.e., elements in $\mathscr{P}^A$, the KPN tracing flow described in Sect. 6.2 is used. For nuclei blocks, i.e., elements in $\mathscr{N}^A$, the functional code can be traced with the same approach. Note however, that the timing estimation provided for these blocks is only relevant if no matching flavor is found by the tool. In the case a flavor is selected, artificial traces must be generated. These traces are created on-demand, depending on the specific flavors selected during nucleus matching. The role of the trace generator in Fig. 7.1a is to create such artificial traces.

Flavor trace generation is trivial for *regular* nuclei, i.e., nuclei with an SDF-like behavior. Since the behavior of the algorithmic block is fully specified by the input and output rates, no instrumentation is required. The trace generator creates a sequence of *firings*, separated by the time provided by the latency property. *Irregular* flavors, in turn, feature data-dependent behavior, so that instrumentation is needed to capture sequences of events. The timing estimation is however taken from the flavor library.

As an example, consider the SDR application in Fig. 7.2a, which represents a simplified iterative receiver. In the example, the source and sink processes model input and output data interfaces. Apart from those processes, the application includes an FFT transformation (*fft1* and *fft2* in the figure), a MIMO demapper (*demap*), a channel decoder (*decod*) and a control process (*ctrl*). The latter is a typical example

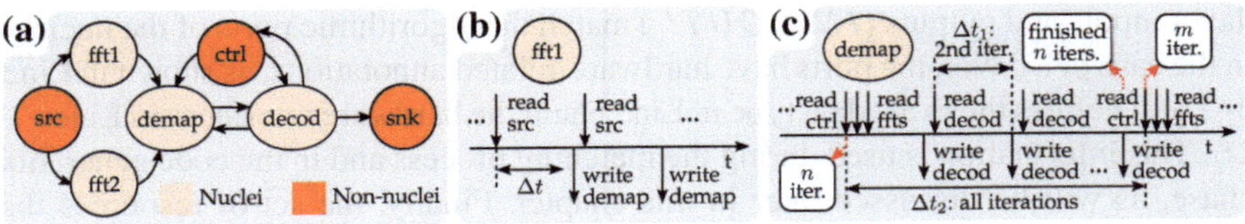

Fig. 7.2 Example of flavor traces. **a** Sample SDR application. **b** Trace for a regular FFT flavor. **c** Trace for an irregular demapping flavor

of a non-nucleus. It takes data-dependent decisions to steer the execution of other processes. In this example, the controller decides how many iterations the demapper and the decoder should perform, before outputting a decoded sample to the sink. Usually, this kind of behavior does not represent a computationally intensive task and is therefore not a nucleus.

The FFT block in Fig. 7.2a is a good example of a regular nucleus. The trace of every flavor would have the form shown in Fig. 7.2b. Traces from two distinct flavors of the same nucleus would differ in the computation latencies (Δt in Fig. 7.2b) and in the communication delays. The latter is influenced by the data representation, the type of the interface and other properties of the flavor model.

The demapper block in Fig. 7.2a is a good example of an irregular nucleus. As shown in Fig. 7.2c, its trace features data-dependent behavior. The demapper first reads from the controller how many iterations to perform (n in the figure). Then, it reads the inputs from both FFT blocks and starts computing. Only after n iterations with the decoder, the demapper reads again the input from the controller. Recall that process traces are obtained in *isolation*, hence time needed by the decoder block is not represented in the trace of Fig. 7.2c (between a write and a read from the decoder).

A summary of the different tracing mechanisms and their characteristics is given in Table 7.1. Note that in this flow, a functional CPN specification of the whole network may be required even if there is only one irregular block or a non-nucleus. This is because the KPN tracing flow requires the token history in all input channels. This means that token logging has to be performed on all the predecessors of the irregular block. The predecessors of these blocks must then also be instrumented. This process may ripple through the entire application graph.

7.2.2 Sequential Performance Estimation Revisited

So far in this book, the source code of an application or a process has been used to provide a performance estimate. For flavors, an annotation-based approach is followed instead. The annotations are not directly added to the application specification, but are computed from the latency equations contained in the flavor description. These equations can be extracted from vendor specifications of hardware blocks and optimized software routines, e.g., FFT implementations on TI DSPs in [185] and on

Table 7.1 Trace generation in the SDR flow

	Non-nuclei	Irregular nuclei	Regular nuclei
Instrumentation required	✓	✓	✗
Timing source	see Sect. 6.2.2	flavor library	flavor library
Block semantics	KPN	KPN	SDF

Xilinx FPGAs in [294]. However, they cannot describe the timing of more complex (IP) blocks. Consider, as an example, the timing model of the OSIP processor presented in Sect. 4.2.3.2.

The latency property of a flavor can be an arbitrary complex equation. Internally, the flavor trace generator uses the GiNaC library [19] to evaluate these functions. Consider an SDR implementation $\mathscr{S}\mathscr{I}^A$ resulting from a nucleus matching $NC^A = (\mu_n, \mu_f)$, where $\mu_n : \mathscr{N}^A \rightarrow \mathscr{F}^{SOC}$ determines the flavors to be used, and μ_f fixes the values of all the parameters of all the flavors. The mapping and scheduling phase uses this information to retrieve time-annotated traces for the flavors in $\mathscr{I}(\mu_n)$ from the flavor trace generator, as suggested by the bi-directional arrow in Fig. 7.1.

For regular nuclei, the time annotation process is straightforward. As an example, suppose that the nucleus matching fixed the points of the FFT to be 64 from the options in Line 3 of Listing 7.2. As a consequence, the latency of the flavor evaluates to $16{\cdot}64{+}100 = 1124$ cycles[1] (see Line 4 of Listing 7.2). This latency value is then used to describe the time between an input and an output of the FFT block, illustrated as Δt in Fig. 7.2b. The time elapsed between a write and a read is considered negligible.

It is more complex to annotate time to the traces of irregular nuclei. Consider the demapper example in Fig. 7.2c. As the trace shows, there are two *relevant processing times*. The first one refers to the latency between a read from and a write to the decoder (Δt_1 in the figure). The second one is the time between consecutive reads from the controller (Δt_2 in the figure). In reality, these values may vary depending on the situation. For example, in a noisy environment the demapper may need more time to compute, so that Δt_1 increases. At the same time, more iterations between the demapper and the decoder may be needed, so that Δt_2 increases as well. Additionally, Δt_1 may vary from iteration to iteration. As can be seen in this example, such a timing characterization cannot be achieved with a single equation. Additionally, it is not possible to generalize to situations with more *relevant processing times* $\Delta t_1, \ldots, \Delta t_n$.

For the reasons above, a pragmatic approach is followed in the SDR flow to annotate timing to irregular traces. Irregular nuclei are forced to include a time checkpoint event in their functional specification. A single equation is used in the flavor description to define the latency between time checkpoints. The flavor trace generator then evenly distributes this latency among the read and write events that happen between the time checkpoints. In the example in Fig. 7.2c, if the time checkpoint is defined exactly before reading the controller's output, the latency equation would provide an

[1] Recall from Sect. 2.3.1.3 that time is measured in terms of cycles of the main clock

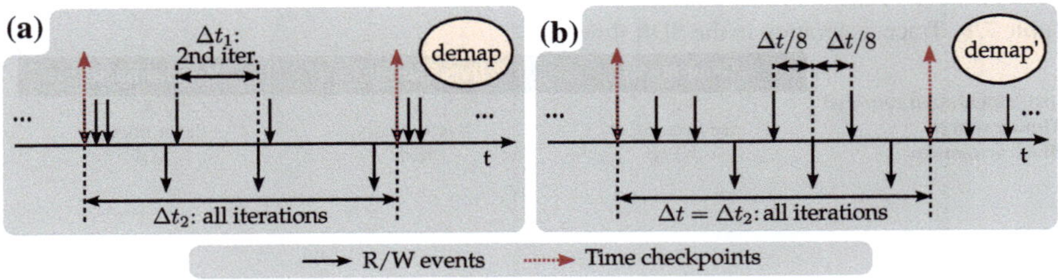

Fig. 7.3 Example of time annotation for an irregular nuclei. **a** Original *real* trace (from Fig. 7.2c). **b** Synthetic trace with the total latency distributed among all segments

average value for Δt_2. An example of this process is shown in Fig. 7.3. The original trace from Fig. 7.2c is shown in Fig. 7.3a. The resulting synthetic trace produced by the flavor trace generator would be similar to the trace shown in Fig. 7.3b. This approach changes completely the detailed synchronization between segments of the application. At a coarser level, the effects may be less noticeable.

7.3 Configuration Matching

Until now, this chapter described the flavor and nucleus libraries as well as the trace generation process that allows to reuse the trace-based mapping and scheduling phase of the parallel flow. After the construction phase in Fig. 7.1a, an SDR application model is available which contains a combination of *normal* processes and nuclei from the nucleus library. This section describes how nuclei are replaced by flavors in Sect. 7.3.1 and how an implementable specification is finally created in Sect. 7.3.2.

7.3.1 From Nucleus to Flavors

Given an SDR application graph $KPN^A = (\mathscr{P}^A \cup \mathscr{N}^A, \mathscr{C}^A, \text{var}_{\text{size}})$ and a target flavor library $\mathscr{F}^{SOC}$, there are many potential implementations, i.e., many possible nucleus mappings $NC^A = (\mu_n, \mu_f)$ in the sense of Def. 2.53. Let $\mathscr{F}^N$ denote the set of all possible flavors that can be used to implement a given nucleus $N \in \mathscr{N}^A$ that match the parameter values specified by the user, i.e.,

$$\mathscr{F}^N = \{F \in \mathscr{F}^{SOC}, (F = (N, V^F, \mathscr{K}^F, \dots) \wedge \forall v \in V^F, D_v^F \subseteq D_{v'}^N)\} \quad (7.1)$$

with v' the variable in the nucleus specification that corresponds to v in the flavor. As an example, consider two instantiations of the nucleus from Listing 7.1, one for a 32-point FFT and another for a 1024-point FFT. The flavor in Listing 7.2 would only be contained in the set for the 32-bit-FFT nucleus. The amount of potential implementations grows rapidly with the problem size. The total number of options can be computed as:

$$n_{\mathcal{SI}} = \prod_{N \in \mathcal{N}^A} (|\mathcal{F}^N| + 1) \tag{7.2}$$

To avoid this exponential complexity, simple tests are performed that allow to prune several options without having to check all of them explicitly. In the following, simple tests are described with a more detailed discussion of interface tests in Sect. 7.3.2.

Given a partial nucleus mapping $NC^A = (\mu_n : \mathcal{N}^A \rightarrow \mathcal{F}^{SOC}, \mu_f)$, if one of the following tests fails, all possible mappings that can be derived from it need not be evaluated.

- Multi-tasking: The number of flavors sharing a processing element cannot exceed the maximum allowed by the underlying resource.
- Redundancy: In platforms with several instances of the same processing element, options may be equivalent even if using a different flavor selection. The redundancy test prunes such equivalent nucleus mappings.
- Local tests: These tests check all local constraints that apply to a single block, e.g., user-defined logical bounds or fixed mappings (see Sect. 6.1.3.2).

7.3.2 Interface Matching

As mentioned in Sect. 2.6.2, flavors that are connected in the application specification must be able to communicate in the target platform. Given a nucleus mapping, this test checks that all interconnected flavors have a matching interface in the sense of Def. 2.54. After a successful matching, the parameters of the interfaces are configured. This configuration is passed to the code generator (see *Flavor cfgs.* in Fig. 7.1).

To enable communication between hardware blocks, the interfaces have to use the same data representation and use the same method to exchange data. For example, two flavors could exchange data using a buffer in shared memory together with a synchronization flag. A flavor could also write its output data directly to the register interface of another one. Actually enabling communication requires that both flavor interfaces use a matching configuration. For a shared memory buffer, both interfaces have to use the same addresses and the same data item size, access stride and item count.

The flavor library contains a detailed interface description for each flavor port, as shown in Lines `7–15` of Listing 7.2. The major part of this description is a specific *type* that defines the type of the interface being used, for example a shared buffer or a register interface. If the type of two connected flavors do not match, the option is not feasible and is discarded. Depending on the type of the interface, several configuration values are determined by matching the allowed ranges for each value described in the flavor library. Special care is taken for memory addresses in order

Table 7.2 Configuration matching for buffer/flag interfaces

	fft1.out	demap.in	matched
type	buffer/flag	buffer/flag	buffer/flag
address	0x10000-0x3FFF0	0x00000-0x1FFF0	0x10000
size	4	4	4
cnt	64-256	64	64
stride	*	8	8
ag-addr	0x10000-0x3FFF0	0x00000-0x1FFF0	0x10200

Table 7.3 Configuration matching for a register interface

	demap.out	decod.in	matched
type	regIFwrite	regIFwrite	regIFwrite
reg-addr	*	0x81400	0x81400
reg-size	2/4	4	4

to avoid overlapping buffers. If the matcher is not able to find values acceptable for both flavors, the nucleus mapping is not feasible and is thus discarded.

Table 7.2 shows a simplified example of the matching process. The output of the FFT block expects both the buffer and the flag within the address range 0x10000-0x3FFF0 and will write 64–256 values of size 4 bytes with any stride. The input port of the demapper needs the buffer and flag in the range 0x00000-0x1FFF0 and is fixed to 64 values of size 4 bytes with a stride of 8 bytes. The results of the matching process are shown in the last column of Table 7.2. These values fulfill the constraints of both ports and prevent that the buffer addresses and the synchronization flag overlap.

An example for the case of a register interface is shown in Table 7.3. The demapper can write its output data to a register interface at an arbitrary address in form of 2 or 4 byte values. The decoder flavor provides a register interface for its input that is located at the address 0x81400 and expects 4 byte values. The matching of the values in this case is only a check if the fixed values provided by the decoder are acceptable for the demapper flavor. It also has to set the configuration values of the demapper to those of the decoder.

The matching process illustrated with the two examples above is performed systematically on every two-flavor interface connected via a KPN channel. Recall the sets of variables $V_c^{F_i}$ and $V_c^{F_j}$ which contain all the interface variables of the flavors F_i and F_j associated with the connection $c \in \mathscr{C}^A$ (see Def. 2.54). The matching process selects a value that is common to the domain of both variables, i.e., $\mu_f(v) = \mu_f(v') = x \in (D_v^{F_i} \cap D_{v'}^{F_j})$. If $D_v^{F_i} \cap D_{v'}^{F_j} = \emptyset$, the option is discarded. If the variables refer to memory locations or other shared resources in the platform, the locations are marked so that they are not used more than once. This is the reason for the definition of address pools in the flavor library (see Line 21 of Listing 7.2).

7.4 Mapping and Code Generation

The previous section described how to find a set of matching nucleus mappings $S_{NC} = \{NC_1^A, \ldots, NC_n^A\}$, corresponding to a set of SDR implementations $S_{\mathscr{S}\mathscr{I}} = \{\mathscr{S}\mathscr{I}_1^A, \ldots, \mathscr{S}\mathscr{I}_n^A\}$. Note that due to the simple checks and the interface matching process, the amount of options left can be much smaller than the maximum amount of options in Eq. 7.2, i.e., $n \ll n_{\mathscr{S}\mathscr{I}}$. In this section, the number of valid options is further reduced, by analyzing the timing behavior in Sect. 7.4.1. Section 7.4.2 presents the extensions to the code generator for SDR applications.

7.4.1 Testing Timing Constraints

From the point of view of the parallel flow, each implementation option is a KPN application in its own, $\mathscr{S}\mathscr{I}^A = KPN^A = (\mathscr{P}^A, \mathscr{C}^A, \text{var}_{\text{size}})$ from Def. 2.56. The test for timing constraints is therefore carried out by using the mapping and scheduling phase of the parallel flow, see Fig. 6.1c.

Algorithm 7.1 Option pruning algorithm for SDR

1: **procedure** SDRPRUNING($A = (\mathscr{M}^A = (\mathscr{P}\mathscr{A}\mathscr{E}^A, KPN^A), V^A, \mathscr{K}^A), \mathscr{F}^{SOC}, SOC$)
 // $A \in \mathscr{A}^{\text{sdr}} \Rightarrow KPN^A = (\mathscr{P}^A \cup \mathscr{N}^A, \mathscr{C}^A, \text{var}_{\text{size}})$
2: $S_{NC} \leftarrow$ GETMATCHS($KPN^A, \{(N \in \mathscr{N}^A, \mathscr{F}^N)\}$) // *tests in Sect. 7.3, $\mathscr{F}^N$ from Eq. 7.1*
3: $\mathscr{T}_{\text{proc}}^A \leftarrow \emptyset$, $S_{\text{out}} \leftarrow \emptyset$ // *initialization*
4: **for** $P \in \mathscr{P}^A$ **do** $\mathscr{T}_{\text{proc}}^A \leftarrow \mathscr{T}_{\text{proc}}^A \cup T^P$ // normal *KPN tracing from Sect. 6.2.1*
5: **end for**
6: **for** $NC = (\mu_n, \mu_f) \in S_{NC}$ **do**
7: $\mathscr{S}\mathscr{I} \leftarrow$ GETKPN(KPN^A, NC)
8: $\mathscr{T}_{\text{flav}}^A \leftarrow$ FLAVORTRACER($\mathscr{I}(\mu_n), \mu_f$) // *from Sect. 7.2*
9: $\mathscr{T}^A \leftarrow \mathscr{T}_{\text{proc}}^A \cup \mathscr{T}_{\text{flav}}^A$
10: $\mathscr{K}_{\text{sdr}}^A \leftarrow \mathscr{K}^A \cup$ GETSDRCONSTRAINTS(NC)
11: $(f_c, RC^A) \leftarrow$ ITERATIVEMAPPING($((\mathscr{P}\mathscr{A}\mathscr{E}^A, \mathscr{S}\mathscr{I}), V^A, \mathscr{K}_{\text{sdr}}^A), \mathscr{T}^A, SOC$)
 // *from Algorithm 6.3*
12: **if** $f_c =$ `True` **then** $S_{\text{out}} \leftarrow S_{\text{out}} \cup (NC, RC^A)$
13: **end if**
14: **end for**
15: **return** S_{out}
16: **end procedure**

The pseudocode of the whole matching and mapping process is shown in Algorithm 7.1. It receives the model of the SDR application, the flavor library and the architecture model. The function GETMATCHS processes all possible options and performs the checks described in Sect. 7.3. The code in Lines 3–5 initializes the output set S_{out} and retrieves the traces for the non-nuclei blocks. The loop in Lines 6–14 tests the timing constraints for every matching implementation. Inside this loop, the function GETKPN replaces nuclei blocks with the respective flavor models, according to the configuration NC. The function FLAVORTRACER represents the flavor tracing component described in Sect. 7.2. It produces traces for the selected flavors, i.e., the flavors in the image of the nucleus mapping function μ_n. Additionally, the flavor tracer needs the values of the flavor parameters (μ_f) to evaluate the latency equations. Once the flavor traces are generated, the trace set of the *new* application is created by merging the traces obtained with the instrumentation phase ($\mathscr{T}^A_{\text{proc}}$) and the flavor traces ($\mathscr{T}^A_{\text{flav}}$), as shown in Line `9`. Before actually running the mapping algorithm of the parallel flow in Line `11`, the function GETSDRCONSTRAINTS extends the application constraints. This is needed in order not to violate assignments made by the matching phase. Two types of additional constraints result from a nucleus mapping, i.e., *fixed mapping* and *logical channel bounds* (see Sect. 6.1.3.2). A fixed mapping is induced by the nucleus-to-flavor mapping. For example, if the FFT nucleus is mapped to a hardware accelerator, the mapping phase of the parallel flow must not override this decision. Logical channel bounds are added due to flavor interface restrictions. For example, a register based communication interface is modeled as a single token logical channel. With the information collected up to Line `11`, it is possible to use the iterative mapping heuristic from Sect. 6.5. If the iterative mapping algorithm succeeds, the implementation option is kept together with its runtime configuration, as shown in Line `12`. Otherwise, the implementation option is discarded.

7.4.2 Code Generator Extensions

After the mapping and scheduling phase, a set of valid implementations is provided to the user (S_{out} in Algorithm 7.1). The user can then select a single implementation based on the results from the TRM and further evaluate it on the target platform.

The code generation process for a single implementation option is illustrated in Fig. 7.4. Apart from the CPN code and the schedule descriptor used in the parallel flow, the SDR code generator receives the nucleus mapping and the flavor configuration files. The nucleus mapping (NC^A) specifies which processes are to be compiled using a traditional compilation approach. The C code for blocks that were replaced by flavors is not compiled with *cpn-cc*. The flavor configuration file contains the values selected during the matching process. These values are required to configure hardware accelerators or optimized software routines.

Code generation for flavors is different for hardware and for software flavors. For hardware flavors, the code generator inserts C code into the main function of the host processor. This code initializes the hardware block and configures it according to the matching configuration (μ_f). Apart from the configuration, the code generator also inserts code that controls the execution of the block. The generated C file is then compiled with the target tool-chains. This process is illustrated in the middle of Fig. 7.4. The flavor library contains code templates for every hardware flavor which are used by the code generator to produce the control and configuration code.

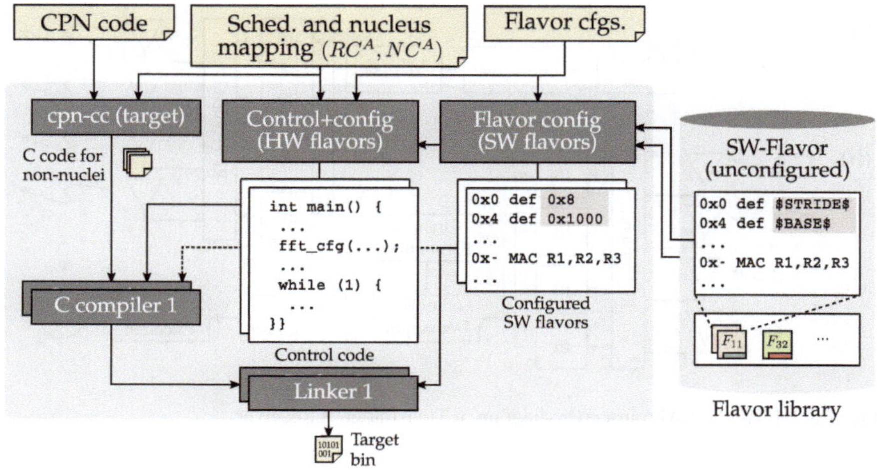

Fig. 7.4 Code generation flow SDR applications

Software flavors do not require control code generation. For these flavors, template code is also stored in the flavor library. It contains placeholders that are replaced by the code generator according to the matching configuration. An example of this process is shown on the right-hand side of Fig. 7.4. In the example, the assembly template code for a flavor is shown which contains placeholders for the `BASE` address and the `STRIDE` used to access the data. The code generator simply replaces these placeholders with the actual values computed by the tool (`0x8` and `0x1000` in the example). For a configurable software flavor specified in C, the code generator would use the target tool-chain compiler, as suggested by the dashed line in the figure.

The low-level code generation flow described in this section applies to bare-metal systems. This is the reason why the code is directly modified by the tool, instead of using high-level programming APIs. This level of detail is required to efficiently execute time-critical SDR applications.

The backend presented in this section can be used to produce an executable for every implementation option. In this way, the programmer can verify more than one option on a virtual platform or on the real target. This is sometimes required, since the accuracy of the TRM module of the parallel flow may not be enough for very time-critical applications. However, note that executing several options on a cycle or instruction-accurate virtual platform may be a time consuming task. For this reason, the SDR flow also includes a backend that generates a high-level SystemC simulation model that uses Synopsys MCO technology [7]. Unlike the TRM, the MCO simulator includes detailed models for the communication and the memory architecture. For this reason, although slower than the TRM, the MCO simulator can provide more accurate results. Besides, the MCO model allows to verify not only the timing but also the functional behavior. Processing elements in an MCO simulator are modeled as *Virtual Processing Units* (VPU) [147]. These units natively execute

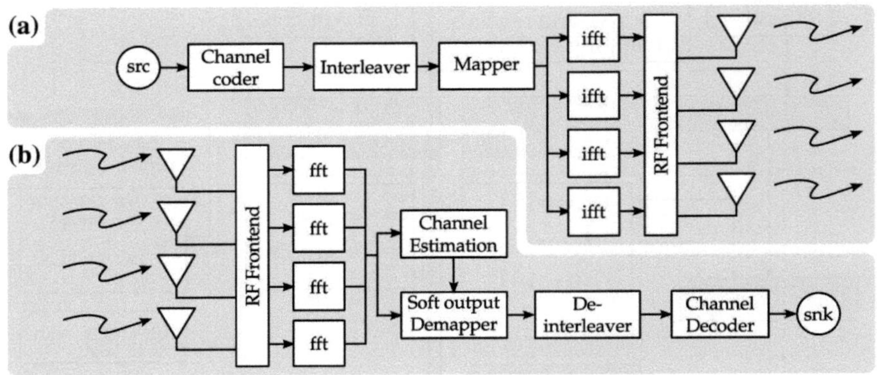

Fig. 7.5 MIMO OFDM transceiver diagram. **a** Transmitter. **b** Receiver

code on the host and use timing annotations to emulate time, e.g., using the SystemC `wait` function. The code for the VPUs is retrieved from the CPN specification and the timing annotations are taken from those produced by the flavor trace generator.

7.5 Case Study

Section 3.4 introduced *cross-platform portability* and *efficiency* as the two key drivers for SDR CBSE methodologies (see also Table 3.4). This section presents a case study in which these two properties are assessed. Efficiency is analyzed by comparing the performance of automatically generated SDR implementations against manual designs. Portability is analyzed by using the same application specification for two heterogeneous MPSoCs. The case study also provides a quantitative analysis of the performance difference of solutions using (1) a traditional compilation approach, (2) optimized software routines and (3) hardware acceleration.

Section 7.5.1 provides details about the SDR application used in this case study. The target platforms and the flavor libraries are introduced in Sect. 7.5.2. Finally, Sects. 7.5.3–7.5.4 discuss the experimental results.

7.5.1 Target Application: MIMO OFDM Transceiver

To test the SDR flow, the MIMO OFDM transceiver introduced in Sect. 2.1 is used. A detailed algorithmic structure of the generic non-iterative transceiver is shown in Fig. 7.5. It is composed of two subsystems: the transmitter in Fig. 7.5a and the receiver in Fig. 7.5b. Both subsystems are usually found in end-user devices for the down-link and the up-link. For simulation purposes, the two subsystems are interconnected with a model of an additive white Gaussian noise channel.

The transmitter (TX) chain of the system includes a traffic generator (*src* in the figure) that simulates incoming data from the upper network layers. The bitstream is coded with a convolutional code of rate $1/2$, interleaved to prevent burst errors and later mapped into a *Quadrature Phase Shift Keying* (QPSK) constellation. Four streams, one for each antenna, are then converted by means of an Inverse Fast Fourier Transform (*ifft*), and are thereafter transmitted. On the receiver (RX) side, the streams are converted back by the *fft* blocks. The channel estimation provides an estimate of the channel matrix to a simple zero forcing soft-output MIMO demapper. This, in turn, computes likelihood values for the Viterbi decoder, which provides the decoded bits to the sink block. The sink block represents data consumption by the MAC layer. The OFDM transmission scheme uses 64 sub-carriers, with 48 containing actual payload and the remaining used for channel estimation. The *(i)fft* blocks compute a 64 point (Inverse) Fast Fourier Transform.

The CPN description of the transceiver comes from an in-house C/C++ code base. This code was mainly intended for simulation and algorithmic research and has therefore no target-dependent code. For the purpose of testing, each of the blocks described above is considered a Nucleus. The case study in this section analyzes the receiver (RX-alone), the transmitter (TX-alone) and the whole transceiver (TX-RX).

7.5.2 Target Platforms and Flavor Libraries

Two target platforms were created for this case study, a *software-dominated platform* (SDP) and a *hardware-dominated platform* (HDP), shown in Fig. 7.6. Platform SDP, in Fig. 7.6a, is a shared memory MPSoC that contains one ARM926EJ-S and three LTVLIW processors interconnected by an AMBA bus. Platform HDP, in Fig. 7.6b, is an MPSoC with hardware accelerators for most of the operations required by the MIMO OFDM transceiver. It contains an ARM926EJ-S, an LTVLIW, two accelerators for 64-point FFTs, one for demapping and one for Viterbi decoding. The components in platform HDP are interconnected by a multi-layer AMBA bus and have access to several shared memories. The accelerators in the platforms were all modeled in SystemC.

The source module in both platforms (*src*) is used to model data sources. For the RX path, it provides 4 input streams at a configurable rate, emulating incoming data from the *Analog-to-Digital Converters* (ADCs). For the TX path, it models data coming from another processor in the system in charge of running higher layers of the protocol stack. Similarly, the sink block (*snk*) models data consumption from other layers for the RX path and models consumption by the RF frontend for the TX path.

Several flavors were created for the algorithmic blocks introduced in Sect. 7.5.1. Table 7.4 shows the cycle count for the flavors on the processing elements (ARM

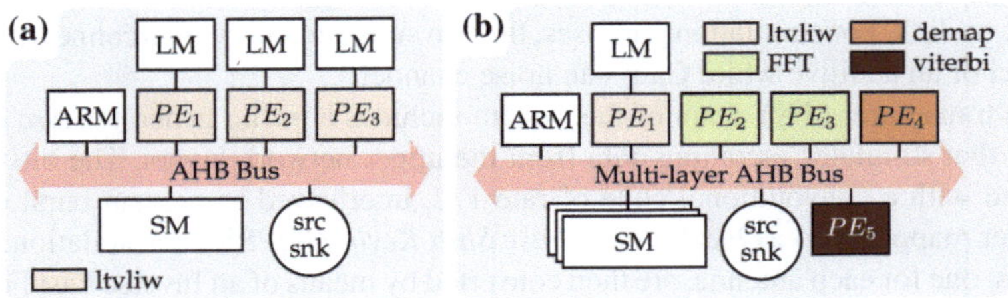

Fig. 7.6 SDR target heterogeneous platforms. **a** *Software-dominated platform* (SDP). **b** *Hardware-dominated platform* (HDP)

Table 7.4 Overview of latency values (in cycles) for all flavors used in the case study

	ARM	LTVLIW	LTVLIW-Opt	HW
Encoder	30306	4635	2074	–
Interleaver	28727	35574	523	–
Mapper	31262	9464	1047	–
fft/ifft	5067749	802580	45000[a]	1124 [c]
Ch. Estimator	808562	-	75000 [b]	–
Demapper	1350190	446730	25048 [b]	6816 [d]
De-interleaver	28727	35574	4121	–
Decoder	3749339	1245598	70020 [a]	1545 [c]

[a] Values estimated from available libraries for commercial VLIW processors.
[b] Values scaled from an available implementation on a vector processor to match the ILP available in the LTVLIW.
[c] Values extracted from off-the-shelf commercially available IP cores. The cycle counts include input and output delay.
[d] Value, for both channel estimation and demapping, scaled from [103]

and LTVLIW) and on the hardware accelerators. A cell with the entry "–" expresses that the given Nucleus has no such flavor.

The C implementations of the nuclei were derived directly from the CPN description (by removing simulation-relevant code). This general-purpose C code was compiled for the ARM and the LTVLIW processors (columns labeled ARM and LTVLIW in Table 7.4). The cycle counts are equivalent to the numbers expected from CBSE SDR approaches based on traditional compilation flows. The LTVLIW-Opt column, instead, contains performance data of optimized assembly routines for the LTVLIW for most of the blocks in the transceiver. An expert low-level programmer provided the implementations for the encoder, the interleaver, the mapper and the de-interleaver. For the rest of the blocks, reference numbers were selected from similar solutions. Finally, the cycle counts under the HW column in Table 7.4 are taken from data sheets of off-the-shelf commercially available IP cores [295], external references and in-house designs. The numbers in Table 7.4 reflect typical implementations. However, note that the accuracy does not affect the tool evaluation performed in this case study.

For the case study, three flavor libraries were created with the implementations described in Table 7.4. The first library (FL1) contains the un-optimized software flavors. The second library (FL2) extends FL1 by including the flavors corresponding to the optimized software routines of the LTVLIW. Finally, the third library (FL3) includes the hardware flavors. The MPSoC model of platform SDP can be extended

Table 7.5 SDR flow: Results summary

	Flavor library	Expected latency (cycles)	Obtained latency (cycles)	Expected throughput (bps)	Obtained throughput (bps)	Throughput deviation (%)	Runtime (s)
RX-alone	FL1	4134777	4141790	7955	7953	0.015	22.7
	FL2	264189	264716	116366	116222	0.13	33.1
	FL3	14730	14888	2816901	2782608	1.21	8.8
TX-alone	FL1	1647986	1654870	11961	11601	3.0	5.8
	FL2	93644	93867	213333	210757	1.2	4.5
	FL3	5892	6037	5268935	5217391	0.98	29.5
TX-RX	FL1	TX: 3210320 RX: 4134777	TX: 3210900 RX: 4135010	5980	5962	0.30	98.2
	FL2	TX: 180000 RX: 288795	TX: 180200 RX: 290133	94843	93439	1.47	33.7
	FL3	TX: 10013 RX: 14730	TX: 10500 RX: 15861	2472634	2461538	0.45	75.6

with the first two flavor libraries (FL1 and FL2), while platform HDP can be extended with any of the libraries.

7.5.3 Results of the SDR Flow

The three waveforms (RX-alone, TX-alone, and TX-RX) were analyzed by the SDR flow for each flavor library (FL1, FL2, FL3). For each test case, a latency and a throughput constraint were provided to the tool. The constraints were set to match the performance of the best implementation for each case. These reference values were provided by an experienced designer using a manual *spreadsheet* design approach. The results of the nine different combinations are summarized in Table 7.5. The values for the first two flavor libraries (FL1 and FL2) were obtained in platform SDP, while platform HDP was used for the third library (FL3). The reference latency and throughput values are shown in the second and fourth columns of Table 7.5 (*expected latency* and *expected throughput*). The results obtained by the SDR flow are contained in the third and the fifth columns.

In Table 7.5, the application throughput θ in bits per second (bps) is computed as

$$\theta = \frac{n_{\text{ant}} \cdot n_{\text{sc}} \cdot s \cdot c}{P}$$

where n_{ant} is the number of antennas ($n_{\text{ant}} = 4$), n_{sc} is the number of sub-carriers with payload ($n_{\text{sc}} = 48$), s is the number of bits per symbol ($s = 2$), c is the code rate ($c = 1/2$) and P is the average time between consecutive executions of the sink block, i.e., the average rate with which the Viterbi decoder produces data. The period P is provided by the TRM of the parallel flow. As the table shows, the results of the SDR flow are close to the values obtained manually. In fact, the average throughput deviation is only of +1.0 % (the average latency deviation, not shown in the table,

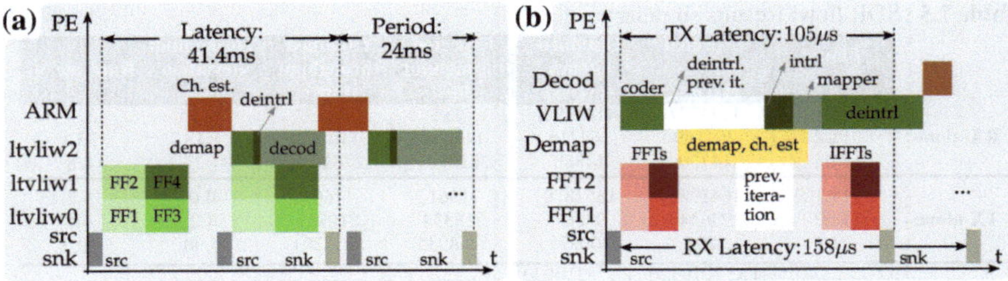

Fig. 7.7 Gantt charts obtained with the MCO simulator (100 MHz clock). **a** Two iterations of RX-alone using FL1. **b** One iteration of TX-RX using FL3

is +1.5 %). This negligible deviation shows that the implementation chosen by the mapper matches the one selected manually. The observed deviations are due to small variations in the communication estimation within the SDR flow. In the case of the TX-RX application, the latencies of both TX and RX chains are reported in Table 7.5. Naturally, these latencies are higher than the latencies obtained when each chain was processed separately. For example, the latency of the RX chain in the TX-RX case with FL3 was of 15861. In the RX-alone case, the latency was of 14888 cycles. The latency of the TX chain in the FL3 case increased from 5892 to 10013 cycles. This increase is a consequence of the throughput constraint, which forces a higher resource utilization that leads to overlapping iterations, i.e., a pipelined execution. The last column in Table 7.5 contains the time it took to run the SDR flow for every test case on an AMD Phenom host processor running at 3.2 GHz with 8 GHz of RAM. The average runtime was slightly above 30s, which is comparable to tolerable traditional compilation times.

For every case, the matching phase analyzed between hundreds and thousands of options and exported one to ten options depending on the case. As an example, for the RX-alone waveform the process started with 638 options when using FL1. The options were reduced to 328 after the matching check from Sect. 7.3, then to 33 after local latency checks and finally 6 met the throughput constraints and were returned by Algorithm 7.1. For the TX-RX waveform on platform HDP, the SDR flow reduced 4096 options to 12.

The best schedules computed by the SDR flow for the two test cases were exported and verified using the MCO simulator. A simplified version of the Gantt charts obtained from the simulation are shown in Fig. 7.7. The figure shows how the latency and the period were measured. The values were compared against the TRM estimation with no considerable variation. From the figure, one can see how tight throughput constraints cause the processing of different iterations to overlap. This contributes to the increase in the latency of the RX and TX chains for the TX-RX case (Fig. 7.7b).

The best rates obtained for the different test cases are plotted in Fig. 7.8. The figure shows how the SDR flow produced efficient implementations of the waveforms, according to the available flavors. With the same input specification, the flow produced implementations with performance differences of a couple of orders

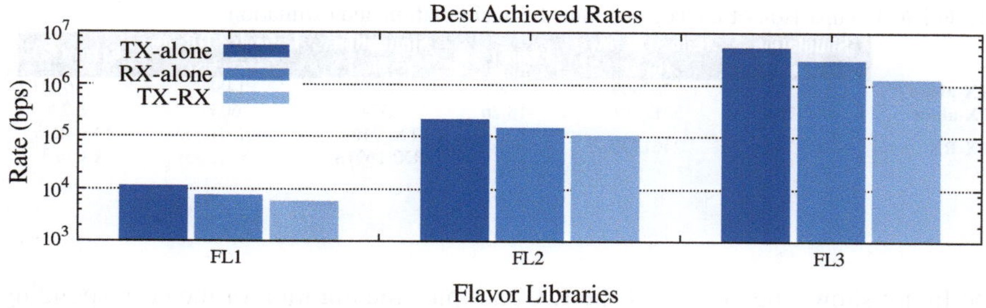

Fig. 7.8 Best rates obtained by the SDR flow for the different waveforms

of magnitude. Optimized software routines reported an average speedup of 16 x compared to un-optimized ones. Hardware accelerated flavors obtained an average speedup of 25 x compared to optimized routines. The performances obtained for the FL1 case would be close to the performances expected by a traditional compilation flow for both platforms (SDP and HDP). The results also corroborate that in order to achieve high data rates with high-end radios, heterogeneous platforms with accelerators are needed.

7.5.4 Execution on the Virtual Platform

For each case, the best option graph found by the SDR flow was used to generate code for the virtual platform. The overall process, including compilation of initialization and control code for hardware flavors, configuration of the software flavors and linking the parts to firmware images for the ARM and LTVLIW processors took 2.95 s for the RX-alone case, 3.14 s for the TX-alone case, and 3.67 s for the TX-RX case. Note that the TX-alone compilation was longer, due to the higher number of software flavors, resulting in more code to be compiled, assembled and linked.

The actual simulation was performed for 10 consecutive frames, using recorded data from functional simulation in the source blocks. The data received by the sink blocks was recorded to trace files and compared to reference data obtained from functional simulation without detecting any difference. The simulations took 6.12 s for TX-alone, 5.65 s for RX-alone and 6.23 s for TX-RX.

To validate the schedules produced by the TRM and the MCO simulator, the SystemC Explorer tool of Synopsys PA was used, producing the traces shown in Fig. 7.9. The bars in the figure represent memory accesses produced by each processing element. The same color scheme of Fig. 7.7b was added to the traces to make it easier to compare the actual simulation results with the predicted ones. The two cursors visible in the figure were used to measure the start and the end of the frames and thus to determine the latency and the throughput. The placement of the cursors in

Table 7.6 Comparison of latency and rates from simulation and estimation

	Simulation (bps)	Rate SDR ow (bps)	Error (%)	Simulation (cycles)	Latency SDR ow (cycles)	Error (%)
RX-alone	2678571	2782609	-3.88	17721	14888	16
TX-alone	4411764	5217391	-18.26	6574	6037	8.2
TX-RX	2419354	2461538	-1.74	TX: 7168 RX: 16918	TX: 10500 RX: 15861	TX: -46.5 RX: 6.2

the figure shows the latency of the RX path plus the duration of the corresponding source and sink (roughly 204 μs).

Table 7.6 shows the rates and latencies that were obtained on the virtual platform and compares the values measured from the simulation with the estimates calculated by the SDR flow. Note that the latency of the RX path on the TX-RX application was of 16918 cycles, whereas the latency of the same path in the RX-alone application was of 17721 cycles. This unexpected timing behavior was caused by timing delays introduced by the bus arbitration which affected the polling behavior on synchronization flags. These kinds of effects are not modeled by the TRM, which explains the relatively high deviation of 16 % for the RX-alone case.

The average absolute error in the latency predicted by the SDR flow was of 19 %, while that of the throughput was of 7.9 %. In almost all cases, the estimate was optimistic, i.e., positive latency errors and negative throughput errors. This is due to the abstract timing model of the TRM which does not account for some of the effects in the virtual platform, e.g., bus contention and the aforementioned polling overhead. The only exception was observed in the TX chain of the TX-RX application, in which the prediction was 46 % worse than what was actually observed in the platform. The reason for this high deviation was tracked down to a slight difference in the time sharing of the LTVLIW resource. The FIFO policy of the TRM schedules the de-interleaver block after the encoder (see Fig. 7.7b). This adds 4121 cycles to the TX path, which explains the deviation. In the platform, the polling mechanism gave preference to the interleaver, thereby keeping the TX latency almost unaltered (see Fig. 7.9).

7.6 Limitations and Outlook

The SDR flow is a proof of concept of a methodology for programming applications with stringent constraints. For this methodology to be widely applicable, the following items should be improved:

- Irregular nuclei: The current solution for the timing characterization of irregular nuclei described in Sect. 7.2.2 has to be extended to improve the accuracy. An option would be to add access pattern descriptions to hardware accelerators and optimized routines, similar to the approach in [96].

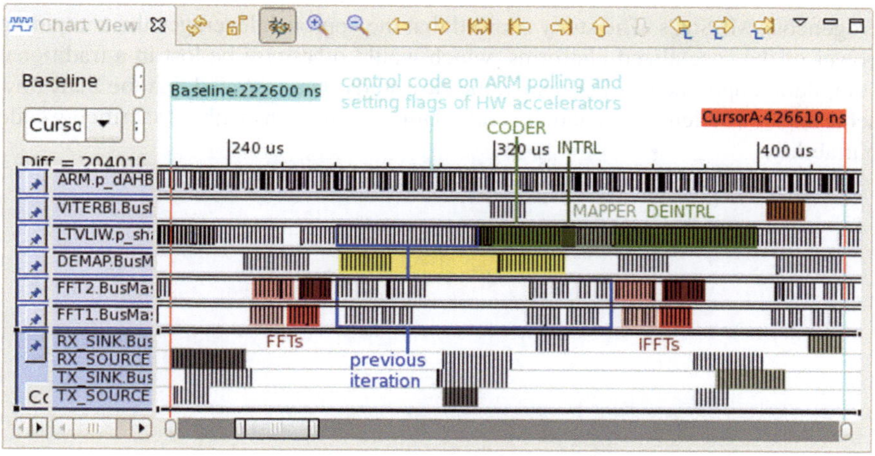

Fig. 7.9 Execution traces for TX-RX obtained from VP (compare to Fig. 7.7b)

- Algorithmic properties: In addition to timing properties, algorithm designers are also interested in algorithmic performance measures, such as the *Bit Error Rate* (BER). How to model these properties as functions of the flavor parameters and how to perform a waveform-wide evaluation remain an open problem.
- Nuclei and flavor libraries: The case study presented in this book uses *ad-hoc* formats for the tool flow libraries. The formats and the characterization of nuclei and flavors should be driven by a standard that enables true portability of waveforms.
- Code generation: The code generation process for flavors must be generalized, based on a more thorough analysis of hardware and software interfaces.
- Timing accuracy: The timing accuracy of the TRM proved to be enough for multimedia applications in Chap. 6. For the more demanding applications analyzed in this chapter, the TRM estimate deviated up to 46 %. For this reason, a cycle/instruction-accurate simulator is still unavoidable in the SDR flow.
- Adapter code: The interface matching presented in Sect. 7.3.2 is very strict. This could be relaxed by allowing some interfaces to be matched with *adapter code*. The code would ensure that the communication is successfully implemented. This must be done carefully in order not to spoil performance.

7.7 Synopsis

This chapter presented a tool flow to solve the SDR problem stated in Sect. 2.6. The flow is an extension to the parallel flow of Chap. 6 that makes it possible to use optimized software routines and hardware accelerators. The SDR flow was showcased on a case study that included a demanding communication algorithm and two heterogeneous MPSoCs. The study showed that the approach leverages the computing power of the specialized platforms, which would otherwise be lost in a traditional compilation approach. Additionally, the case study demonstrated that the SDR flow achieves the efficiency of manually crafted implementations without sacrificing code portability.

Chapter 8
Multi-Application Flow

Until now, this book presented three programming flows for applications of different natures, namely sequential, parallel and SDR. However, as mentioned in Chap. 1, today's embedded systems are no longer designed to serve a single task. As a consequence, the results of a single application flow may not hold true in a real system where other applications share the hardware resources. This chapter presents a tool flow that computes different valid runtime configurations for each application, depending on the execution scenario. The flow is therefore a solution to the multi-application problem stated in Def. 2.18.

This chapter is organized as follows. Sect. 8.1 provides an overview of the tool flow, which uses the single application flows from Chaps. 5–7. The core of the multi-application flow is a scenario analysis phase, which is described in Sect. 8.3 presents a case study that validates the approach. Thereafter, Sect. 8.4 lists potential improvements to the flow, followed by a summary of the chapter in Sect. 8.5.

8.1 Tool Flow Overview

An overview of the multi-application flow is shown in Fig. 8.1. This flow is a detailed version of the MAPS flow presented in Fig. 1.7. It receives as inputs, a set of applications, an architecture file, a set of constraints and configuration options, as well as a multi-application description. The latter describes how applications are expected to interact at runtime (see Defs. 2.7–2.9). The main output of the flow is a set of jointly valid use case runtime configurations in the sense of Def. 2.17. The multi-application flow is divided into three phases. (1) The single application phase, in Fig. 8.1a, produces a parallel application model for all the applications. (2) The scenario analysis phase, in Fig. 8.1b, produces a valid runtime configuration for each of the use cases in the multi-application description. (3) Finally, the backend phase, in Fig. 8.1c stores the configurations for later binary images. An overview of these

J. Castrillón Mazo and R. Leupers, *Programming Heterogeneous MPSoCs*, DOI: 10.1007/978-3-319-00675-8_8,

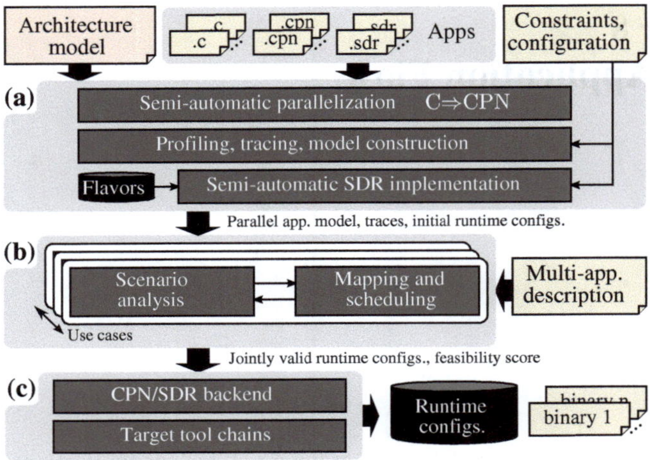

Fig. 8.1 Multi-application tool flow. **a** Single application phase. **b** Scenario analysis phase. **c** Backend phase

phases is presented in Sect. 8.1.1, followed by a description of the new input and output models in Sect. 8.1.2.

8.1.1 Tool Flow Components

The multi-application flow in Fig. 8.1 is divided into three main phases. The purpose and functionality of the components in each phase are described in the following sections.

8.1.1.1 Single Application Phase

The single application phase in Fig. 8.1a contains components from the three single application flows described in Chaps. 5–7. The purpose of this phase is to provide the multi-application flow with a uniform representation of the different applications (A). This representation is composed of a parallel application model and application constraints (see Def. 2.6). The model is extended with execution traces for both processes and flavors, as described in Sect. 6.2 and Sect. 7.2.

Obtaining the parallel model for CPN applications is a straightforward process, as described in Sect. 6.3.1. For sequential applications, the parallel model is obtained by using the semi-automatic parallelization flow described in Chap. 5. In the case of SDR applications, an SDR implementation is produced as described in Chap. 7.

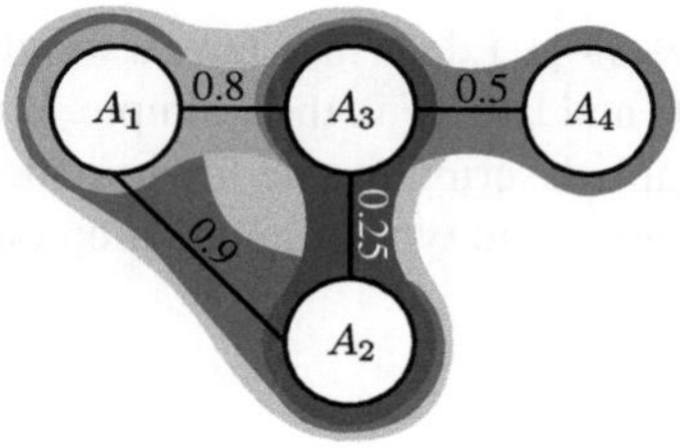

$$UC_1 = (\{A_1, A_2\}, 0.9)$$
$$UC_2 = (\{A_1, A_3\}, 0.8)$$
$$UC_3 = (\{A_2, A_3\}, 0.25)$$
$$UC_4 = (\{A_3, A_4\}, 0.5)$$
$$UC_5 = (\{A_1, A_2, A_3\}, 0.18)$$

Fig. 8.2 Example of multi-application description

8.1.1.2 Scenario Analysis Phase

The purpose of this phase is to analyze each use case and produce a set of jointly valid runtime configurations for each of the applications in the use case. As shown in Fig. 8.1b, the mapping and scheduling phase of the parallel flow is used to produce single-application configurations. In this phase, a relaxed composability approach is used, as discussed in Sect. 3.5. It is therefore not possible to provide hard guarantees about constraint compliance. To give the programmer an idea of the safety of a multi-application configuration, the scenario analysis phase exports a *feasibility score*.

8.1.1.3 Backend Phase

Currently the target MPSoCs do not support switching configurations at runtime, nor loading different versions of the application binary. Therefore, each use case is executed in isolation. The backend phase of the multi-application flow reuses the backends of the parallel and the SDR flows.

8.1.2 Input and Output Modeling

In addition to the input and output files discussed in the previous chapters, the multi-application flow includes a multi-application input description and a runtime configuration output database. The formats of these files are described in the following.

8.1.2.1 Multi-Application Description

The multi-application description provides information about all the possible use cases that may appear at runtime. This file provides the information as a graph or as a list, as introduced by Def. 2.7 and Def. 2.9. An example of a graph representation is shown in Listing 8.1. Nodes in the graph represent applications, `"JPEG MJPEG LP-AF"` in Line `2`. Nodes are connected with edges that express that two applications may run simultaneously (see Lines 3–4). Edges may have a weight annotation, like 0.75 on the edge that connect JPEG and LP-AF in the example. If not specified, a weight of 1 is assumed. After the graph description, the file provides extra information about the applications, e.g., location and type of application (see Lines 6–7 in Listing 8.1).

```
1 <GraphDescription>
2  <Applications List="JPEG MJPEG LP-AF"/>
3  <ConcurrencyEdge SourceApp="JPEG" TargetApp="MJPEG"/>
4  <ConcurrencyEdge SourceApp="JPEG" TargetApp="LP-AF" Weight="0.75"/>
5 </GraphDescription>
6 <Application Name="JPEG"
7              Location="../apps/cpn/" Type="KPN" RTClass="best-effort"/>
8 <!-- ... Other applications -->
```

Listing 8.1: Example of an application concurrency graph (see Def. 2.7)

Recall, that every clique in the graph represents a use case whose weight is determined by the product of the weights on all the edges (see Eq. 2.1). The list of use cases is computed by a recursive backtracking algorithm [151]. If the use cases are described as a list, no further processing of this input file is required. An example of a graph representation and the associated use cases is shown in Fig. 8.2.

8.1.2.2 Runtime Configurations

This file is a simple collection of configurations for each use case, i.e., a file representation of the solution to the multi-application problem in Def. 2.18. An example of this file for the input file in Listing 8.1 is shown in Listing 8.2. For every use case, the file specifies which schedule descriptor to use for every application. In the example, the second schedule descriptor is used in the first use case for JPEG (`jpeg_2.scheddesc`) whereas the first one is used in the second use case (`jpeg_1.scheddesc`). Each of the files referenced in Listing 8.2 is a descriptor file, as introduced in Sect. 6.1.3.4. Note also that for every solution, the file includes the *score* (see Line 2 and Line 5). As mentioned before, the score gives an idea of the certainty of the computed configuration. The bigger the score, the better the configuration is. In the example, the second use case seems to have more slack than the first one. Additionally, the whole multi-application problem is also given a score, as shown in Line 1.

```
1 <MultiappSolution TotalScore="-0.3075">
2  <UseCase List="JPEG MJPEG" Weight="1.0" Score="-0.3">
3    <AppCfg App="JPEG"   Cfg="output/jpeg_2.scheddesc"/>
4    <AppCfg App="MJPEG" Cfg="output/mjpeg_1.scheddesc"/></UseCase>
5  <UseCase List="JPEG LP-AF" Weight="0.75" Score="-0.01">
6    <AppCfg App="JPEG"   Cfg="output/jpeg_1.scheddesc"/>
7    <AppCfg App="LP-AF"  Cfg="output/lp-af_3.scheddesc"/></UseCase>
8 <MultiappSolution>
```

Listing 8.2: Example of the output file of the multi-application flow

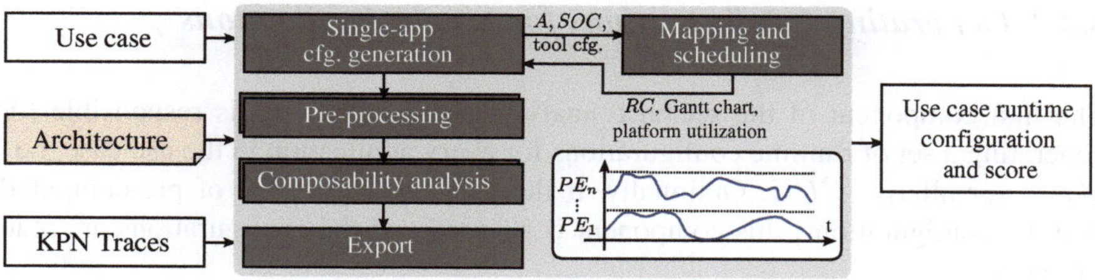

Fig. 8.3 Scenario analysis flow

8.2 Scenario Analysis

In the scenario analysis phase, each use case is analyzed separately. A straightforward approach to verify a use case consists in merging all applications in a single large KPN and running the single application flow on it. However, this approach results in an exponential amount of applications to analyze, which makes the approach unfeasible for practical situations. In this book a relaxed composability approach is followed (see Sect. 3.5.1.3), in which single application schedules are tested in a multi-application context. A *composability function* is defined that allows to judge the quality of a composition of individual schedule configurations.

The flow of the scenario analysis phase is shown in Fig. 8.3. It receives the architecture description, a use case $UC = (S^{UC} = \{A_1, \ldots, A_m\}, p^{UC})$ (see Def. 2.8) and a set of traces for all the processes of the applications in the use case, i.e., $\mathcal{T}^{UC} = \bigcup_{A \in S^{UC}} \mathcal{T}^A$. As output, the flow produces a use case configuration $\mathcal{RC}^{UC} = \{RC^{A_1}, \ldots, RC^{A_m}\}$ (see Def. 2.17) with RC^{A_i} a valid runtime configuration for A_i. This phase also produces a *feasibility score* that expresses with what certainty the configuration is jointly valid. The score is based on the composability function, which is defined on the platform utilization profiles of the scheduled traces. For this purpose, the mapping and scheduling phase of the parallel flow is used to produce several runtime configurations for every application. For every runtime configuration, the TRM returns the platform utilization profile, as shown in the upper part of Fig. 8.3.

Given a set of runtime configurations for every single application in the use case, $\mathcal{RC}^{A_i} = \{RC_1^{A_i}, \ldots, RC_{m_i}^{A_i}\}$, the scenario analysis phase can be seen as a filter that quickly discards combinations of configurations that clearly lead to a processor overutilization. Such combinations would most likely violate the application constraints. Finally, note that the scenario analysis flow has no automated feedback. This means that if the scores for some of the use cases are low, it is the user responsibility to modify the applications in order to make them fit better on the platform. The four components of the scenario analysis phase are described in the following.

8.2.1 Generating Configurations for Single Applications

The first component of the scenario analysis phase in Fig. 8.3 is responsible for generating a set of runtime configurations for every application in the use case, i.e., $\mathscr{RC}^{A_i}$ for all $A_i \in UC$. Optionally, if the user provides a list of pre-computed runtime configurations, this component is skipped and the configurations are read from disk.

The simplest implementation of this component would iterate over all the applications in the use case and call the mapping heuristic in Algorithm 6.3. However, having a single runtime configuration for every application means that there is only one possible use case configuration $\mathscr{RC}^{UC}$. This reduces the scenario analysis to a decision whether $\mathscr{RC}^{UC}$ is jointly valid or not.

In order to provide more candidates to the scenario analysis, this component keeps adding platform resources to the mapping process even after a valid runtime configuration is found. This is done by removing the termination criterion in Line 24 of Algorithm 6.3. Alternatively, the user is allowed to pass a set of different constraints for every application. By tightening the timing constraints or by restricting the mapping to different platform subsets, the iterative mapping algorithm would produce different runtime configurations. Naturally, every runtime configuration produced by the parallel flow is stored. This makes it possible to reuse configuration across use cases, instead of recomputing them every time.

As suggested by Fig. 8.3, every runtime configuration is associated with a Gantt chart, generated by the TRM as disccussed in Sect. 6.3.2. The chart can be seen as a set of functions that model the utilization of each of the platform resources. More formally,

Definition 1 A **utilization function** ($\vartheta_{PE}^{RC^A}$) is a time-discrete function $\vartheta_{PE}^{RC^A} : D_\vartheta \subset \mathbb{N} \rightarrow [0, 1]$ that represents the utilization of $PE \in \mathscr{PE}$ due to application A according to a runtime configuration RC^A.

Definition 2 A **platform utilization set** ($\mathscr{U}_{SOC}^{RC^A}$) is a set with the utilization functions of every $PE \in \mathscr{PE}$ for application A according to a runtime configuration RC^A, i.e., $\mathscr{U}_{SOC}^{RC^A} = \{\vartheta_{PE_1}^{RC^A}, \ldots, \vartheta_{PE_n}^{RC^A}\}$. This set is a representation of the Gantt chart produced by the TRM.

The output of the first component of the scenario analysis flow is then a set of valid runtime configurations for every application and the corresponding platform utilization sets. That is, $\mathscr{RC}_{\text{all}}^{UC} = \bigcup_{A_i \in UC} \mathscr{RC}^{A_i}$, with $\mathscr{RC}^{A_i} = \{RC_1^{A_i}, \ldots, RC_m^{A_i}\}$, and $\mathscr{U}_{\text{all}}^{UC} = \bigcup_{A_i \in UC} \mathscr{U}_{SOC}^{A_i}$, with $\mathscr{U}_{SOC}^{A_i} = \{\mathscr{U}_{SOC}^{RC_1^{A_i}}, \ldots, \mathscr{U}_{SOC}^{RC_m^{A_i}}\}$.

The pre-processing and the composability analysis steps analyze different use case configurations in the sense of Def. 2.17. The use case configurations $\mathscr{RC}^{UC} = \{RC^{A_1}, \ldots, RC^{A_m}\}$ are created as a combination of configurations from $\mathscr{RC}_{\text{all}}^{UC}$, with $RC^{A_i} \in \mathscr{RC}^{A_i}$. For every application A_i in the use case, there is a platform utilization set $\mathscr{U}_{SOC}^{RC^{A_i}} \in \mathscr{U}_{SOC}^{A_i}$ that corresponds to the candidate runtime configuration

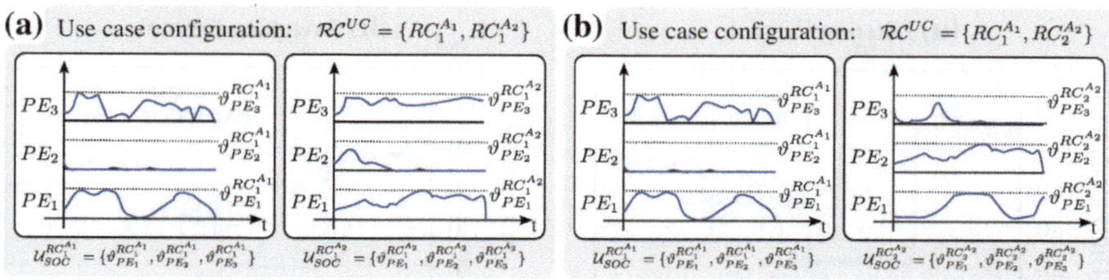

Fig. 8.4 Utilization functions and platform utilization sets. **a** *Bad* configuration. **b** *Good* configuration

RC^{A_i}. The set of all utilization functions associated with a use case configuration is then $\mathcal{U}^{UC} = \bigcup_{A_i \in UC} \mathcal{U}_{SOC}^{RC^{A_i}}$. The purpose of the next two components is to analyze the feasibility of the candidate configuration based on the utilization functions. An intuitive example is shown in Fig. 8.4 for a use case with two applications on a platform with three processors. The first use case configuration in Fig. 8.4a would most likely violate some of the application constraints (if any), since both applications are intensively using PE_1 and PE_3. The second candidate in Fig. 8.4b is a more promising configuration. In the second runtime configuration for A_2 ($RC_2^{A_2}$), the processor PE_2 is used, so that the load in the other processors is not as high as with $RC_1^{A_2}$. Additionally, the high utilization periods on PE_1 and PE_3 seem to happen on low utilization periods of the first application on the same processors.

The next sections explain how this intuitive analysis of the utilization functions is automated in order to select a use case configuration that is likely to be jointly valid.

8.2.2 Pre-Processing

Before the actual composability analysis, the pre-processing component of the scenario analysis phase in Fig. 8.3 performs *signal conditioning* of the utilization functions produced by the previous component. Consider a candidate use case configuration $\mathcal{RC}^{UC}$ with associated platform utilization sets $\mathcal{U}^{UC}$. Pre-processing includes:

Filtering: This process applies a sliding window filter to the utilization functions in $\mathcal{U}^{UC}$ that serves to smooth the functions. By suppressing *details* of the functions, the runtime of the composability analysis is reduced. An example of the filtering process is shown in Fig. 8.5a. In this example, the blue solid curve has four times less samples that the original utilization curve.

Time scale: The filtered utilization functions from different applications may have a different time scale. This may be caused by a different configuration of the TRM or by a different window size during filtering. This conditioning process synchronizes the utilization samples so that they all refer to the same time.

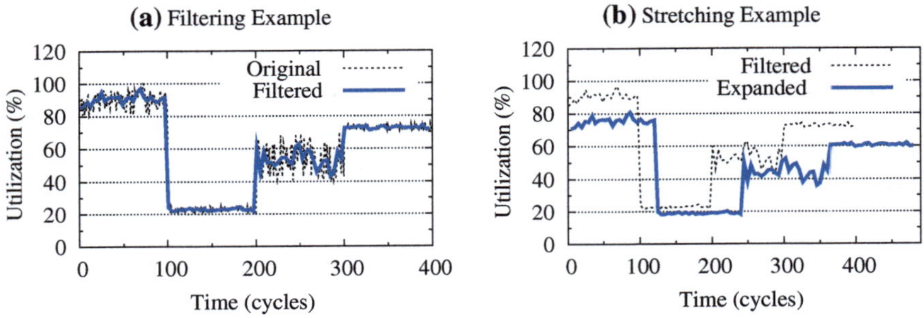

Fig. 8.5 Pre-processing example. **a** Filtering example. **b** Stretching example

Length: The composability analysis sometimes requires the utilization functions to have exactly the same amount of samples, i.e., same domain $D_\vartheta \subset \mathbb{N}$ in Def. 1. This length equalization is performed by *stretching* the functions while keeping the area constant. This step requires re-sampling and interpolation. For applications with real time constraints, the iterations are stretched only until the deadline. This applications are treated as periodic by the *crop* conditioning process. An example of this process is shown in Fig. 8.5b. In this example, the blue solid curve was extended from 400 to 480 cycles.

Crop: For real time applications, the utilization functions between time checkpoints are extracted. When analyzing the composability of applications with periodic schedules, a *hyper-period* has to be computed. Ideally, the hyper-period would be the *Least Common Multiple* (LCM) of the periods of the individual applications. This however may produce long traces that would slow down the composability analysis. To avoid this, the utilization functions are cropped so that a smaller hyper-period can be found. This is done by finding the greatest period among all the applications and using a multiple of it as hyper-period. The amount of iterations of the hyper-period is selected, so that the time that is cropped from the other applications is minimized. More formally, let $\{t_{A_1}, \ldots, t_{A_m}\}$ be the periods of the periodic applications in the use case and let $t_{\max} = \max_{A \in UC}(t_A)$ be the greatest period. The hyper-period is defined as $t_{\text{hyper}} = K^{\text{crop}} \cdot t_{\max}$, with

$$K^{\text{crop}} = \underset{k \in \{1,\ldots,N_{\max}^{\text{crop}}\}}{\operatorname{argmin}} \left(\sum_{A \in UC} \left\lceil \frac{k \cdot t_{\max}}{t_A} \right\rceil \cdot t_A - k \cdot t_{\max} \right) \tag{8.1}$$

$N_{\max}^{\text{crop}}$ is an internal parameter of the tool that constrains the search in order to reduce the computation time. Note that if t_{hyper} is really the LCM of the other periods, the argument of the argmin operator in Eq. 8.1 evaluates to zero. After determining the hyper period, the utilization functions of every application A are replicated as many times as $\lceil t_{\text{hyper}}/t_A \rceil$ and then cropped to the length given by t_{hyper}.

8.2.3 Composability Analysis

The purpose of the composability analysis component in Fig. 8.3 is to provide information about the timing behavior of different applications running concurrently. In order for this analysis to be fast, the schedules are not run on the actual platform nor in the TRM. Instead, the pre-processed utilization functions are used. Recall that this component processes all candidate use case configurations for a given use case UC produced by the first component (in Sect. 8.2.1).

Let $\hat{\mathscr{U}}^{UC} = \bigcup_{A_i \in UC} \hat{\mathscr{U}}_{SOC}^{RC^{A_i}} = \{\hat{\vartheta}_{PE_1}^{RC^{A_1}}, \ldots, \hat{\vartheta}_{PE_j}^{RC^{A_i}}, \ldots, \hat{\vartheta}_{PE_n}^{RC^{A_m}}\}$ denote the pre-processed utilization functions for a candidate use case runtime configuration $\mathscr{RC}^{UC}$. Two composability functions are analyzed in this book. The first one is a *mean-criterion*, similar to the one in [155]. The second one, is a *displacement-criterion* which provides a more thorough analysis by testing different application starting times.

Mean-criterion Composability: This is a simple approach in which the mean utilization of every application on every processor is computed. The combined utilization of every processor is then compared against a threshold. More formally, the combined mean utilization of processor PE due to a use case configuration $\mathscr{RC}^{UC}$ is:

$$\bar{\vartheta}_{PE}^{\mathscr{RC}^{UC}} = \sum_{A \in UC} \left(\frac{1}{N_{PE}^{A}} \cdot \sum_{k=0}^{N_{PE}^{A}-1} \hat{\vartheta}_{PE}^{RC^{A}} \right) \tag{8.2}$$

where N_{PE}^{A} is the number of samples in the utilization function of application A on processor PE. Whether the use case is feasible or not is a binary decision, given by: $\bar{\vartheta}_{\max} > \max_{PE \in \mathscr{PE}}(\bar{\vartheta}_{PE}^{\mathscr{RC}^{UC}})$. $\bar{\vartheta}_{\max} \in [0, 1]$ is a threshold provided by the user.

Note that the lower the mean utilization on each processor, the better the configuration is, i.e., the more likely it is that the constraints will be met. This is used to define the score of a use case runtime configuration,

$$\omega_{\mathrm{mc}}^{\mathscr{RC}^{UC}} = \bar{\vartheta}_{\max} - \max_{PE \in \mathscr{PE}} (\bar{\vartheta}_{PE}^{\mathscr{RC}^{UC}}) \tag{8.3}$$

Note that $\omega_{\mathrm{mc}}^{\mathscr{RC}^{UC}} \in [\bar{\vartheta}_{\max} - |S^{UC}|, \bar{\vartheta}_{\max}]$, so that the higher the value, the better the configuration is. Configurations with a negative score $\omega_{\mathrm{mc}}^{\mathscr{RC}^{UC}} < 0$ are discarded.

Displacement-criterion Composability: Note that the previous approach provides a coarse guarantee. The mean utilization criterion states that all the computations can be carried out in the available computation time, if the mean utilization is not above 100 %. This would mean that, for example, the mean throughput of a process may be respected. However, this criterion does not say anything about the variation of the throughput along the entire execution. Additionally, instantaneous load situ-

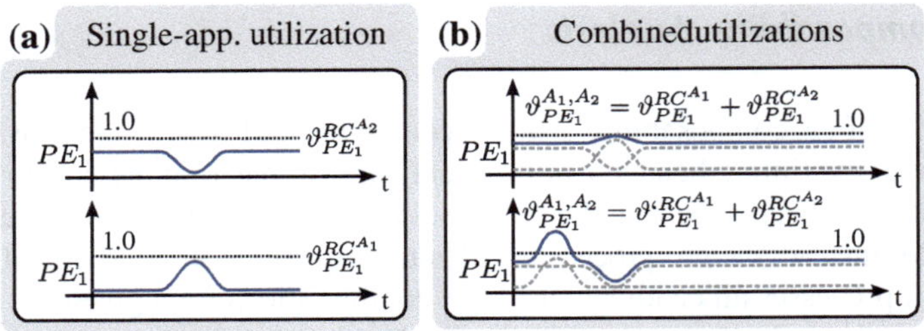

Fig. 8.6 Combined processor utilization. **a** Sample utilization functions on a single processor. **b** Combined utilization functions with different displacements

ations may introduce path latency violations that cannot be detected with the mean criterion. This is illustrated with the example in Fig. 8.6. Figure 8.6a shows the utilization functions of two different applications (A_1, A_2) on a single processor PE_1. Intuitively, it can be seen that the mean computation in Eq. 8.2 would produce a value under 1.0, i.e., less than 100 % utilization. Figure 8.6b shows that depending on the relative starting time of the applications, different load situations are observed. In the upper plot of Fig. 8.6b, it is assumed that both applications start at the same time. As a result, the combined utilization is computed by adding the functions in Fig. 8.6a, as shown by the solid blue line in Fig. 8.6b. In this case, the combined utilization lies below the 100 % mark, indicating that both applications may run correctly on processor PE_1. However, if the second application starts while the first application is running, the combined load situation changes, as shown in the bottom plot of Fig. 8.6b. In this case, the instantaneous load surpasses the 100 % mark, indicating that the timing behavior of the applications may be affected. If the computational peak observed in the utilization function of A_1 is required to meet a path constraint, it is likely that this constraint will be missed.

The displacement-criterion tries to account for these changes in the instantaneous load, by analyzing the combined utilization of every processor, given different, relative application starting times. For two applications A_1 and A_2, with candidate configurations RC^{A_1} and RC^{A_2}, the displaced combined utilization is defined by:

$$\hat{\vartheta}_{PE}^{A_1,A_2}(k,\tau) = \hat{\vartheta}_{PE}^{RC^{A_1}}(k) + \hat{\vartheta}_{PE}^{RC^{A_2}}(k-\tau) \tag{8.4}$$

The displacement-criterion then analyzes how often the instantaneous combined load goes above a user-defined threshold $\tilde{\vartheta}_{\max}$ for every displacement τ. For this analysis a *displacement function* is defined as:

$$\tilde{\vartheta}_{PE}^{RC^{A_1},RC^{A_2}}(\tau) = \sum_{k=0}^{N_{PE}^{A_1}-1} (\hat{\vartheta}_{PE}^{A_1,A_2}(k,\tau) - \tilde{\vartheta}_{\max}) \cdot H(\hat{\vartheta}_{PE}^{A_1,A_2}(k,\tau) - \tilde{\vartheta}_{\max}) \tag{8.5}$$

where H is the Heaviside function and $\tau \in \{0, \ldots, N_{PE}^{A_2}-1\}$. Note that the Heaviside function cancels all the values of $\hat{\vartheta}_{PE}^{A_1,A_2}(k,\tau)$ below the threshold $\tilde{\vartheta}_{\max}$. In the case of periodic applications, the time shift in Eq. 8.4 is replaced by a circular shift.

The displacement analysis for two applications given by Eq. 8.5 can be extended to more applications. For three applications, the worst-case displacement of the first two applications is determined first. The worst-case is determined by the displacement for which the instantaneous load surpasses the threshold the most, i.e., $\tau^* = \underset{\tau}{\text{argmax}}(\tilde{\vartheta}_{PE}^{RC^{A_1},RC^{A_2}}(\tau))$. Thereafter, the displacement function in Eq. 8.5 is computed for the functions $\hat{\vartheta}_{PE}^{RC^{A_3}}$ and $\hat{\vartheta}_{PE}^{A_1,A_2}|_{\tau=\tau*}$ (from Eq. 8.4). The resulting function $\tilde{\vartheta}_{PE}^{RC^{A_1},RC^{A_2},RC^{A_3}}$ can be then used to compute the displacement function for more applications. The process can be repeated until all applications have been analyzed and a joint displacement function $\tilde{\vartheta}_{PE}^{\mathscr{RC}^{UC}}$ has been computed for every processor.

The discrete function $\tilde{\vartheta}_{PE}^{\mathscr{RC}^{UC}}$ gives a qualitative measure of the feasibility of a use case. The higher the values of the function, the more unlikely it is for the two applications to run simultaneously and yet meet the constraints. In order to provide a single score for the current candidate use case configuration, the maximum average value across all the processors is used. This value is then scaled and inverted, i.e.,

$$\omega_{\text{dc}}^{\mathscr{RC}^{UC}} = \frac{-1}{N_{PE^*}^{A_m}} \cdot \max_{PE \in \mathscr{PE}} \left(\frac{1}{N_{PE}^{A_{m-1}}} \sum_{\tau=0}^{N_{PE}^{A_{m-1}}-1} \tilde{\vartheta}_{PE}^{\mathscr{RC}^{UC}}(\tau) \right) \tag{8.6}$$

where PE^* is the processor with the maximum average displacement function. The outer normalization ($N_{PE^*}^{A_m}$) in Eq. 8.6 is inserted in order to remove the effect of the summation in Eq. 8.5. As with the mean-criterion score from Eq. 8.3, the higher the displacement-criterion score, the better the configuration is. Note that $\omega_{\text{dc}}^{\mathscr{RC}^{UC}} \in [\bar{\vartheta}_{\max} - |S^{UC}|, 0]$. A hard decision criterion would discard all configurations for which $\omega_{\text{dc}}^{\mathscr{RC}^{UC}} < 0$. Note however, that the score does not carry as much information as the individual functions $\tilde{\vartheta}_{PE}^{\mathscr{RC}^{UC}}$. For this reason, the functions are also exported to provide additional information to the programmer about use case configurations.

In this section, the platform utilization profiles are used to draw conclusions on whether it is possible for two or more applications to run simultaneously while still meeting their constraints. In the case of different application classes, e.g., real and non-real time, a hierarchical scheduler is used in which the best effort applications are executed only when the real time applications are blocked (recall Fig. 4.1). This ensures that the execution of real time applications is not affected by the presence of best-effort applications. For applications of the same class, the composability analysis evaluates the jointly required processing bandwidth. How the bandwidth relates to the real time schedulability of the underlying scheduling algorithm is not addressed in the multi-application flow.[1]

8.2.4 Results Export

The last component of the scenario analysis flow in Fig. 8.3 is in charge of selecting the best use case runtime configuration for every use case in the multi-application description. The selected configurations are exported in a file with the format shown in Listing 8.2.

Choosing the best use case configuration can be done interactively or automatically. In the interactive approach, the user is presented with the different displacement functions of the scenario analysis. Based on the shape of the functions, the user can decide which configuration to finally choose.

The automatic approach simply selects the configuration with the highest score among all possible configurations. Recall the set of valid runtime configurations for every application $A_i \in UC$, $\mathscr{RC}^{A_i} = \{RC_1^{A_i}, \ldots, RC_{k_i}^{A_i}\}$. The final use case configuration is then determined by:

$$\mathscr{RC}^{*UC} = \{RC^{*A_1}, \ldots, RC^{*A_m}\} = \underset{\mathscr{RC}^{UC} \in S_{\mathscr{RC}}}{\text{argmax}} \; (\omega_{\text{dc}}^{\mathscr{RC}^{UC}}) \tag{8.7}$$

with $S_{\mathscr{RC}} = \{\{RC^{A_1}, \ldots, RC^{A_m}\} \subset \mathscr{RC}_{\text{all}}^{UC}, (RC^{A_1}, \ldots, RC^{A_m}) \in \times_{i=0}^{m} \mathscr{RC}^{A_i}\}$ and $\omega_{\text{dc}}^{\mathscr{RC}^{UC}}$ the score from Eq. 8.6. The total multi-application score mentioned in Sect. 8.1.2.2 is then computed as a weighted sum of the individual scores, i.e.,

$$\omega^{ACS} = \sum_{UC \in ACS} \omega_{\text{dc}}^{*UC} \cdot p^{UC} \tag{8.8}$$

with ω_{dc}^{*UC} the best score for each use case, i.e., $\omega_{\text{dc}}^{*UC} = \max_{\mathscr{RC}^{UC} \in S_{\mathscr{RC}}} (\omega_{\text{dc}}^{\mathscr{RC}^{UC}})$.

8.3 Case Study

This section presents results of the multi-application flow in Fig. 8.1 on an experimental setup that is described in Sect. 8.3.1. The purpose of this case study is to assess how accurately the scores produced by the composability analysis phase reflect constraint compliance. This is tested on the TRM in Sect. 8.3.2.

[1] For some algorithms and problem statements, it is possible to find a theoretical bound on the bandwidth that guarantees that an application will meet its real time constraints. Consider, for (Footnote 1 continued)
example, the unity bound ($U = 1$) for Earliest Deadline First (EDF) or the bound $U \approx 0.693$ for Rate Monotonic (RM) with infinite tasks [179].

8.3.1 Experimental Setup

This case study provides a solution to the motivational example introduced in Sect. 2.1 (see also Fig. 2.1). As target MPSoC, the platform DCP defined in Sect. 6.7.1.1 is used (see also Fig. 6.12a), extended with the hardware accelerators needed for the MIMO-OFDM benchmark, shown in Fig. 7.6b.

In this case study, it is assumed that the single application phase in Fig. 8.1a has been already run for every application. This means that the LP-AF application from the motivational example has already been parallelized, as described in Sect. 5.5.2. It also means that an SDR implementation with a valid runtime configuration has been found for the MIMO-OFDM benchmark, as described in Sect. 7.5. Note that this benchmark mostly uses the hardware accelerators in the platform, so that its execution does not interfere with the other applications. For this reason, this case study focuses on the resource sharing among the LP-AF, the JPEG and the MJPEG applications.

8.3.2 Scenario Analysis Results

This section analyzes the results of the scenario analysis phase in Fig. 8.3. As multi-application description, a fully connected graph is provided to the tool. The analysis focuses on the worst-case scenario in which all applications are running concurrently.

The first step in the scenario analysis phase consists in generating valid runtime configurations for every application. As explained in Sect. 8.2.1, configurations are generated by varying the available processors for the mapping process. As discussed in Sect. 6.7.3, the minimum configuration for MJPEG has one RISC and 3 VLIW processors while the minimum configuration for LP-AF has 2 RISCs and 2 VLIWs processors.

The results of the scenario analysis phase for selected combinations are summarized in Table 8.1. The first column contains an identifier of the candidate use case runtime configuration ($\mathscr{RC}_i^{UC_j}$). In the table, UC_1 refers to the three-application use case, i.e., $S^{UC_1} = \mathscr{A}$, while UC_2 contains the two real-time applications LP-AF and MJPEG. The numbers in the next three columns refer to the processors used in each single-application runtime configuration. As an example, the first row describes a multi-application configuration in which both RISCs and the first two VLIWs are used for LP-AF, all processors are used for JPEG and a RISC and three VLIWs are used for MJPEG. This configuration corresponds to the best-effort mapping of JPEG and the real time mappings for LP-AF and MJPEG obtained in Sect. 6.7 (see Fig. 6.15).

The fifth and the sixth columns in Table 8.1 contain the scores computed by the mean-criterion from Eq. 8.3 and the displacement-criterion in Eq. 8.6 respectively. For all the test cases, the thresholds were both set to a 90 % utilization bound, i.e., $\bar{\vartheta}_{\max} = \tilde{\vartheta}_{\max} = 0.9$. Note that all the scores in the table are negative, which means

Table 8.1 Results of the scenario analysis phase

	LP-AF	JPEG	MJPEG	Mean score	Displacement score ($\times 10^{-3}$)	Constraints	TRM runtime (s)
$\mathcal{RC}_1^{UC_1}$	1-2-3-4	1→6	2-3-4-5	-1.32	-2.44	✗	99.09
$\mathcal{RC}_2^{UC_1}$	1-2-4-5	1→6	2-3-4-6	-0.929	-2.00	✓	101.03
$\mathcal{RC}_3^{UC_1}$	1-4-5-6	1→6	1-2-3-4-5	-0.912	-3.01	✓	94.35
$\mathcal{RC}_4^{UC_1}$	3-4-5-6	1→6	3-4-5-6	-1.440	-3.15	✗	97.13
$\mathcal{RC}_5^{UC_1}$	1-2-5-6	1→6	1-2-5-6	-1.27	-3.53	✗	97.18
$\mathcal{RC}_6^{UC_1}$	1→6	1→6	1→6	-1.146	-2.94	✗	97.44
$\mathcal{RC}_1^{UC_2}$	2-3-4-6	–	1-3-4-5	-0.24	-0.45	✓	63.95
$\mathcal{RC}_2^{UC_2}$	1-2-3-5	–	1-2-3-4	-0.57	-1.01	✗	63.95

that the tool could not find an *absolutely* valid configuration. This is due to the tight real time constraints of both applications, more notably of MJPEG as can be observed in Fig. 6.16b. The best configurations proposed by the tool for the two use cases are then the ones with the highest score (see Eq. 8.7), i.e., $\mathcal{RC}_2^{UC_1}$ and $\mathcal{RC}_1^{UC_2}$.

In order to verify if the configurations are actually valid, the traces of the three applications were merged and replayed together on the TRM by using the individually valid runtime configurations. Whether or not the constraints were met is shown in the last but one column in Table 8.1. Finally, the time employed by the TRM to verify the configuration is reported in the last column of the table.

The total execution time of the scenario analysis phase of Fig. 8.3 was of 81 s (on a AMD Phenom running at 3.2 GHz with 8 GB of RAM). This time includes around 4 s for pre-processing, half a second for exporting the results and 76s for the composability analysis. During this time, a total of 3080 different options were analyzed. Note that to test the same amount of configurations on the TRM would require 80.67 h (three orders of magnitude more time). On the cycle-accurate simulator the evaluation would have required one or two orders of magnitude more time.

Table 8.1 shows six different configurations for the three-application use case (UC_1) and two configurations for the two-application use case (UC_2). These configurations were among the best solutions found by the multi-application flow. In all of them, the best effort mapping of JPEG is used. Notice that directly using the results of the single application flow would not work in a multi-application context, i.e., configuration $\mathcal{RC}_1^{UC_1}$ misses the constraints. Two valid configurations were found for UC_1. The first one corresponds to the configuration with the best displacement score ($\mathcal{RC}_2^{UC_1}$) while the second one corresponds to the best configuration according to the mean criterion ($\mathcal{RC}_3^{UC_1}$). As shown in Table 8.1, the single-application constraints were met when running on the TRM for both solutions. The configuration obtained with the displacement criterion displayed a better load distribution which translated in a higher slack for the constrained processes. This is due to a higher transient load on the third VLIW (PE_5). Similar load conditions are the reason why the remaining three configurations for UC_1 failed. The last two rows in Table 8.1 contain the best and the worst results for the two-application use case. As expected, the best configu-

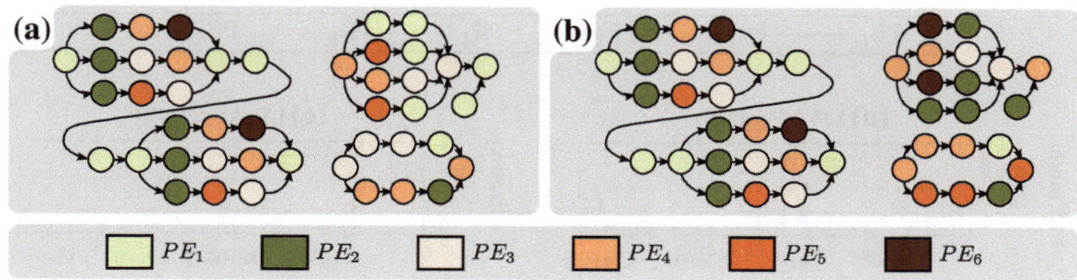

Fig. 8.7 Multi-application configurations. **a** Best configuration for each application in isolation ($\mathscr{RC}_1^{UC_1}$). **b** Best configuration selected for the use case ($\mathscr{RC}_2^{UC_1}$)

ration for this case does not uses the single-application runtime configurations used in $\mathscr{RC}_2^{UC_1}$.

The actual mappings for $\mathscr{RC}_1^{UC_1}$ and $\mathscr{RC}_2^{UC_1}$ are shown in Fig. 8.7a. For this use case, the solution of the composability analysis is quite logical. The mapping of LP-AF and MJPEG are almost equivalent to the mappings obtained in isolation. The only real change lies on a better distribution of the VLIWs among the applications, so that only one VLIW is shared by the applications (PE_4).

Figure 8.8 shows the results of the scenario analysis for the second and the sixth configuration in Table 8.1 ($\mathscr{RC}_2^{UC_1}$, $\mathscr{RC}_6^{UC_1}$). The pre-processed utilization vectors for the different single application configurations are shown in Fig. 8.8a–c for $\mathscr{RC}_6^{UC_1}$ and in Fig. 8.8e–g for $\mathscr{RC}_2^{UC_1}$. The displacement functions from Eq. 8.5 for both configurations are shown in Fig. 8.8d, h. As can be seen from the plots, the curves obtained for $\mathscr{RC}_6^{UC_1}$ have higher values than those of the second configuration. In particular, configuration $\mathscr{RC}_6^{UC_1}$ makes heavy use of PE_6 and PE_4. The better load distribution of $\mathscr{RC}_2^{UC_1}$ is reflected in the low values observed in Fig. 8.8h.

Note that $\mathscr{RC}_6^{UC_1}$ corresponds to the best-effort configurations for the three applications. That is, the configuration in which each application uses all the resources in the platform. The plots in Fig. 8.8 therefore illustrate that the best mapping for individual applications may not be the best choice in a multi-application scenario.

8.4 Limitations and Outlook

This chapter presented a general solution approach to the multi-application problem, based on the analysis of the platform utilization. Several aspects can be improved in the programming flow, including:

- Config generation: The first component of the flow generates single-application configurations in a batch mode. Due to the small amount of applications in the case study, the *brute force* generation described in Sect. 8.2.1 was applicable. However, for more complex multi-application problems, a better generation is needed. The

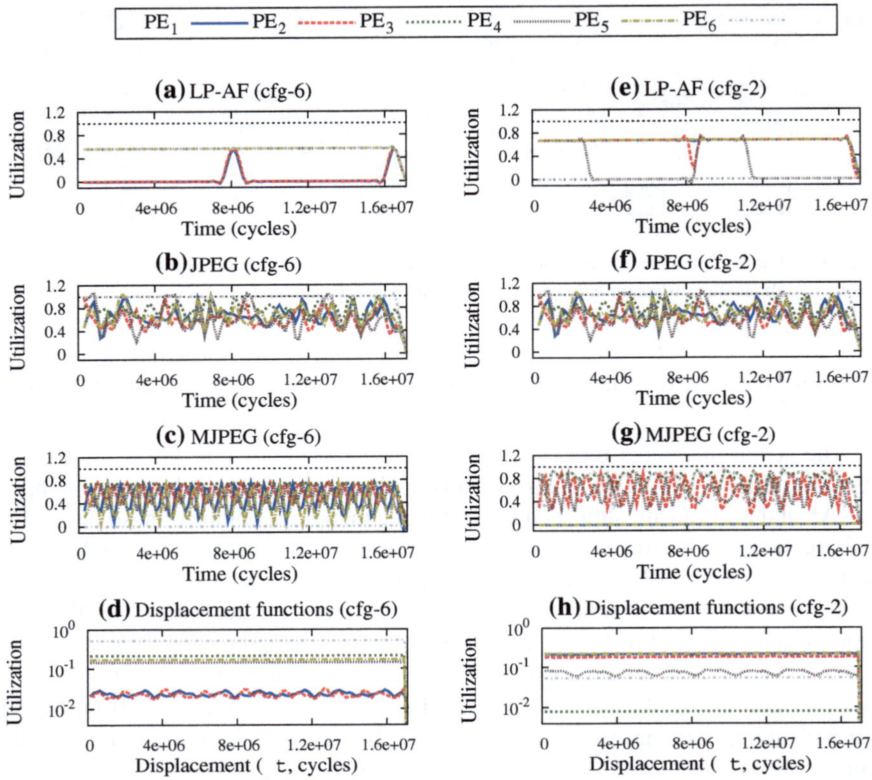

Fig. 8.8 Composability results. **a–c** Utilizations for LP-AF, JPEG and MJPEG in $\mathscr{RC}_6^{UC_1}$. **e–g** Utilizations for $\mathscr{RC}_2^{UC_1}$. **d** Displacement analysis for $\mathscr{RC}_6^{UC_1}$. **h** Displacement analysis for $\mathscr{RC}_2^{UC_1}$

batch generation strategy can be improved with a selection of combinations, so that the rest of the flow converges faster. Alternatively, the configurations could be generated on demand, by observing the results of the scenario analysis phase.

- Reliability: The multi-application flow is also based on execution traces. Therefore, it shares the reliability issues discussed for the parallel flow in Sect. 6.8.
- Complexity: The amount of use cases grows exponentially with the number of edges in the application concurrency graph. This complexity cannot be circumvented, since it comes from the problem specification. However, the design space can be searched in a more efficient way, similar to how it was done in the SDR flow (see Sect. 7.3). Use cases can be described in a tree data structure that represent the set inclusion relation. If configurations fail in a use case, they also fail in parent nodes.
- Switching costs: The flow presented in this chapter does not consider the cost associated with a change in the use case configuration during runtime. If the cost

is too high, some configurations should not be switched, even if there is a better configuration in the data base. A similar tradeoff analysis was presented in [21]. How to implement dynamic switching of scenarios at runtime and how to model the associated costs is outside the scope of this book.
- Communication resources: The multi-application flow must be made aware of communication resources. The current version supposes that the memory constraints set for the single application analysis ensure that the available memory is not surpassed in multi-application scenarios.

8.5 Synopsis

This chapter presented a solution to the multi-application problem stated in Sect. 2.3. Due to the lack of formal properties of KPN applications, a pragmatic solution approach was followed, in which estimated platform utilization profiles are used. By *merging* utilization functions, the schedule of the KPN applications is modified. This however, does not invalidate the tracing approach, since KPNs are determinate (see Sect. 2.5.1.3). The multi-application flow was used to find a solution to the motivational example presented in Sect. 2.1. Jointly valid runtime configurations were found in around a minute. Note that, a manual approach based on simulation would require several tens of minutes per simulation. In order to test more configurations, a programmer would also need to re-code the application and manually change the mapping. All in all, the productivity boost provided by the multi-application flow in Fig. 8.1 is of at least three orders of magnitude, effectively reducing the software productivity gap mentioned in Sect. 1.1.3.

Chapter 9
Conclusions and Outlook

The software component of embedded devices has become the dominant differentiation factor and, at the same time, the dominant design cost factor. Even before this software boom, embedded hardware platforms had already undergone a migration to parallel, multi-processor architectures. Surprisingly, today's embedded systems are still programmed with fragmented, single-processor compilers with no true support for heterogeneity or multi-processing. This makes parallelization and application mapping a daunting, error-prone task, contributing to further open the software productivity gap. This book presented a set of tools and methodologies that aim at narrowing the gap, namely (1) a sequential flow that helps to port legacy C code to a parallel programming model, (2) a parallel flow for KPN applications that automatically computes a valid runtime configuration, (3) an extension to the parallel flow that allows to use specialized software routines and hardware accelerators in highly heterogeneous platforms, and finally (4) a multi-application flow that reuses the results of the previous flows to quickly filter out combinations of runtime configurations that are not feasible in a given multi-application context. Together with the tool flows, this book proposed several algorithms, e.g., Algorithm 5.6, 6.3, and 7.1, which aim at providing efficient execution of applications on heterogeneous MPSoCs. The case studies presented in this book revealed a potential improvement of around three orders of magnitude in design-time effort, which can effectively improve programmers' productivity. Besides the tooling aspects, this book also described a thin runtime system, accelerated by the OSIP customized processor. This lightweight multi-tasking support makes it possible to execute fine-grained tasks more efficiently, thereby increasing the optimization potential of the mapping algorithms described in this book.

Chapter 4 described the design decisions that led to the final OSIP hardware and software architecture. With an average scheduling latency of around 1000 cycles, OSIP outperforms traditional software solutions by at least an order of magnitude. When running the OSIP scheduling approach on an off-the-shelf ARM processor, the range of operation is clearly restricted, as shown in Fig. 4.7. This shows how the ASIP paradigm improves efficiency while retaining programmability. The cost of

J. Castrillón Mazo and R. Leupers, *Programming Heterogeneous MPSoCs*, DOI: 10.1007/978-3-319-00675-8_9,

having a programmable solution is paid by a reduced efficiency while compared to hardware solutions, as discussed in Sect. 3.1.2.

Chapter 5 described the parallelism extraction flow for sequential C programs. This flow is based on dynamic data flow analysis, enabled by application tracing. The proposed heuristic in Algorithm 5.6 extends previous graph-based clustering techniques with a pattern discovery phase that allows to exploit DLP and PLP in addition to TLP. The new algorithms were tested on synthetic examples and on a multimedia application to validate the approach. Generating the hints for the parallel implementation took around 3 min and obtaining the actual parallel implementation about half a day.

Chapter 6 described the mapping and scheduling flow for KPN applications. This flow also uses dynamic analysis, based on application tracing, to gain an insight into the behavior of KPN processes. A new heuristic, namely GBM in Algorithm 6.2, was proposed for jointly mapping KPN processes and KPN channels onto heterogeneous MPSoCs. The case studies showed that this algorithm outperforms typical mapping heuristics, such as load balancing, by 17 % for random KPNs and 40 % for real applications. Furthermore, the algorithm succeeded even in platforms with heterogeneous, asymmetric interconnect. The tool flow was validated on the TI Keyston (CC6678) commercial platform. Additionally, an iterative mapping heuristic was proposed in Algorithm 6.3 for applications with real-time constraints. This heuristic managed to find valid runtime configurations in a few minutes.

Chapter 7 described the extensions to the parallel flow for demanding applications that require hardware acceleration. The chapter presented a proof of concept of the methodology for applications in the SDR domain. In the SDR flow, different potential matchings for a platform-independent application specification are computed and evaluated in which platform-specific kernels are used. The pruning technique in Algorithm 7.1 serves to retrieve the most promising implementation options, quickly reducing the size of the search space. The methodology was applied to a MIMO-OFDM transceiver in a case study that showed the applicability and the importance of such a programming flow. With the SDR flow, it was possible to match the performance of implementations obtained manually by an expert. Compared to a traditional compilation flow, performance gaps of two orders of magnitude were observed.

Finally, Chap. 8 described the proposed approach for dealing with multiple applications. The core of this multi-application flow is a scenario analysis phase in which the platform utilization profiles of applications in isolation are combined. The combined profiles are analyzed for potential over-utilization periods that could endanger constraint compliance. A new measure was introduced with the displacement-criterion in Eq. 8.5, which provides a safer analysis than previous criteria based on the mean utilization. This was verified with a case study in which 3 applications were analyzed and mapped to a single heterogeneous MPSoC in less than 2 min.

All together, the programming flows presented in this book contribute towards closing the software productivity gap, by allowing to test and obtain valid configurations for single and multiple applications within reasonable time.The heuristics that underlie the proposed programming flows are relatively simple and hence fast, which enables integration into interactive programming environments such as the MAPS IDE. Although possibly sub-optimal, the heuristics provide the programmer with a working implementation, which can be further optimize by hand. The proposed tooling infrastructure can serve as basis for future research along the contributions of this book, including (1) better algorithms for parallelism extraction, (2) better mapping and scheduling policies for KPN applications, (3) better understanding of algorithmic kernels and how to characterize them towards a better tooling support for hardware acceleration and (4) better *scores* and analysis for multi-application configurations. The methodologies have all room for improvement as discussed in Sects. 5.6, 6.8, 7.6 and 8.4. Additionally, the following items are promising directions for future work:

- Power/energy awareness: The goal of the algorithms is to generate *efficient* implementations. To achieve this goal, the algorithms search a configuration that meets the constraints while using as few resources as possible. This works under the assumption that by using less resources the energy consumption is reduced, since idle processors can be turned off. Future work should focus on making the algorithms energy-aware in order to truly strive for energy efficiency. However, this will only be practical once power and energy estimation techniques are available at the electronic system level.
- Design-time safety: The dynamic nature of the programming paradigms addressed in this book made it necessary to make decisions from information collected during profiling runs.The tool flows rely on the programmer to select suitable application inputs and other representative stimuli. As a consequence, the results may not be safe if the execution conditions change. Several techniques could contribute to improve the design-time safety, including code coverage and a joint analysis of multiple execution traces. This techniques would also serve to give the programmer an idea of the safety of the results.
- Runtime safety: In general, design-time and runtime deviations are inevitable. Therefore, new runtime strategies are needed to provide control for situations where the design-time solution fails. The interaction between design-time and runtime safety measures could open many new research problems.
- Applications: This book focused on the KPN model in order to provide the expressiveness required by today's and upcoming applications. The analysis of new applications and benchmarks would help to further understand dynamic effects. This understanding could enrich the heuristics presented in this work.

References

1. ACE Associated Compiler Experts (1998) DSP-C An extension to ISO/IEC IS 9899:1990
2. Adl-Tabatabai AR, Kozyrakis C, Saha B (2006) Unlocking concurrency. Queue 4(10):24–33. doi:10.1145/1189276.1189288
3. Aho AV, Sethi R, Ullman JD (1986) Compilers: principles, techniques, and tools. Addison-Wesley Longman Publishing Co, Boston
4. Akesson B, Molnos A, Hansson A, Angelo JA, Goossens K (2010) Composability and predictability for independent application development, verification, and execution. In: Huebner M, Becker J (eds) Multiprocessor system-on-chip-hardware design and tool integration circuits and systems, chap 2. Springer, Berlin, pp 25–56
5. Amdahl GM (1967) Validity of the single processor approach to achieving large scale computing capabilities. In: Proceedings of the April 18–20, 1967, spring joint computer conference (AFIPS '67). ACM, (Springer), New York, pp 483–485, doi:10.1145/1465482.1465560. http://doi.acm.org/10.1145/1465482.1465560
6. Anderson E, Bai Z, Bischof C, Blackford S, Demmel J, Dongarra J, Du Croz J, Greenbaum A, Hammarling S, McKenney A, Sorensen D (1999) LAPACK users' guide, 3rd edn. Society for industrial and applied mathematics, Philadelphia. http://www.netlib.org/lapack/lug/lapack_lug.html
7. Anne-Francoise G (2011) Synopsys unveils multicore optimization technology. http://www.edn.com/article/519849-Synopsys_unveils_multicore_optimization_technology.php
8. ARM (2010) AHB CPU wrappers technicalr reference manual. http://infocenter.arm.com/help/index.jsp?topic=/com.arm.doc.set.amba/index.html
9. ARM (2013) ARM926 processor. http://www.arm.com/products/processors/classic/arm9/arm926.php
10. Arnold O, Fettweis G (2010) Power aware heterogeneous MPSoC with dynamic task scheduling and increased data locality for multiple applications. In: 2010 International conference on embedded computer saystems (SAMOS), pp 110–117. doi:10.1109/ICSAMOS.2010.5642075
11. Asanovic K, Bodik R, Catanzaro BC, Gebis JJ, Husbands P, Keutzer K, Patterson DA, Plishker WL, Shalf J, Williams SW, Yelick KA (2006) The landscape of parallel computing research: a view from Berkeley technical report, EECS Department. University of California, Berkeley. http://www.eecs.berkeley.edu/Pubs/TechRpts/2006/EECS-2006-183.pdf
12. Atmel (2011) Diopsis, D940. http://www.atmelroma.it
13. Baert R, Brockmeyer E, Wuytack S, Ashby TJ (2009) Exploring parallelizations of applications for MPSoC platforms using MPA. In: DATE '09: proceedings of the conference on design, automation and test in europe
14. Balakrishnan S, Rajwar R, Upton M, Lai K (2005) The impact of performance asymmetry in emerging multicore architectures. In: Proceedings of the 32nd international symposium on computer architecture (ISCA '05), pp 506–517. doi:10.1109/ISCA.2005.51

J. Castrillón Mazo and R. Leupers, *Programming Heterogeneous MPSoCs*, DOI: 10.1007/978-3-319-00675-8,

15. Balarin F, Watanabe Y, Hsieh H, Lavagno L, Passerone C, Sangiovanni-Vincentelli A (2003) Metropolis: an integrated electronic system design environment. Computer 36(4):45–52. http://dx.doi.org/10.1109/MC.2003.1193228
16. Bamakhrama MA, Zhai JT, Nikolov H, Stefanov T (2012) A methodology for automated design of hard-real-time embedded streaming systems. In: Proceedings of the 15th design, automation, and test in europe conference (DATE 2012)
17. Banerjee U, Eigenmann R, Nicolau A, Padua DA (1993) Automatic program parallelization. Proc IEEE 81(2):211–243
18. Basten T, Hoogerbrugge J (2001) Efficient execution of process networks. In: Chalmers A, Mirmehdi M, Muller H (eds) Communicating process architectures—2001. IOS Press, Amsterdam, pp 1–14
19. Bauer C, Frink A, Kreckel R (2002) Introduction to the GiNaC framework for symbolic computation within the C++ programming language. J Symb Comput 33:1–12. http://dx.doi.org/10.1006/jsco.2001.0494,
20. Bekooij M, Hoes R, Moreira O, Poplavko P, Pastrnak M, Mesman B, Mol JD, Stuijk S, Gheorghita V, van Meerbergen J (2005) Dataflow analysis for real-time embedded multiprocessor system design. In: van der Stok P (ed) Dynamic and robust streaming in and between connected consumer-electronic devices. Springer, Berlin, pp 81–108
21. Benini L, Bertozzi D, Milano M (2008) Resource management policy handling multiple use-cases in MPSoC platforms using constraint programming. In: Proceedings of the 24th international conference on logic programming (ICLP '08). Springer-Verlag, Berlin, pp 470–484, doi:10.1007/978-3-540-89982-2_41. http://dx.doi.org/10.1007/978-3-540-89982-2_41
22. van Berkel CHK (2009) Multi-core for mobile phones. In: Proceedings of the conference on design, automation and test in europe, european design and automation association, 3001 (DATE '09), Belgium, pp 1260–1265. http://www.dl.acm.org/citation.cfm?id=1874620.1874924
23. Bhattacharyya SS, Murthy PK, Lee EA (1999) Synthesis of embedded software from synchronous dataflow specifications. J VLSI Signal Process Syst 21(2):151–166. http://www.dx.doi.org/10.1023/A:1008052406396
24. Bhattacharyya SS, Deprettere EF, Keinert J (2010) Dynamic and multidimensional dataflow graphs. In: Bhattacharyya SS, Deprettere EF, Leupers R, Takala J (eds) Handbook of signal processing systems. Springer, Berlin, pp 899–930
25. Bhattacharyya SS, Deprettere EF, Leupers R, Takala J (2010a) Handbook of signal processing systems, Part II. Springer, Berlin
26. Bhattacharyya SS, Deprettere EF, Leupers R, Takala J (2010b) Handbook of signal processing systems, Part IV. Springer, Berlin
27. Bijlsma T, Bekooij MJG, Smit GJM, Jansen PG (2007) Omphale: streamlining the communication for jobs in a multi processor system on chip. Technical Report TR-CTIT-07-44. Centre for Telematics and Information Technology University of Twente, Enschede
28. Bilsen G, Engels M, Lauwereins R, Peperstraete J (1996) Cyclo-static dataflow. IEEE Trans Signal Process 44(2):397–408. doi:10.1109/78.485935
29. Biscondi E, Flanagan T, Fruth F, Lin Z, Moerman F (2012) Maximizing multicore efficiency with navigator runtime. White Pap. www.ti.com/lit/wp/spry190/spry190.pdf
30. Blair G, Coupaye T, Stefani JB (2009) Component-based architecture: the fractal initiative. Ann Telecommun 64:1–4, doi:10.1007/s12243-009-0086-1. http://dx.doi.org/10.1007/s12243-009-0086-1
31. Borlenghi F, Witte EM, Ascheid G, Meyr H, Burg AP (2011) A 772 Mbit/s 8.81 bit/nJ 90 nm CMOS soft-input soft-output sphere decoder. In: IEEE Asian solid state circuits conference (A-SSCC)
32. Bridges MJ, Vachharajani N, Zhang Y, Jablin T, August DI (2007) Revisiting the sequential programming model for multi-core. In: Proceedings of the 40th annual IEEE/ACM international symposium on microarchitecture. IEEE computer society, Washington, DC, USA, MICRO 40, pp 69–84, Doi:10.1109/MICRO.2007.35. http://liberty.princeton.edu/Publications/micro40_scale.pdf

33. Buck JT (1993) Scheduling dynamic dataflow graphs with bounded memory using the token flow model. PhD thesis, EECS Department. University of California, Berkeley
34. Burger D, Austin TM (1997) The simplescalar tool set, Version 2.0. SIGARCH Comput Archit News 25(3):13–25, doi:10.1145/268806.268810. http://doi.acm.org/10.1145/268806.268810
35. C Meenderinck, A Azevedo, M Alvarez, B Juurlink, and A Ramirez (2008) Parallel scalability of H.264. In: Workshop on programmability issues for multi-core computers (MULTIPROG)
36. Canada CRC (2010) Scari software suite. http://www.crc.gc.ca/en/html/crc/home/research/satcom/rars/sdr/products/scari_suite/scari_suite
37. Carloni L, Bernardinis F, Pinello C, Sangiovanni-Vincentelli A, Sgroi M (2005) Platform-based design for embedded systems. In: Zurawski R (ed) The embedded systems handbook. CRC Press, Boca Raton, p 36
38. Carro L, Rutzig MB (2012) Multi-core systems on chip. In: Bhattacharyya SS, Deprettere EF, Leupers R, Takala J (eds) Handbook of signal processing systems, 2nd edn. Springer, Berlin
39. Carta S, Alimonda A, Pisano A, Acquaviva A, Benini L (2007) A control theoretic approach to energy-efficient pipelined computation in MPSoCs. ACM Trans Embed Comput Syst 6, doi:10.1145/1274858.1274865. url:http://doi.acm.org/10.1145/1274858.1274865
40. Castrillon J, Velasquez R, Stulova A, Sheng W, Ceng J, Leupers R, Ascheid G, Meyr H (2010) Trace-based KPN composability analysis for mapping simultaneous applications to MPSoC platforms. In: Proceedings of the design, automation and test in europe conference and exhibition, Dresden, pp 753–758
41. Castrillon J, Schürmans S, Stulova A, Sheng W, Kempf T, Leupers R, Ascheid G, Meyr H (2011) Component-based waveform development: the nucleus tool flow for efficient and portable software defined radio. Analog Integr Circuits Signal Process 69(2):173–190, doi:10.1007/s10470-011-9670-1. http://dx.doi.org/10.1007/s10470-011-9670-1
42. Castrillon J, Shah A, Murillo L, Leupers R, Ascheid G (2011) Backend for virtual platforms with hardware scheduler in the MAPS framework. In: Proceedings of the 2nd IEEE Latin American symposium on circuits and systems LASCAS'11, IEEE, pp 1–4
43. Castrillon J, Sheng W, Leupers R (2011) Trends in embedded software synthesis. In: Carro L, Pimentel AD (eds) International conference On embedded computer systems: architecture, modeling, and, simulation (SAMOS'11), pp 347–354
44. Castrillon J, Tretter A, Leupers R, Ascheid G (2012) Communication-aware mapping of KPN applications onto heterogeneous MPSoCs. In: Proceedings of the 49th annual conference on design automation (DAC '12)
45. Castrillon J, Leupers R, Ascheid G (2013) MAPS: mapping concurrent dataflow applications to heterogeneous MPSoCs. IEEE Trans. Industr Inf 9(1):527–545, doi:10.1109/TII.2011.2173941. (Online since 10.2011)
46. Castrillon J, Schürmans S, Stulova A, Sheng W, Kempf T, Ishaque A, Leupers R, Ascheid G, Meyr H, (2010) Component-based waveform development: the nucleus tool flow for efficient and portable SDR. In; (2010) Wireless innovation conference and product exposition (SDR'10). Wireless innovation forum, Washington, DC, USA
47. Ceng J (2011) A methodology for efficient multiprocessor system-on-chip software development. PhD thesis, RWTH Aachen Univeristy. Institute for Communication Technologies and Embedded Systems (ICE), Aachen
48. Ceng J, Castrillon J, Sheng W, Scharwächter H, Leupers R, Ascheid G, Meyr H, Isshiki T, Kunieda H (2008) MAPS: an integrated framework for MPSoC application parallelization. In: Proceedings of the 45th annual conference on design automation (DAC '08). ACM, New York, pp 754–759. http://doi.acm.org/10.1145/1391469.1391663
49. Ceng J, Sheng W, Castrillon J, Stulova A, Leupers R, Ascheid G, Meyr H (2009) A high-level virtual platform for early MPSoC software development. In: Proceedings of the 7th IEEE/ACM international conference on hardware/software codesign and system synthesis (CODES+ISSS '09). ACM, New York, pp 11–20. http://doi.acm.org/10.1145/1629435.1629438

50. Chakraborty S, Kunzli S, Thiele L (2003) A general framework for analysing system properties in platform-based embedded system designs. In: Proceedings of the conference on design, automation and test in europe (DATE '03). IEEE Computer Society, Washington, DC, USA, pp 190–195. doi:10.1109/DATE.2003.1253607
51. Chandraiah P, Domer R (2008) Code and data structure partitioning for parallel and flexible MPSoC specification using designer-controlled recoding. IEEE Trans Comput Aided Des Integr Circuits Syst 27(6):1078–1090. doi:10.1109/TCAD.2008.923244
52. Cheung E, Hsieh H, Balarin F (2007) Automatic buffer sizing for rate-constrained KPN applications on multiprocessor system-on-chip. In: Proceedings of the 2007 IEEE international high level design validation and test workshop. IEEE Computer Society, Washington, DC, USA, pp 37–44, doi:10.1109/HLDVT.2007.4392782. http://portal.acm.org/citation.cfm?id=1546679.1546842
53. Choi J, Oh H, Kim S, Ha S (2012) Executing synchronous dataflow graphs on a SPM-based multicore architecture. In: Proceedings of the 49th annual design automation conference (DAC '12). ACM, New York, pp 664–671, doi:10.1145/2228360.2228480. http://doi.acm.org/10.1145/2228360.2228480
54. Clarke P (2011) IHS: embedded wireless market to grow 35 %. http://www.eetimes.com/electronics-news/4215209/IHS-2011-embedded-wireless-market-to-grow-35-, in EETimes
55. Coherent Logix (2013) HyperX development system. http://www.coherentlogix.com/
56. Compaan Design BV (2012) HotSpot parallelizer for C. http://www.compaandesign.com/
57. Cordes D, Marwedel P (2012) Multi-objective aware extraction of task-level parallelism using genetic algorithms. In: Proceedings of design, automation and test in europe (DATE 2012), Dresden
58. Cordes D, Marwedel P, Mallik A (2010) Automatic parallelization of embedded software using hierarchical task graphs and integer linear programming. In: Proceedings of the eighth IEEE/ACM/IFIP international conference on hardware/software codesign and system synthesis (CODES/ISSS '10). ACM, New York, pp 267–276, doi:10.1145/1878961.1879009. http://doi.acm.org/10.1145/1878961.1879009
59. Coussy P, Gajski D, Meredith M, Takach A (2009) An introduction to high-level synthesis. IEEE Des Test Comput 26(4):8–17. doi:10.1109/MDT.2009.69
60. CriticalBlue (2012) Prism: simplifying multicore programming. http://criticalblue.com/prism/index.php
61. Cui Y, Wang Y, Chen Y, Shi Y (2011) Experience on comparison of operating systems scalability on the multi-core architecture. In: 2011 IEEE international conference on cluster computing (CLUSTER), pp 205–215. doi:10.1109/CLUSTER.2011.31
62. Cumming P (2003) The TI OMAP platform approach to SoC. In: Martin G, Chang H (eds) Winning the SoC revolution: experiences in real design, Kluwer, chap 5
63. Davare A (2007) Automated mapping for heterogeneous multiprocessor embedded systems. PhD thesis, EECS Department. University of California, Berkeley. http://www.eecs.berkeley.edu/Pubs/TechRpts/2007/EECS-2007-115.html
64. Davis RI, Burns A (2011) A survey of hard real-time scheduling for multiprocessor systems. ACM Comput Surv (CSUR) 43(4):35:1–35:44, doi:10.1145/1978802.1978814. http://doi.acm.org/10.1145/1978802.1978814
65. Demers A, Keshav S, Shenker S (1989) Analysis and simulation of a fair queueing algorithm. SIGCOMM Comput Commun Rev 19(4):1–12, doi:10.1145/75247.75248. http://doi.acm.org/10.1145/75247.75248
66. Derin O, Diken E, Fiorin L (2011) A Middleware approach to achieving fault-tolerance of Kahn process networks on networks-on-chips. Int J Reconfigurable Comput 2011(Article ID 295385):15pages http://www.downloads.hindawi.com/journals/ijrc/2011/295385.pdf, selected papers from the international workshop on reconfigurable communication-centric systems on chips (ReCoSoC' 2010)
67. Design and Reuse (2010) Blue wonder communications' BWC200 passed first IOT tests against ZTE environment. http://www.designreuse.com/news/23192/lte-ip-iot-tests.html

68. Dongarra JJ, Du Croz J, Hammarling S, Duff IS (1990) A set of level 3 basic linear algebra subprograms. ACM Trans Math Softw 16(1):1–17, doi:10.1145/77626.79170. http://doi.acm.org/10.1145/77626.79170
69. van Dongen SM (2000) Graph clustering by flow simulation. PhD thesis, Faculteit Wiskunde en Informatica. Universiteit Utrecht, Utrecht
70. Dubey P (2005) A platform 2015 workload model: recognition, mining and synthesis moves computers to the Era of Tera. White Pap
71. Dunkels A, Schmidt O, Voigt T, Ali M (2006) Protothreads: simplifying event-driven programming of memory-constrained embedded systems. In: Proceedings of the 4th international conference on embedded networked sensor systems (SenSys '06). ACM, New York, pp 29–42. http://doi.acm.org/10.1145/1182807.1182811
72. Ecker W, Müller W, Dömer R (2008) Hardware-dependent software—principles and practice. In: Ecker W, Müller W, Dömer R (eds) Hardware-dependent software—introduction and overview. Springer, Berlin
73. Eclipse Foundation (2012) Eclipse— an open development platform. http://www.eclipse.org
74. Edwards S, Tardieu O (2006) SHIM: a deterministic model for heterogeneous embedded systems. IEEE Trans VLSI Syst 14(8):854–867. doi:10.1109/TVLSI.2006.878473
75. EE Times (2007) Embedded software Stuck at C. http://www.eetimes.com/news/design/$
76. Eker J, Janneck J, Lee E, Liu J, Liu X, Ludvig J, Neuendorffer S, Sachs S, Xiong Y (2003) Taming heterogeneity—the ptolemy approach. Proc IEEE 91(1):127–144. doi:10.1109/JPROC.2002.805829
77. Electronic Systems Group (2012) CoMPSoC project. http://compsoc.eu/
78. Electronic Systems Group (2013) SDF^3: SDF for free. http://www.es.ele.tue.nl/sdf3/
79. Erbas C, Cerav-Erbas S, Pimentel A (2006) Multiobjective optimization and evolutionary algorithms for the application mapping problem in multiprocessor system-on-chip design. IEEE Trans Evol Comput 10(3):358–374. doi:10.1109/TEVC.2005.860766
80. Ester M, peter Kriegel H, S J, Xu X (1996) A density-based algorithm for discovering clusters in large spatial databases with noise. In: Proceedings of 2nd international conference on knowledge discovery and data mining. AAAI Press, New York, pp 226–231
81. EURETILE Consortium (2013) European reference tiled architecture experiment (euretile). http://euretile.roma1.infn.it/mediawiki/index.php/Main_Page
82. Falk J, Keinert J, Haubelt C, Teich J, Bhattacharyya SS (2008) A generalized static data flow clustering algorithm for MPSoC scheduling of multimedia applications. In: Proceedings of the 8th ACM international conference on embedded software (EMSOFT '08). ACM, New York, pp 189–198, doi:10.1145/1450058.1450084. http://doi.acm.org/10.1145/1450058.1450084
83. Falk J, Zebelein C, Haubelt C, Teich J (2011) A rule-based static dataflow clustering algorithm for efficient embedded software synthesis. In: Proceedings of IEEE computer society (DATE'11), pp 14–18
84. Fiorin L, Ferrante A, Padarnitsas K, Carucci S (2010) Hardware-assisted security enhanced linux in embedded systems: a Proposal. In: Proceedings of the 5th workshop on embedded systems security (WESS '10). ACM, New York, pp 3:1–3:7, doi:10.1145/1873548.1873551. http://doi.acm.org/10.1145/1873548.1873551
85. Fisher J, Farabosch P, Young C (2005) Embedded computing: a VLIW approach to architecture compilers and tools. Morgan-Kaufmann (Elsevier), Burlington
86. Fisher JA, Faraboschi P, Young C (2004) Embedded computing : a VLIW approach to architecture compilers and tools. Morgan Kaufmann, Burlington
87. Flynn MJ (1972) Some computer organizations and their effectiveness. IEEE Trans Comput C-21(9):948–960. doi:10.1109/TC.1972.5009071
88. Gao L, Huang J, Ceng J, Leupers R, Ascheid G, Meyr H (2009) TotalProf: a fast and accurate retargetable source code profiler. In: Proceedings of the 7th IEEE/ACM international conference on hardware/software codesign and system synthesis (CODES+ISSS '09). ACM, New York, pp 305–314. http://doi.acm.org/10.1145/1629435.1629477
89. Garey MR, Johnson DS (1990) Computers and intractability. A guide to the theory of NP-completeness. W. H. Freeman and Co. New York

90. Gartner (2011) Gartner says western europe PC market declined 19 percent in second quarter of 2011. http://www.gartner.com/it/page.jsp?id=1769215
91. Geilen M, Basten T (2003) Requirements on the execution of Kahn process networks. In: Proceedings of the 12th european symposium on programming (ESOP 2003). Springer Verlag, Berlin, pp 319–334
92. Geilen M, Tripakis S, Wiggers M (2010) The earlier the better: a theory of timed actor interfaces. Technical Report UCB/EECS-2010-130, EECS Department. University of California, Berkeley. http://www.eecs.berkeley.edu/Pubs/TechRpts/2010/EECS-2010-130.html
93. Gelernter D, Carriero N (1992) Coordination languages and their significance. Commun ACM 35(2):97–107, doi:10.1145/129630.129635. http://doi.acm.org/10.1145/129630.129635
94. Geuns S, Bekooij M, Bijlsma T, Corporaal H (2011) Parallelization of while loops in nested loop programs for shared-memory multiprocessor systems. In: Design automation test in europe conference exhibition (DATE), pp 1–6
95. Gheorghita S, T Basten Corpoaal H (2006) An overview of application scenario usage in streaming-oriented embedded system design
96. Ghosal A, Limaye R, Ravindran K, Tripakis S, Prasad A, Wang G, Tran TN, Andrade H (2012) Static dataflow with access patterns: semantics and analysis. In: Proceedings of the 49th annual design automation conference (DAC '12). ACM, New York, pp 656–663, doi:10.1145/2228360.2228479. http://doi.acm.org/10.1145/2228360.2228479
97. Girkar M, Polychronopoulos C (1992) Automatic extraction of functional parallelism from ordinary programs. IEEE Trans Parallel Distrib Syst 3(2):166–178. doi:10.1109/71.127258
98. Girkar M, Polychronopoulos CD (1995) Extracting task-level parallelism. ACM Trans Program Lang Syst 17(4):600–634. http://doi.acm.org/10.1145/210184.210189
99. GNU Project (2012) GDB: The GNU project debugger. http://www.gnu.org/software/gdb/
100. Goyal P, Guo X, Vin HM (1996) A hierarchical CPU scheduler for multimedia operating systems. In: Proceedings, 2nd symposium on operating systems design and implementations (OSDI'96). Seattle, Washington, pp 107–122
101. Graham SL, Kessler PB, Mckusick MK (1982) Gprof: a call graph execution profiler. SIGPLAN Not 17(6):120–126, doi:10.1145/872726.806987. http://doi.acm.org/10.1145/872726.806987
102. Group OS (2013) Open64. http://www.open64.net/
103. Haene S (2008) VLSI circuits for MIMO-OFDM physical layer. ETH Zurich, Zurich
104. Haid W, Keller M, Huang K, Bacivarov I, Thiele L (2009) Generation and calibration of compositional performance analysis models for multi-processor systems. In: Proceedings of SAMOS'09. IEEE Press, Thessaloniki, pp 92–99
105. Hall MH, Amarasinghe SP, Murphy BR, Liao SW, Lam MS (1995) Detecting coarse-grain parallelism using an interprocedural parallelizing compiler. In: Proceedings of the 1995 ACM/IEEE conference on Supercomputing (CDROM) Supercomputing '95. ACM, New York, doi:10.1145/224170.224337. http://doi.acm.org/10.1145/224170.224337
106. Hankins RA, Chinya GN, Collins JD, Wang PH, Rakvic R, Wang H, Shen JP (2006) Multiple instruction stream processor. SIGARCH Comput Archit News 34(2):114–127. http://doi.acm.org/10.1145/1150019.1136495
107. Hansson A, Goossens K, Bekooij M, Huisken J (2009) CoMPSoC: a template for composable and predictable multi-processor system on chips. ACM Trans Des Autom Electron Syst 14(1):1–24. http://doi.acm.org/10.1145/1455229.1455231
108. Harriss T, Walke R, Kienhuis B, Deprettere E (2002) Compilation from matlab to process networks realized in FPGA. Des Autom Embed Syst 7(4):656-679. http://doi.acm.org/10.1145/1146909.1146981
109. Haubelt C, Falk J, Keinert J, Schlichter T, Streubühr M, Deyhle A, Hadert A, Teich J (2007) A systemC-based design methodology for digital signal processing systems. EURASIP J Embed Syst 2007(1):22. http://dx.doi.org/10.1155/2007/47580
110. Hennessy J (1999) The future of systems research. Computer 32(8):27–33. doi:10.1109/2.781631

111. Hind M (2001) Pointer analysis: haven't we solved this problem Yet? In: Proceedings of the 2001 ACM SIGPLAN-SIGSOFT workshop on program analysis for software tools and engineering (PASTE '01). ACM, New York, pp 54–61. http://doi.acm.org/10.1145/379605.379665
112. Hoare CAR (1978) Communicating sequential processes. Commun ACM 21(8):666–677. doi:10.1145/359576.359585
113. Hoare CAR (2004) Communicating sequential processes. Prentice Hall Int, Upper Saddle River
114. Howard J, Dighe S, Hoskote Y, Vangal S, Finan D, Ruhl G, Jenkins D, Wilson H, Borkar N, Schrom G, Pailet F, Jain S, Jacob T, Yada S, Marella S, Salihundam P, Erraguntla V, Konow M, Riepen M, Droege G, Lindemann J, Gries M, Apel T, Henriss K, Lund-Larsen T, Steibl S, Borkar S, De V, Van Der Wijngaart R, Mattson T (2010) A 48-core IA-32 message-passing processor with DVFS in 45nm CMOS. In: 2010 IEEE international on solid-state circuits conference digest of technical papers (ISSCC), pp 108–109. doi:10.1109/ISSCC.2010.5434077
115. Huang K, Bacivarov I, Liu J, Haid W (2009) A modular fast simulation framework for stream-oriented MPSoC. In: IEEE symposium on industrial embedded systems (SIES), IEEE, Ecole Polytechnique Fédérale de Lausanne. Switzerland, pp 74–81. doi:10.1109/SIES.2009.5196198
116. Huang K, Haid W, Bacivarov I, Keller M, Thiele L (2012) Embedding formal performance analysis into the design cycle of MPSoCs for real-time streaming applications. ACM Transactions in Embedded Computing Systems (TECS), New York
117. Hwu W-M, Ryoo S, Ueng SZ, Kelm JH, Gelado I, Stone SS, Kidd RE, Baghsorkhi SS, Mahesri AA, Tsao SC, Navarro N, Lumetta SS, Frank MI, Patel SJ (2007) Implicitly parallel programming models for thousand-core microprocessors. In: Proceedings of the 44th annual conference on design automation (DAC '07). ACM, New York, pp 754–759. http://doi.acm.org/10.1145/1278480.1278669
118. IEEE (2005) Open systemC language reference manual
119. IEEE (2010) IEEE standard for IP-XACT, standard structure for packaging, integrating, and reusing IP within tool flows. In: IEEE computer society and the IEEE standards association corporate advisory group
120. Ienne P, Leupers R (2006) Customizable embedded processors: design technologies and applications (Systems on silicon). Morgan Kaufmann Publishers Inc, Burlington
121. IMEC (2012) CleanC analysis tools. http://www.imec.be/cleanc/Welcome.html
122. International Technology Roadmap for Semiconductors (ITRS) (2005) Design. http://www.itrs.net/
123. International Technology Roadmap for Semiconductors (ITRS) (2011) Design. http://www.itrs.net/
124. International Technology Roadmap for Semiconductors (ITRS) (2011) System drivers. http://www.itrs.net/
125. International Technology Roadmap for Semiconductors (ITRS) (2013). http://www.itrs.net/
126. ISO/IEC (1996) Information technology—open systems interconnection—basic reference model: the basic model. International standard ISO-IEC 7498–1
127. Jain AK, Murty MN, Flynn PJ (1999) Data clustering: a review. ACM Comput Surv 31(3):264–323. http://doi.acm.org/10.1145/331499.331504
128. Jantsch A, Sander I (2005) Models of computation and languages for embedded system design. IEE Proc Comput Digital Tech 152(2):114–129. doi:10.1049/ip-cdt:20045098
129. Java Community Process (2013) Real-time specification for Java. http://www.rtsj.org/
130. Javaid H, Shafique M, Parameswaran S, Henkel J (2011) Low-power adaptive pipelined MPSoCs for multimedia: an H.264 video encoder case study. In: 48th ACM/EDAC/IEEE design automation conference (DAC), pp 1032–sa1037
131. Johnson R, Pingali K (1993) Dependence-based program analysis. In: Conference on programming language design and implementation, pp 78–89. http://citeseer.ist.psu.edu/johnson93dependencebased.html

132. Joint Tactical Networking Center (JTNC) (2013) JTRS standards. http://jtnc.mil/sca/Pages/sca1.aspx
133. Jones D, Topham N (2009) High speed CPU simulation using LTU dynamic binary translation. In: Proceedings of the 4th international conference on high performance embedded architectures and compilers (HiPEAC '09). Springer-Verlag, Berlin, pp 50–64, doi:10.1007/978-3-540-92990-1_6. http://dx.doi.org/10.1007/978-3-540-92990-1_6
134. JTRS Standards Joint Program Executive Office (JPEO) Joint Tactical Radio System (JTRS) (2011) Software communications architecture specification. Version Next (Draft), San Diego. http://www.public.navy.mil/jpeojtrs/sca/Pages/scanext.aspx
135. Kahn G (1974) The semantics of a simple language for parallel programming. In: Rosenfeld JL (ed) Information processing '74: proceedings of the IFIP congress, New York, pp 471–475
136. Kanda W, Yumura Y, Kinebuchi Y, Makijima K, Nakajima T (2008) SPUMONE: lightweight CPU virtualization layer for embedded systems. In: IEEE/IFIP international conference on embedded and ubiquitous computing (EUC '08), vol 1. pp 144–151. doi:10.1109/EUC.2008.157
137. Karam L, AlKamal I, Gatherer A, Frantz G, Anderson D, Evans B (2009) Trends in multicore DSP platforms. IEEE Signal Process Mag 26(6):38–49. doi:10.1109/MSP.2009.934113
138. Karkowski I, Corporaal H (1997) Design of heterogenous multi-processor embedded systems: applying functional pipelining. In: Proceedings of the 1997 international conference on parallel architectures and compilation techniques (PACT '97). IEEE Computer Society, Washington, DC, USA, p 156
139. Karp RM, Miller RE (1966) Properties of a model for parallel computations: determinacy, termination, queuing. SIAM J Appl Math 14(6)
140. Karuri K, Al Faruque MA, Kraemer S, Leupers R, Ascheid G, Meyr H (2005) Fine-grained application source code profiling for ASIP design. In: Proceedings of the 42nd annual design automation conference. ACM, New York, pp 329–334. http://doi.acm.org/10.1145/1065579.1065666
141. Keinert J, Streubühr M, Schlichter T, Falk J, Gladigau J, Haubelt C, Teich J, Meredith M (2009) SystemCodesigner—an automatic ESL synthesis approach by design space exploration and behavioral synthesis for streaming applications. ACM Trans Des Autom Electron Syst 14:1:1–1:23, doi:10.1145/1455229.1455230. http://www.doi.acm.org/10.1145/1455229.1455230
142. Kempf T, Guenther D, Ishaque A, Ascheid G (2011) MIMO OFDM transceiver for a many-core computing fabric—a nucleus based implementation. In: The wireless innovation forum conference on communications technologies and software defined radio (SDR'11). Washington, DC, USA
143. Kempf T, Witte EM, Ramakrishnan V, Ascheid G, Adrat M, Antweiler M, (2008) A practical view on SDR baseband processing portability. In: 2008 Software defined radio technical conference (SDR'08). Washington DC, USA
144. Kennedy K, Allen JR (2002) Optimizing compilers for modern architectures: a dependence-based approach. Morgan Kaufmann Publishers Inc. San Francisco
145. Khronos Group (2013) OpenCL—the open standard for parallel programming of heterogeneous systems. http://www.khronos.org/opencl/
146. Kienhuis B, Deprettere E, Vissers K, Van Der Wolf P (1997) An approach for quantitative analysis of application-specific dataflow architectures. In: Proceedings IEEE international conference on application-specific systems, architectures and processors, pp 338–349. doi:10.1109/ASAP.1997.606839
147. Kogel T, Leupers R, Meyr H (2006) Integrated system-level modeling of network-on-chip enabled multi-processor platforms. Springer-Verlag New York Inc. Secaucus
148. Kogel T, Meyr H (2004) Heterogeneous MP-SoC—the solution to energy-efficient signal processing. In: Design automation conference (DAC). San Diego
149. Kohout P, Ganesh B, Jacob B (2003) Hardware support for real-time operating systems. In: Proceedings of the 1st IEEE/ACM/IFIP international conference on Hardware/software codesign and system synthesis (CODES+ISSS '03). ACM, New York, pp 45–51. http://www.doi.acm.org/10.1145/944645.944656

150. Kopetz H (2011) Real-time systems: design principles for distributed embedded applications, 2nd edn. Springer, New York
151. Kreher DL, Stinson DR (1998) Combinatorial algorithms: generation, enumeration, and search (Discrete Mathematics and Its Applications). 1 edn. CRC Press, ISBN-13: 978–0849339882
152. Krishnan V, Torrellas J (1999) A chip-multiprocessor architecture with speculative multithreading. IEEE Trans Comput 48(9):866–880
153. Kulkarni M, Pingali K, Walter B, Ramanarayanan G, Bala K, Chew LP (2007) Optimistic parallelism requires abstractions. SIGPLAN Not 42:211–222, doi:10.1145/1273442.1250759. http://www.doi.acm.org/10.1145/1273442.1250759
154. Kumar A, Fernando S, Ha Y, Mesman B, Corporaal H (2008) Multiprocessor systems synthesis for multiple use-cases of multiple applications on FPGA. ACM Trans Des Autom Electron Syst 13(3):1–27. http://doi.acm.org/10.1145/1367045.1367049
155. Kumar A, Mesman B, Theelen B, Corporaal H, Ha Y (2008) Analyzing composability of applications on MPSoC platforms. J Syst Archit 54(3–4):369–383. http://dx.doi.org/10.1016/j.sysarc.2007.10.002
156. Kumar S, Hughes CJ, Nguyen A (2007) Carbon: architectural support for fine-grained parallelism on chip multiprocessors. SIGARCH Comput Archit News 35(2):162–173. http://doi.acm.org/10.1145/1273440.1250683
157. Kwok YK, Ahmad I (1999) Static scheduling algorithms for allocating directed task graphs to multiprocessors. ACM Comput Surv 31(4):406–471. http://doi.acm.org/10.1145/344588.344618
158. Lab K (2013) METIS—serial graph partitioning and fill-reducing matrix ordering. http://glaros.dtc.umn.edu/gkhome/metis/metis/publications
159. Lam M (1988) Software pipelining: an effective scheduling technique for VLIW machines. SIGPLAN Not 23(7):318–328. http://doi.acm.org/10.1145/960116.54022
160. Lattner C (2008) LLVM and clang: next generation compiler technology. The BSD Conference, Ottawa
161. Lattner C, Adve V (2004) LLVM: a compilation framework for lifelong program analysis and transformation. In: Proceedings of the international symposium on code generation and optimization (CGO '04). IEEE Computer Society, Washington, DC, USA, p 75
162. Lattner C, Lenharth A, Adve V (2007) Making context-sensitive points-to analysis withheap cloning practicalf for the real world. In: Proceedings of the 2007 ACM SIGPLANconference on programming language design and implementation (PLDI'07), San Diego
163. Lee E, Parks T (1995) Dataflow process networks. Proc IEEE 83(5):773–801. doi:10.1109/5.381846
164. Lee EA (1991) Consistency in dataflow graphs. IEEE Trans Parallel Distrib Syst 2(2):223–235
165. Lee EA (2006) The problem with threads. Computer 39(5):33–42. http://portal.acm.org/citation.cfm?id=1137232.1137289
166. Lee EA, Messerschmitt DG (1987) Synchronous data flow. Proc IEEE 75(9):1235–1245
167. Lee EA, Sangiovanni-Vincentelli A (1996) Comparing models of computation. In: Proceedings of the IEEE/ACM international conference on computer-aided design (ICCAD '96). IEEE Computer Society, Washington, DC, USA, pp 234–241
168. Lee EA, Sangiovanni-Vincentelli A (1998) A framework for comparing models of computation. IEEE Trans Comput Aided Design Integr Circuits Syst 17(12):1217–1229
169. Lee J, Mooney VJ III, Daleby A, Ingström K, Klevin T, Lindh L (2003) A Comparison of the RTU hardware RTOS with a hardware/software RTOS. In: Proceedings of the 2003 conference on Asia south pacific design automation (ASPDAC). ACM, New York, pp 683–688. http://doi.acm.org/10.1145/1119772.1119925
170. Lehtoranta O, Hamalainen T (2003) Complexity analysis of spatially scalable MPEG-4 encoder. In: Proceedings international symposium on system-on-chip, pp 57–60. doi:10.1109/ISSOC.2003.1267717
171. Leupers R (2000) Code selection for media processors with simd instructions. In: DATE '00. ACM, New York, pp 4–8. http://www.doi.acm.org/10.1145/343647.343679

172. Leupers R (2011) LANCE retargetable C compiler. http://www.lancecompiler.com
173. Leupers R, Castrillon J (2010) MPSoC programming using the MAPS compiler. In: Proceedings of the 15th Asia and south pacific design automation conference (ASP-DAC '10), pp 897–902
174. Levy HM (1984) Capability-based computer systems. Butterworth-Heinemann, Chap Intel iAPX 432:159–186
175. Liao CH, Lai IW, Nikitopoulos K, Borlenghi F, Kammler D, Witte M, Zhang D, Chiueh TD, Ascheid G, Meyr H (2009) Combining orthogonalized partial metrics: efficient enumeration for soft-input sphere decoder. In: IEEE 20th international symposium on personal, indoor and mobile radio communications, pp 1287–1291. doi:10.1109/PIMRC.2009.5450104
176. Limberg T, Ristau B, Fettweis G (2008) A real-time programming model for heterogeneous MPSoCs. Chap embedded computer systems: architectures, modeling, and, simulation. Springer, Berlin, pp 75–84
177. Limberg T, Winter M, Bimberg M, Klemm R, Matus E, Tavares M, Fettweis G, Ahlendorf H, Robelly P (2008) A fully programmable 40 GOPS SDR single chip baseband for LTE/WiMAX terminals. In: 34th European solid-state circuits conference (ESSCIRC 2008), pp 466–469. doi:10.1109/ESSCIRC.2008.4681893
178. Lippett M (2004) An IP core based approach to the on-chip management of heterogeneous SoCs. IP/SOC 2004
179. Liu CL, Layland JW (1973) Scheduling algorithms for multiprogramming in a hard-real-time environment. J ACM 20:46–61
180. Lyrtech Inc (2013) SFF SDR DP API guide. 1st edn. http://www.ceanet.com.au/Products/Lyrtech/SFFSDRDevelopment.aspx
181. Ma Z, Marchal P, Scarpazza DP, Yang P, Wong C, Gmez JI, Himpe S, Ykman-Couvreur C, Catthoor F (2007) Systematic methodology for real-time cost-effective mapping of dynamic concurrent task-based systems on heterogenous platforms. Springer Publishing Company Incorporated, New York
182. Martin G (2005) ESL requirements for configurable processor-based embedded system design. In: IP-SoC 2005, pp 15–20
183. Martin G (2006) Overview of the MPSoC design challenge. In: 43rd ACM/IEEE design automation conference, pp 274–279. doi:10.1109/DAC.2006.229245
184. MathWorks (2013) Simulink—simulation and model-based design. http://www.mathworks.com/products/simulink/
185. McKeown M (2010) FFT implementation on the TMS320VC5505, TMS320C5505, and TMS320C5515 DSPs. Application report. Texas instruments, (SPRABB6A)
186. Meijer S, Nikolov H, Stefanov T (2010) Combining process splitting and merging transformations for polyhedral process networks. In: 8th IEEE workshop on embedded systems for real-time multimedia (ESTIMedia), pp 97–106. doi:10.1109/ESTMED.2010.5666985
187. Meijer S, Nikolov H, Stefanov T (2010) Throughput modeling to evaluate process merging transformations in polyhedral process networks. In: 13th International conference design, automation and test in, europe (DATE'10), pp 747–752
188. Meyr H, Moeneclaey M, Fechtel S (1997) Digital communication receivers: synchronization, channel estimation, and signal processing. Wiley, New York
189. Mignolet JY, Baert R, Ashby TJ, Avasare P, Jang HO, Son JC (2009) MPA: parallelizing an application onto a multicore platform made easy. IEEE Micro 29(3):31–39. http://www.doi.ieeecomputersociety.org/10.1109/MM.2009.46
190. Miyamori T (2007) MPSoC architecture trade-offs for multimedia applications. In: 7th International forum on application-specific multi-processor SoC (MPSoC'07), pp 336–337
191. Moorby P, Kelf D, Lin Y, (2010) Open vector radio, a C dialect standard proposal for high performance software baseband coding. In: 2010 Wireless innovation conference and product exposition (SDR'10). Wireless Innovation Forum, Washington, DC, USA
192. Moreira O, Valente F, Bekooij M (2007) Scheduling multiple independent hard-real-time jobs on a heterogeneous multiprocessor. In: Proceedings of the 7th ACM and IEEE international conference on embedded software (EMSOFT '07). ACM, New York, pp 57–66. http://www.doi.acm.org/10.1145/1289927.1289941

193. Muchnick SS (1997) Advanced compiler design and implementation. Morgan Kaufmann Publishers Inc, San Francisco
194. Murtaza Z, Khan S, Rafique A, Bajwa K, Zaman U (2006) Silicon real time operating system for embedded DSPs. In: Proceedings of the international conference on emerging technologies (ICET '06), pp 188–191. doi:10.1109/ICET.2006.336032
195. Nácul AC, Regazzoni F, Lajolo M (2007) Hardware scheduling support in SMP architectures. In Proceedings of the conference on design, automation and test in Europe, EDA consortium (DATE '07). San Jose, pp 642–647
196. Nadezhkin D, Stefanov T (2011) Automatic derivation of polyhedral process networks from while-loop affine programs. In: 9th IEEE symposium on embedded systems for real-time multimedia (ESTIMedia), pp 102–111. doi:10.1109/ESTIMedia.2011.6088516
197. Nadezhkin D, Nikolov H, Stefanov T (2010) Translating affine nested-loop programs with dynamic loop bounds into polyhedral process networks. In: 8th IEEE Workshop on embedded systems for real-time multimedia (ESTIMedia), pp 21–30. doi:10.1109/ESTMED.2010.5666977
198. Nakano T, Utama A, Itabashi M, Shiomi A, Imai M (1995) Hardware implementation of a real-time operating system. In: Proceedings of the 12th TRON project international symposium (TRON '95). IEEE Computer Society, Washington, DC, USA, p 34
199. Nass R (2008) An Insider's view of the 2008 embedded market study. http://eetimes.com/design/embedded/4007664/An-insider-s-view-of-the-2008-Embedd ed-Market-Study, in EETimes
200. National Instruments (2013) LabVIEW system design software. http://www.ni.com/labview/
201. Nichols B, Buttlar D, Farrell JP (1996) Pthreads programming. O'Reilly and Associates Inc. Sebastopol
202. Nikolov H (2009) System-level design methodology for streaming multi-processor embedded systems. PhD thesis Universiteit Leiden
203. Nohl A, Braun G, Schliebusch O, Leupers R, Meyr H, Hoffmann A (2002) A universal technique for fast and flexible instruction-set architecture simulation. In: Proceedings of the 39th annual design automation conference (DAC '02). ACM, New York, pp 22–27, doi:10.1145/513918.513927. http://doi.acm.org/10.1145/513918.513927
204. NVIDIA (2013) CUDA: parallel programming made easey. http://www.nvidia.com/object/cuda_home_new.html
205. Odendahl M, Sheng W, Aguilar M, Leupers R, Ascheid G (2012) Automated code generation of streaming applications for C6000 multicore DSPs. In: 5th European DSP education and research conference
206. Olivier BE (2010) Embedded C for Digital Signal Processing. In: Bhattacharyya SS, Deprettere EF, Leupers R, Takala J (eds) . Handbook of signal processing systems. Springer, Berlin, pp 769–787
207. Ottoni G, Rangan R, Stoler A, August DI (2005) Automatic thread extraction with decoupled software pipelining. In: Proceedings of the 38th annual IEEE/ACM international symposium on microarchitecture (MICRO 38). IEEE Computer Society, Washington, DC, USA, pp 105–118
208. Owens JD, Luebke D, Govindaraju N, Harris M, Krüger J, Lefohn AE, Purcell TJ (2007) A survey of general-purpose computation on graphics hardware. Comput Graph Forum 26(1):80–113
209. Park S, Hong DS, Chae SI (2008) A hardware operating system kernel for multi-processor systems. IEICE Electrons Express 5(9):296–302
210. Parks TM (1995) Bounded scheduling of process networks. PhD thesis, EECS Department, University of California, Berkeley
211. Patel NK, Shelby KA, Dalio BA, (2010) Radio waveform development system providing an integrated approach to SDR waveform design and implementation. In: Wireless innovation conference and product exposition (SDR'10). Wireless Innovation Forum, Washington, DC, USA

212. Paulin P (2011) Programming challenges and solutions for multi-processor SoCs: an industrial perspective. In: 48th ACM/EDAC/IEEE design automation conference (DAC), pp 262–267
213. Paulin P, Pilkington C, Langevin M, Bensoudane E, Lyonnard D, Benny O, Lavigueur B, Lo D, Beltrame G, Gagne V, Nicolescu G (2006) Parallel programming models for a multiprocessor SoC platform applied to networking and multimedia. Very large scale integration (VLSI) systems. IEEE Trans 14(7):667–680. doi:10.1109/TVLSI.2006.878259
214. Pham D, Asano S, Bolliger M, Day M, Hofstee H, Johns C, Kahle J, Kameyama A, Keaty J, Masubuchi Y, Riley M, Shippy D, Stasiak D, Suzuoki M, Wang M, Warnock J, Weitzel S, Wendel D, Yamazaki T, Yazawa K (2005) The design and implementation of a first-generation CELL processor. In: IEEE international solid-state circuits conference on digest of technical papers (ISSCC), vol 1. pp 184–592. doi:10.1109/ISSCC.2005.1493930
215. Pimentel A, Erbas C, Polstra S (2006) A systematic approach to exploring embedded system architectures at multiple abstraction levels. IEEE Trans Comput 55(2):99–112. http://dx.doi.org/10.1109/TC.2006.16
216. Plishker W, Sane N, Kiemb M, Anand K, Bhattacharyya S (2008) Functional DIF for rapid prototyping. In: The 19th IEEE/IFIP international symposium on rapid system prototyping (RSP '08), pp 17–23. doi:10.1109/RSP.2008.32
217. Plishker W, Sane N, Bhattacharyya SS (2009) A generalized scheduling approach for dynamic dataflow applications. In: Proceedings of DATE'09, pp 111–116
218. PrismTech Ltd (2011) Spectra CX—the SCA development tool. www.prismtechnologies.com
219. Püschel M, Moura J, Johnson J, Padua D, Veloso M, Singer B, Xiong J, Franchetti F, Gacic A, Voronenko Y, Chen K, Johnson R, Rizzolo N (2005) SPIRAL: code generation for DSP transforms. Proc IEEE 93(2):232–275. doi:10.1109/JPROC.2004.840306
220. Qualcomm (2011) Snapdragon S4 processors: system on chip solutions for a new mobile age. White Pap. https://developer.qualcomm.com/download
221. Qualcomm (2013) Qualcomm snapdragon processors. http://www.qualcomm.eu/products/snapdragon
222. Ramakrishnan, V, Witte, EM, Kempf, T, Kammler, D, Ascheid, G and H Meyr, Adrat, M and M Antweiler (2009) Efficient and portable SDR waveform development: the nucleus concept. In: IEEE military communications conference (MILCOM 2009). Boston
223. Ramey C (2011) TILE-Gx100 manyCore processor: acceleration interfaces and architecture. Presented at HotChips 23
224. Reid AD, Flautner K, Grimley-Evans E, Lin Y (2008) SoC-C: efficient programming abstractions for heterogeneous multicore systems on chip. In: Proceedings of the 2008 international conference on compilers, architectures and synthesis for embedded systems (CASES '08). ACM, New York, pp 95–104, doi:10.1145/1450095.1450112. http://www.doi.acm.org/10.1145/1450095.1450112
225. Rik Myslewski (2011) ARM vet: the CPU's future is threatened. http://www.theregister.co.uk/2011/08/20/microprocessors_may_face_trouble_ahead/
226. Robin Saxby (2006) Don't just survive, thrive. Presentation: 9th Annual Semico Summit, Scottsdale
227. Rul S, Vandierendonck H, De Bosschere K (2010) A profile-based tool for finding pipeline parallelism in sequential programs. Parallel Comput 36:531–551, doi:10.1016/j.parco.2010.05.006. http://www.dx.doi.org/10.1016/j.parco.2010.05.006
228. Sangiovanni-Vincentelli A, Martin G (2001) Platform-based design and software design methodology for embedded systems. IEEE Design Test Comput 18(6):23–33. doi:10.1109/54.970421
229. Sarkar V (1991) Automatic partitioning of a program dependence graph into parallel tasks. IBM J Res Dev 35(5.6):779–804. doi:10.1147/rd.355.0779
230. Schaeffer SE (2007) Graph clustering. Comput Sci Rev 1(1):27–64, doi:10.1016/j.cosrev.2007.05.001. http://www.sciencedirect.com/science/article/pii/S1574013707000020
231. Seidel H (2006) A task-level programmable processor. PhD thesis, Duisburg
232. Shabbir A, Kumar A, Stuijk S, Mesman B, Corporaal H (2010) CA-MPSoC: an automated design flow for predictable multi-processor architectures for multiple applications. J Syst

Archit EUROMICRO J 56(7):265–277, doi:10.1016/j.sysarc.2010.03.007. http://dx.doi.org/10.1016/j.sysarc.2010.03.007

233. Shen CC, Wu HH, Sane N, Plishker W, Bhattacharyya SS (2011) A design tool for efficient mapping of multimedia applications onto heterogeneous platforms. In: IEEE International conference on multimedia and expo (ICME), pp 1–6. doi:10.1109/ICME.2011.6011952
234. Shen CC, Plishker W, Wu HH, Bhattacharyya SS, (2010) A lightweight dataflow approach for design and implementation of SDR systems. In Wireless innovation conference and product exposition (SDR'10), Washington, DC, USA
235. Sheng W, Wiebe A, Stulova A, Leupers R, Kienhuis B, Walters J, Ascheid G (2012) FIFO exploration in mapping streaming applications onto the TI OMAP3530 platform: case study and optimizations. In: IEEE 6th international symposium on embedded eulticore SoCs (MCSoC-12)
236. Sheng W, Schürmans S, Odendahl M, Leupers R, Ascheid G (2013) Automatic calibration of streaming applications for software mapping exploration. IEEE Des Test Comput
237. Sih GC, Lee EA (1993) A compile-time scheduling heuristic for interconnection-constrained heterogeneous processor architectures. IEEE Trans Parallel Distrib Syst 4(2):175–187. http://dx.doi.org/10.1109/71.207593
238. Snir M, Otto S (1998) MPI-the complete reference: the MPI core. MIT Press, Cambridge
239. Sriram S, Bhattacharyya SS (2009) Embedded multiprocessors: scheduling and synchronization, 2nd edn. Marcel Dekker, Inc, New York
240. Standard for information technology—portable operating system interface (POSIX) Shell and utilities IEEE Std 10031–2004, The Open Group Base Specifications Issue 6, section 29 (2004) IEEE and The Open Group
241. Stefanov T (2004) Converting weakly dynamic programs to equivalent process network specifications. PhD thesis, Leiden University, The Netherlands
242. STMicroelectronics and CEA (2010) Platform 2012: a many-core programmable accelerator for ultra-efficient embedded computing in nanometer technology. White Pap. http://www.2parma.eu/documents/publications.html
243. Straumann T (2001) Open source real time operating systems overview. CoRR cs.OS/0111035
244. Stuijk S, Geilen M, Basten T (2006) SDF^3: SDF for free. In: Proceedings of the 6th international conference on application of concurrency to system design (ACSD 2006). IEEE Computer Society Press, Los Alamitos, pp 276–278, doi:10.1109/ACSD.2006.23. http://www.es.ele.tue.nl/sdf3
245. Stuijk S, Basten T, Geilen MCW, Corporaal H (2007) Multiprocessor resource allocation for throughput-constrained synchronous dataflow graphs. In: Proceedings of the 44th annual design automation conference (DAC '07). ACM, New York, pp 777–782. http://doi.acm.org/10.1145/1278480.1278674
246. Stulova A, Leupers R, Ascheid G (2012) Throughput driven transformations of synchronous data flows for mapping to heterogeneous MPSoCs. In: International conference on embedded computer systems: architectures, modeling and simulation (SAMOS XII)
247. Sun Y, Amiri K, Brogioli M, Cavallaro JR (2012) Application-specific accelerators for communications. In: Bhattacharyya SS, Deprettere EF, Leupers R, Takala J (eds) Handbook of signal processing systems, 2nd edn. Springer, Berlin
248. Suzanne Deffree (2009) ASIC design starts to drop 22reports http://www.edn.com/article/459021-ASIC_design_starts_to_drop_22_in_2009_Gartner_reports.php
249. Synopsys (2012) Design compiler graphical. http://www.synopsys.com/Tools/Implementation/RTLSynthesis/DCGraphical
250. Synopsys (2012) Platform architect. http://www.synopsys.com/Tools/SLD/VirtualPrototyping/Pages/PlatformArchitect.aspx
251. Synopsys (2012) Prime time. http://www.synopsys.com/tools/implementation/signoff/pages/primetime.aspx
252. Synopsys (2012) Processor designer. http://www.synopsys.com/Tools/SLD/ProcessorDev/Pages/default.aspx

253. Synopsys (2012) Signal processing worksystem (SPW). http://www.synopsys.com/systems/blockdesign/digitalsignalprocessing/pages/signa l-processing.aspx
254. Synopsys (2012) System studio. http://www.synopsys.com/Systems/BlockDesign/DigitalSignalProcessing/Pages/Syste mStudio.aspx
255. Target (2013) IP designer. http://www.retarget.com/index.php
256. Tensilica (2013) Xtensa customizable processors. http://www.tensilica.com/products/xtensa-customizable
257. Texas Instruments (2013) Keystone device architecture. http://processors.wiki.ti.com/index.php/Keystone
258. Texas Instruments (2013) OMAP mobile processors. http://www.ti.com/lsds/ti/omap-applications-processors/features.page
259. The Object Management Group (OMG) (2013) CORBA 3.1. http://www.omg.org/spec/CORBA/3.1/
260. The OpenMP Architecture Review Board (2009) The openMP specification for parallel programming. http://www.openmp.org
261. Thiele L (2013) Distributed Application Layer (DAL). http://www.tik.ee.ethz.ch/euretile/dal.php
262. Thiele L, Wilhelm R (2004) Design for timing predictability. Real Time Syst 28(2–3):157–177, doi:10.1023/B:TIME.0000045316.66276.6e. http://dx.doi.org/10.1023/B:TIME.0000045316.66276.6e
263. Thiele L, Bacivarov I, Haid W, Huang K (2007) Mapping applications to tiled multiprocessor embedded systems. In: Proceedings of the ACSD '07, IEEE Computer Society, New Yor. http://dx.doi.org/10.1109/ACSD.2007.53
264. Thies W, Chandrasekhar V, Amarasinghe S (2007) A practical approach to exploiting coarse-grained pipeline parallelism in C programs. In: Proceedings of the 40th annual IEEE/ACM international symposium on microarchitecture (MICRO 40). IEEE Computer Society, New York, pp 356–369, doi:10.1109/MICRO.2007.7. http://dx.doi.org/10.1109/MICRO.2007.7
265. Tobias Grosser (2012) Polly: polyhedral optimizations for LLVM. http://www.polly.llvm.org/
266. Tournavitis G, Wang Z, Franke B, O'Boyle M (2009) Towards a holistic approach to auto-parallelization—integrating profile-driven parallelism detection and machine-learning based mapping. In: Proceedings of the programming language design and implementation conference (PLDI 09), Dublin, 15–20 June
267. Tsang E (1993) Foundations of constraint satisfaction. Academic press limited, department of computer science. University of Essex Colchester, Essex
268. Tudor D, Macariu G, Jebelean C, Cretu V (2009) Towards a load balancer architecture for multi-core mobile communication systems. In: 5th International symposium on applied computational intelligence and informatics (SACI '09), pp 391–396. doi:10.1109/SACI.2009.5136280
269. UBM Electronics (2012) 2012 Embedded market survey. http://www.embedded.com/electronics-blogs/embedded-market-surveys/4405646/2012- Embedded-Market-Survey
270. Ullman JD (1975) NP-complete scheduling problems. J Comput Syst Sci 10(3):384–393, doi:10.1016/S0022-0000(75),80008--0. http://dx.doi.org/10.1016/S0022-0000(75)80008--0
271. University RA (2012) Ultra-high speed mobile information and communication (UMIC). http://www.umic.rwth-aachen.de
272. Urfianto MZ, Isshiki T, Khan AU, Li D, Kunieda H (2006) A multiprocessor system-on-chip architecture with enhanced compiler support and efficient interconnect. In: IP-SOC 2006, Design and reuse
273. Vasudevan N, Edwards SA (2009) Celling SHIM: compiling deterministic concurrency to a heterogeneous multicore. In: Proceedings of the ACM symposium on applied computing (SAC), Honolulu
274. VectorFabrics (2012) vfEmbedded and vfThreaded-x86. http://www.vectorfabrics.com/
275. Verdoolaege S, Nikolov H, Stefanov T (2007) PN: a tool for improved derivation of process networks. EURASIP J Embed Syst 2007(1):13. http://www.dx.doi.org/10.1155/2007/75947

276. Voronenko Y, Arbatov V, Berger CR, Peng R, Püschel M, Franchetti F, (2010) Computer generation of platform-adapted physical layer software. In: Wireless innovation conference and product exposition (SDR'10). Wireless Innovation Forum, Washington, DC, USA
277. Wallace G (1992) The JPEG still picture compression standard. Consumer electronics, IEEE Trans 38(1):18–34. doi:10.1109/30.125072
278. Wei YC, Cheng CK (1991) Ratio cut partitioning for hierarchical design. IEEE Trans Comput Aided Des 10(7):911–921
279. Weng RN, Wolf T (2009) Analytic modeling of network processors for parallel workload mapping. ACM Trans Embed Comput Syst 8:18:1–18:29
280. Whittier R (1996) Push/Pull: PC technology/End user demand. In: Symposium on VLSI technology. Digest of technical papers, pp 2–5. doi:10.1109/VLSIT.1996.507773
281. Wiggers MH (2009) Aperiodic multiprocessor scheduling for real-time stream processing applications. PhD thesis, University of Twente, 978–90-365-2850-4
282. Wiggers MH, Bekooij MJG, Smit GJM (2008) Buffer capacity computation for throughput constrained streaming applications with data-dependent inter-task communication. In: Proceedings of the 2008 IEEE real-time and embedded technology and applications symposium (RTAS '08). IEEE Computer Society, Washington, DC, USA, pp 183–194. http://dx.doi.org/10.1109/RTAS.2008.10
283. Wiggers MH, Bekooij MJG, Smit GJM (2011) Buffer capacity computation for throughput-constrained modal task graphs. ACM Trans Embed Comput Syst 10(2):17:1–17:59, doi:10.1145/1880050.1880053. http://doi.acm.org/10.1145/1880050.1880053
284. van der Wijngaart RF, Mattson TG, Haas W (2011) Light-weight communications on intel's single-chip cloud computer processor. SIGOPS Oper Syst Rev 45:73–83, doi:10.1145/1945023.1945033. http://doi.acm.org/10.1145/1945023.1945033
285. Wikipedia (2013) Embedded system. http://en.wikipedia.org/wiki/Embedded_system
286. Wikipedia (2013) List of concurrent and parallel programming languages. http://en.wikipedia.org/wiki/List_of_concurrent_and_parallel_programming_langua ges
287. Wikipedia (2013) OMAP. http://en.wikipedia.org/wiki/OMAP
288. Wikipedia (2013) Snapdragon (system on chip). http://en.wikipedia.org/wiki/Snapdragon_(System_on_Chip)
289. Wilhelm R, Engblom J, Ermedahl A, Holsti N, Thesing S, Whalley D, Bernat G, Ferdinand C, Heckmann R, Mitra T, Mueller F, Puaut I, Puschner P, Staschulat J, Stenström P (2008) The worst-case execution-time problem—overview of methods and survey of tools. ACM Trans Embed Comput Syst 7(3):1–53. http://doi.acm.org/10.1145/1347375.1347389
290. Wilson RP, French RS, Wilson CS, Amarasinghe SP, Anderson JM, Tjiang SWK, Liao SW, Tseng CW, Hall MW, Lam MS, Hennessy JL (1994) SUIF: an infrastructure for research on parallelizing and optimizing compilers. SIGPLAN Not 29:31–37, doi:10.1145/193209.193217. http://doi.acm.org/10.1145/193209.193217
291. Witte EM (2012) Efficient and flexibility tradeoff of soft-input soft-output sphere decoding architecture. PhD thesis, RWTH Aachen Univeristy. Institute for Communication Technologies and Embedded Systems (ICE), Aachen
292. Witte EM, Borlenghi F, Ascheid G, Leupers R, Meyr H (2010) A Scalable VLSI-architecture for soft-input soft-output single tree-search sphere decoding. IEEE Trans Circuits Syst Part II: Express Briefs (57):706–710
293. Wolf W (2004) The future of multiprocessor systems-on-chips. In: Proceedings of the 41st annual conference on design automation (DAC '04). ACM Press, New York, pp 681–685. http://doi.acm.org/10.1145/996566.996753
294. Xilinx (2011) LogiCORE IP fast fourier transform v7.1. Product specification. Xilinx Inc, San Jose (DS260)
295. Xilinx (2012) IP documentation. http://www.xilinx.com/support/documentation/
296. Yehia S, Girbal S, Berry H, Temam O (2009) Reconciling specialization and flexibility through compound circuits. In: IEEE 15th International symposium on high performance computer architecture (HPCA 2009), pp 277–288. doi:10.1109/HPCA.2009.4798263

297. Yoon JS, Kim JH, Kim HE, Lee WY, Kim SH, Chung K, Park JS, Kim LS (2010) A graphics and vision unified processor with 0.89 uW/fps pose estimation engine for augmented reality. In: IEEE international solid-state circuits conference digest of technical papers (ISSCC), pp 336–337. doi:10.1109/ISSCC.2010.5433907
298. Zeligsoft (2012) Zeligsoft CE 2.4. http://www.zeligsoft.com/tools/zeligsoft-ce
299. Zhang D, Nikitopoulos K, Lai IW, Ascheid G, Meyr H (2010) Iterative channel estimation control for MIMO-OFDM systems. In: 44th Annual conference on information sciences and systems (CISS), pp 1–6. doi:10.1109/CISS.2010.5464953
300. Zhang D, Zhang H, Castrillon J, Kempf T, Vanthournout B, Ascheid G, Leupers R (2011) Optimized communication architecture of MPSoCs with a hardware scheduler: a system-level analysis. Int J Embed Real Time Commun Syst 2(3):1–20. doi:10.4018/jertcs.2011070101
301. Zhang D, Lu L, Castrillon J, Kempf T, Ascheid G, Leupers R, Vanthournout B (2013) Application-aware spinlock control using a hardware scheduler in MPSoC platforms. Inte J Embed Real Time Commun Syst (to appear)

Glossary

Notation (General, Multi-application Flow)

Of	big O notation for asymptotic runtime complexity order
$\wp(S)$	power set of set S
$\mathscr{A}$	Set of all applications, $\mathscr{A} = \mathscr{A}^{\text{seq}} \sqcup \mathscr{A}^{\text{kpn}} \sqcup \mathscr{A}^{\text{sdr}}$
ACG	application concurrency graph
ACS	application concurrency set
$\mathscr{A}^{\text{hrt}}$	set of applications with hard real-time constraints
$\mathscr{A}^{\text{seq}}$	set of sequential applications
$\mathscr{A}^{\text{kpn}}$	set of parallel KPN applications
$\mathscr{A}^{\text{srt}}$	set applications with soft real time constraints
$\mathscr{A}^{\text{nrt}}$	set of applications with no real time constraints
$\mathscr{A}^{\text{sdr}}$	set of applications for SDR
A	generic application, $A \in \mathscr{A}$
$\mathscr{CM}^{CP}$	cost model of a communication primitive
ζ^{CP}	cost function associated with communication primitive CP
$\mathscr{CP}$	set of all communication channels in the target MPSoC
$\mathscr{CM}^{PT}$	cost model of a processor type
ζ^{PT}	cost function associated with processor type PT
$\zeta^{PT,\text{dy}}$	dynamic cost function associated with processor type PT
$\zeta^{PT,\text{st}}$	static cost function associated with processor type PT
CP	generic communication channel, $CP \in \mathscr{CP}$
$\mathscr{CR}$	set of all communication resources in the target MPSoC
CR	generic communication resource, $CR \in \mathscr{CR}$
D_v^A	domain of variable v of application A
D_v^{PT}	domain of variable v of processor type PT
dst	generic function that returns the target node of an edge within a graph
$\mathscr{K}^A$	application constraints
K_i^A	i-th constraint of application A
$\mathscr{M}^A$	application model
μ_a	assignment of platform and application variables
μ_c	mapping of communication to communication primitives
μ_p	mapping of processes to processing elements
$\mathscr{PE}$	set of all processing elements in the target MPSoC
PE	generic processing element, $PE \in \mathscr{PE}$

J. Castrillón Mazo and R. Leupers, *Programming Heterogeneous MPSoCs*, DOI: 10.1007/978-3-319-00675-8,

$\mathcal{PE}^{v}$	shorthand notation for the set of all processing elements of type v
PE_i^v	shorthand notation for the i-th processing element of processor type v, $PE_i^v \in PE, v \in \mathcal{PT}$
$\mathcal{PT}$	set of all processor type in the target MPSoC
PT	generic processor type, $PT \in \mathcal{PT}$
RC^A	runtime configuration for application A
$\mathcal{RC}^A$	set of runtime configurations for application A
$\mathcal{RC}^{UC}$	set of runtime configurations for each application in use case UC
res	function that maps a communication primitive with its resources, res : $\mathcal{CP} \rightarrow \wp(\mathcal{CR})$
SOC	graph model of a given MPSoC, $SOC = (\mathcal{PE}, \mathcal{CP})$
src	generic function that returns the source node of an edge within a graph
UC	use case, subset of applications that may run concurrently and associated weight
$\vartheta_{PE}^{RC^A}$	utilization function of an application A on a processor PE according to a runtime configuration RC^A
$\mathcal{U}_{SOC}^{RC^A}$	set of all utilization functions of application A on platform SOC according to a runtime configuration RC^A
$\mathrm{var}_{\mathrm{size}}$	function that returns the size of a variable associated with a data flow edge
V^A	application variables
V^{PT}	set of variables of a processor type
v_{SP}^{PT}	variable of a processor type that represents the scheduling policy
$\omega_{\mathrm{dc}}^{\mathcal{RC}^{UC}}$	multi-application score for use case UC with use case runtime configuration $\mathcal{RC}^{UC}$ according to the displacement-criterion
$\omega_{\mathrm{mc}}^{\mathcal{RC}^{UC}}$	multi-application score for use case UC with use case runtime configuration $\mathcal{RC}^{UC}$ according to the mean-criterion
x_{CH}^{CR}	amount of channels that can be implemented over a communication resource CR
x_{MEM}^{CR}	memory size of communication resource CR
X^{PT}	attribute set of a processor type
x_{cs}^{PT}	attribute of a processor type that models the time spent in a context switch
x_{tasks}^{PT}	attribute of a processor type that defines the maximum amount of tasks that can executed
π^{P^A}	sequential profile of a process P^A
BB^A	generic basic block of application A
$\mathcal{BB}^{f^A}$	set of basic blocks of function f^A in application A
$\mathcal{BB}^A$	set of basic blocks of application A
CB^f	coupled block in a function f
$CDFG^{f^A}$	control-data flow graph of function f^A
CFG^{f^A}	control flow graph of function f^A
CG^A	call graph of application A
$\mathcal{C}^G$	clustering of a graph G
$CDFG_{\mathrm{par}}^{f^A}$	parallel annotated graph for function f^A

Notation (Sequential Flow)

CG_{par}^A	call graph of application A with function graphs that are parallel-annotated
CG_{pi}^A	call graph of application A with function graphs for which a parallel implementation has been determined
δ^c	control dependence relation
δ^f	forward data dependence relation
δ^a	anti-data dependence relation

δ^o	output data dependence relation
DFG^{f^A}	data flow graph of function f^A
dom	dominance relationship of nodes in a control flow graph
$D_v^{PA^n}$	domain of variable v of a parallel annotation PA^n
$E_c^{f^A}$	set of control flow edges of function f^A
E_c^A	set of control flow edges of all functions in application$A$
E_{cg}^A	set of call graph edges of application A
$E_d^{f^A}$	set of data flow edges of function f^A
$E_d^{f^A*}$	set of data flow edges of function f^A defined over basic blocks
E_d^A	set of data flow edges of all functions in application A
f^A	generic function of application A
IR^A	intermediate representation of application A
$\mathscr{M}_{\text{par}}^A$	parallel-annotated application model for application A
PA^n	parallel annotation for node n of a given graph
$\mathscr{P}\mathscr{I}^A$	parallel implementation option for a sequential application A
$\mathscr{P}\mathscr{A}^V$	set of parallel annotations for a set of nodes V of a given graph
pdom	post-dominance relationship of nodes in a control flow graph
pred	function that returns the control flow predecessors of a node
π^A	sequential profile of application A
s^A	generic IR-statement of application A
σ^A	function that provides information about function call sites of application A
$\mathscr{S}\mathscr{E}^A$	set of all elements in the model of a sequential application A
S_{f}^A	set of all functions of application A
$S_{\text{stmt}}^{f^A}$	set of all statements in function f^A
S_{stmt}^A	set of all statements of application A
succ	function that returns the control flow successors of a node
V^{PA^n}	set of variables of a parallel annotation for node n
X^{PA^n}	attribute set of a parallel annotation for node n

Notation (Parallel Flow)

$b_{C^A}^A$	variable that represents the size of channel C^A of application A
β	assignment of buffer size variables
C^A	fifo channel of a KPN application A
CG^{P^A}	call graph of process P^A
$\mathscr{C}^A$	set of all fifo channel of KPN application A
$I_i^{P^A}$	i-th process iteration of process P^A
$\mathscr{I}^{P^A}$	sequence of process iterations of process P^A
KPN^A	KPN graph model of an application
$N_{\text{it}}^{P^A}$	number of iterations of process P^A
P^A	process of application A
$\mathscr{P}\mathscr{A}\mathscr{E}^A$	set of all elements in the model of a parallel application A
$\mathscr{P}^A$	set of processes of application A
RE^A	set with all read events of a KPN application A
$S_i^{P^A}$	i-th segment of process P^A
$\mathscr{S}^A$	set of all segments in application A
$\mathscr{S}\mathscr{E}^{P^A}$	set of all elements in the sequential model of a process P^A
$\mathscr{T}\mathscr{G}^A$	trace graph of application A
T^{P^A}	trace of process P^A

$\mathcal{T}^A$	set of all process traces of application A
V_{size}^A	set of channel size variables of application A
WE^A	set with all write events of a KPN application A

Notation (SDR Flow)

$\mathcal{CM}^{F^{SOC}}$	cost model of flavor F^{SOC}
$\zeta^{F^{SOC}}$	cost function associated with a flavor F^{SOC}
$D_v^{F^{SOC}}$	domain of variable v of a flavor F^{SOC}
$D_v^{N^A}$	domain of variable v of a nucleus N^A
F^{SOC}	a flavor in platform SOC
$\mathcal{F}^{SOC}$	set of all flavors in platform SOC
$IN^{F^{SOC}}$	set of of input ports of a flavor F^{SOC}
IN^{N^A}	set of of input ports of a nucleus N^A
$K_i^{F^{SOC}}$	i-th constraint of flavor F^{SOC}
$\mathcal{K}^{F^{SOC}}$	set of constraints of a flavor F^{SOC}
$K_i^{N^A}$	i-th constraint of nucleus N^A
$\mathcal{K}^{N^A}$	set of constraints of a nucleus N^A
μ_f	assignment of flavor variables
μ_n	mapping of nucleus to flavors
$\equiv_{\text{IF}}$	relation that describes that two interfaces match
N^A	a nucleus of application A
$\mathcal{N}^A$	set of all nuclei in application A
NC^A	nucleus configuration for an application A, mapping of nucleus to flavors and matching flavor configuration
$\mathcal{N}$	set of all nuclei (for example defined in a library)
$OUT^{F^{SOC}}$	set of output ports of a flavor F^{SOC}
OUT^{N^A}	set of output ports of a nucleus N^A
$\mathcal{PE}^{F^{SOC}}$	set of PEs that contain the flavor F^{SOC}
P^{N^A}	process description associated with a nucleus N^A
$\mathcal{SI}^A$	SDR implementation of application A
$V_{C^A}^F$	set of variables of a flavor F that describe the interface of the port connected over a channel C^A
$V^{F^{SOC}}$	set of variables of a flavor F^{SOC}
V^{N^A}	set of variables of a nucleus N^A

Index

A
Abstract Syntax Tree, 29, 91, 108
Advanced High-performance Bus, 82, 84, 157
Advanced Microcontroller Bus Architecture, 82, 179
Analog-to-Digital Converter, 179
Application Programming Interface, 9, 12, 22, 67, 68, 81, 85, 87, 88, 92, 124, 155, 156, 177
Application Specific Instruction-set Processor, 4, 13, 55, 75, 76, 205
Application-Specific Integrated Circuit, 3
Architecture Description Language, 7
As Late As Possible, 151
As Soon As Possible, 97, 102

B
Basic Block, 31–34, 45, 91, 94–97, 102, 131
Basic Linear Algebra Subprograms, 48
Board Support Package, 48
Boolean Dataflow, 39, 43, 62

C
C for Process Networks, 11, 44
Call Graph, 33, 34, 45, 92, 118, 119
Central Processing Unit, 4
Common Object Request Broker Architecture, 66, 68
Communicating Sequential Processes, 62
Communication Assist based MPSoC, 71
Compiler Known Functions, 79
Component-Based Software Engineering, 66–68, 178, 180
Composable and predictable Multi-Processor System on Chip, 70, 71
Compute Unified Device Architecture, 61
Constraint Satisfaction Problem, 24, 132
Control Flow Analysis, 30
Control Flow Graph, 29, 31, 44, 45, 92, 130, 131
Control-Data Flow Graph, 33, 36, 37, 46, 91, 96, 100
Core Functional Dataflow, 62
Coupled Block, 100
Cyclo-Static Dataflow, 39, 43, 63, 65, 70, 71, 158

D
Data Flow Analysis, 32, 95, 96, 206
Data Flow Graph, 29, 32, 63
Data Level Parallelism, 35–37, 56, 98, 101, 103, 104
Dataflow Interchange Format, 62, 65, 67
Decoupled Software Pipelining, 57
Digital Signal Processor, Digital Signal Processing, 4, 9, 10, 16, 18, 19, 57, 62, 67, 74, 119, 162
Direct Memory Access, 10
Directed Acyclic Graph, 18, 19, 35, 41, 44, 52, 71, 145, 148, 150
Distributed Application Layer, 71
Distributed Operation Layer, 64, 71
Domain Specific Language, 9
Dynamic Dataflow, 39, 41–43, 62, 63, 65

E
Earliest Finishing Time, 17, 18, 145
Electronic Design Automation, 7
Electronic System Level, 7, 10, 11, 80, 81, 207
EUropean REference TILed architecture Experiment, 71, 113

J. Castrillón Mazo and R. Leupers, *Programming Heterogeneous MPSoCs*, DOI: 10.1007/978-3-319-00675-8,

F

Fast Fourier Transform, 48, 59, 174, 176, 179
Field Programmable Gate Arrays, 59, 63, 65, 171
Finite Impulse Response, 15
Finite State Machine, 63
First Come First Served, 17, 18, 70, 74
First-In-First-Out, 41–44, 46, 58, 125, 127–129, 133, 135, 137, 146, 149, 155, 162, 184

G

General-Purpose Graphics Processing Unit, 61
GNU Project Debugger, 155
Graphics Processing Unit, 4

H

Hardware Operating System Kernel, 54–56, 79
Hardware, 7, 55, 56, 57, 69, 72, 74, 80, 82, 137, 150, 156, 180
Hardware-dependent Software , 9
High Level Synthesis, 57
High Performance Computing, 2, 48, 53, 56–58, 60–62
High-level Design Language, 7
Homogeneous Synchronous Dataflow, 40, 42, 44, 70, 71

I

ICE RISC core, 81, 118, 154, 155, 157, 161
Institute for Communication Technologies and Embedded Systems at the RWTH Aachen University, 83
Instruction Level Parallelism, 29, 35, 56, 58
Instruction Set Architecture, 20, 22, 54–56
Instruction Set Simulator, 20
Integer Linear Programming, 29, 35, 56, 58, 59
Integrated Development Environment, 11, 92, 98, 112, 118, 138, 164, 207
Intel's Single-chip Cloud Computer, 3, 9
Intermediate Representation, 29–31, 33, 60, 90–97, 108
International Technology Roadmap for Semiconductors, 3, 4, 7
Inter-Processor Communication, 23
Interuniversity MicroelEctronics Centre, 57

J

Joint Photographic Experts Group, 15–17, 138, 157, 159, 161, 162, 163, 189, 190, 199, 200

K

Kahn Process Network, 11, 12, 16, 38, 39, 41–47, 51, 52, 58

L

Language for Instruction Set Architectures, 78, 82
Least Common Multiple, 194
Line Of Code, 7, 19
Linear Algebra PACKage, 48
LISA Tek Very Long Instruction Word processor core, 81, 82, 118, 154, 155, 161, 179–181, 184
Long Term Evolution, 5, 55
Low Level Virtual Machine, 11, 90, 91, 94, 97, 108, 124

M

Medium Access Control (layer), 47, 66, 179
Message Passing Interface , 9, 12, 61, 93, 108, 114
Model of Computation, 10, 38, 39, 40, 43, 55, 58, 60, 65, 70, 72, 123
Moving Picture Experts Group, 6
MPSoC Application Programming Studio, 2, 10–13, 33, 38, 43, 58, 60, 65, 68, 88, 90, 92, 93, 98, 100, 105, 114, 118, 124, 162, 164, 187, 207
MultiCore Optimization technology from Synopsys, 11, 177, 182, 183
Multi-Dimensional Synchronous Dataflow, 158, 169
Multiple Instruction Multiple Data, 35, 54
Multiple Instruction Stream Processor, 54
Multiple-Input Multiple-Output, 16, 164, 169, 178, 179, 199, 206
Multi-Processor System-on-Chip, 1–3, 6, 8–11, 15, 17, 20, 22, 55, 57, 71, 73, 80, 83, 84, 87, 91, 122, 154, 157, 165, 166, 178, 180, 186, 199, 205, 206
Multi-Rate Dataflow (same as SDF), 40

N
Nested Loop Programs, 58, 111
Network on Chip, 10, 71, 113
Non Recurring Engineering, 3, 5

O
Omphale Input Language, 63, 65
Open Computing Language, 61
Open Multimedia Application Platform, 4, 5, 54, 88
Operating System application specific Instruction-set Processor, 13, 53, 55, 73, 74, 76, 88, 123, 129, 154–157, 171, 205
Operating System, 2, 13, 54, 86
Operations Per Second, 5, 6
Orthogonal Frequency-Division Multiplexing, 16, 164, 178, 179, 199, 206

P
Personal Computer, 2, 5, 76, 77
PHYsical (layer), 47, 66, 68
Pipeline Level Parallelism, 35, 37, 56, 60, 98, 101, 105, 110, 112, 113, 119, 120, 206
Platform Architect from Synopsys, 11, 20, 82, 183
Polyhedral Process Networks, 64, 65
POSIX (Portable Operating System Interface), threads (Pthreads), 9
Process Network, 39, 43, 62, 64
Processing Element, 3–5, 9, 21, 22, 27, 54, 61, 66, 75, 80, 82, 93, 128, 137, 173, 177, 179
Processor Designer from Synopsys, 20, 75, 78, 81
Processor Support Package, 81
Program Counter, 2, 22, 76, 77

Q
Quadrature Phase Shift Keying, 179
Quasi Dynamic Mapping, 19
Quasi Static Scheduling, 18, 63, 65

R
Radio Frequency, 2, 66, 179
Real-Time Operating System, 46, 54
Real-time Task Manager, 54
Recognition, Mining and Synthesis, 167
Reduced Instruction Set Computer, 9, 18, 54, 74, 77, 81, 90, 115, 199
Register Level Transfer, 78
Round-Robin With Skipping, 70, 129
Round-Robin, 17, 18, 22, 63, 70, 129
Run-Length Encoded, 44, 138

S
Scenario Aware Dataflow, 62
Signal Processing Worksystem, 62
Silicon Real-Time Operating System, 54
Single Instruction Multiple Data, 19, 35
Single-Rate Dataflow, 40
Software Communication Architecture, 68, 69
Software Defined Radio, 1, 5, 12, 13, 15, 23, 26, 28, 47, 49, 51, 57, 66–69, 166, 169, 171, 172, 175, 176, 178, 180, 181, 182, 184
Software for Systems on Silicon, chair at ICE, 10
Software, 1–10, 13, 15, 22, 23, 39, 43, 47, 49, 50, 53–55, 57, 62, 67, 70, 71, 74, 76, 78, 80, 82, 93, 114, 165, 170, 176, 180, 185, 203, 205, 206
Software/Hardware Integration Medium, 64, 65
Stanford University Intermediate Framework, 56
Static Affine Nested Loop Programs, 58, 65
Static Single Assignment, 90
Strongly connected component, 3, 9
Synchronous Dataflow, 39–42, 55, 63–65, 70, 158
System on Chip, 1, 4, 5, 21, 23, 64

T
Tagged Signal Model, 82
Task Level Parallelism, 35, 36, 57, 98, 99, 101, 105–109, 112, 120, 206
Texas Instruments, 4, 22, 54, 55, 67, 88, 156, 161, 164, 170, 206
Thread Level Speculation, 57, 61
Three-Address Code, 31
Tightly Coupled Thread, 12
Time Division Multiplexing, 63, 70
Time-annotated Communication Extended Finite State Machine, 82
Transaction Level Model, 82, 84, 113

U
Ultra-high speed Mobile Information and Communication, 10, 48

V

Variable-rate Phased Dataflow, 63
Very Long Instruction Word, 9, 74, 81, 97, 115, 138, 154, 159, 162, 199, 201
Virtual Platform Analyzer, 155
Virtual Platform, 185
Virtual Processing Unit, 177

W

Whole Program Analysis, 33
Worst-Case Execution Time, 20

MIX
Papier aus verantwortungsvollen Quellen
Paper from responsible sources
FSC® C105338

If you have any concerns about our products,
you can contact us on
ProductSafety@springernature.com

In case Publisher is established outside the EU,
the EU authorized representative is:
Springer Nature Customer Service Center GmbH
Europaplatz 3, 69115 Heidelberg, Germany

Printed by Libri Plureos GmbH
in Hamburg, Germany